CW01376100

Optimization Methods
For Logical Inference

WILEY-INTERSCIENCE
SERIES IN DISCRETE MATHEMATICS AND OPTIMIZATION

ADVISORY EDITORS

RONALD L. GRAHAM
AT & T Laboratories, Florham Park, New Jersey, U.S.A.

JAN KAREL LENSTRA
Department of Mathematics and Computer Science,
Eindhoven University of Technology, Eindhoven, The Netherlands

ROBERT E. TARJAN
Princeton University, New Jersey, and
NEC Research Institute, Princeton, New Jersey, U.S.A.

A complete list of titles in this series appears at the end of this volume.

Optimization Methods
For Logical Inference

Vijay Chandru

John Hooker

A Wiley-Interscience Publication
JOHN WILEY & SONS, INC.
New York • Chichester • Weinheim • Brisbane • Singapore • Toronto

This book is printed on acid-free paper. ∞

Copyright © 1999 by John Wiley & Sons, Inc. All rights reserved.

Published simultaneously in Canada.

No part of this publication may be reproduced, stored in a retrieval system or transmitted in any form or by any means, electronic, mechanical, photocopying, recording, scanning or otherwise, except as permitted under Sections 107 or 108 of the 1976 United States Copyright Act, without either the prior written permission of the Publisher, or authorization through payment of the appropriate per-copy fee to the Copyright Clearance Center, 222 Rosewood Drive, Danvers, MA 01923, (978) 750-8400, fax (978) 750-4744. Requests to the Publisher for permission should be addressed to the Permissions Department, John Wiley & Sons, Inc., 605 Third Avenue, New York, NY 10158-0012, (212) 850-6011, fax (212) 850-6008, E-Mail: PERMREQ@WILEY.COM.

For ordering and customer service, call 1-800-CALL WILEY.

Library of Congress Cataloging in Publication Data:

Chandru, Vijay, 1953–
    Optimization methods for logical inference / Vijay Chandru, John Hooker.
        p. cm. — (Wiley-Interscience series in discrete mathematics and optimization)
    "A Wiley-Interscience publication."
    Includes bibliographical references.
    ISBN 0-471-57035-4 (cloth : alk. paper)
    1. Combinatorial optimization.   2. Logic, Symbolic and mathematical.   I. Hooker, John, 1944– . II. Title. III. Series.
QA402.5.C447   1999
519.3—dc21   98-44622
                                                       CIP

Printed in the United States of America

10 9 8 7 6 5 4 3 2 1

To our parents

Contents

	Preface	xiii
1	**Introduction**	**1**
1.1	Logic and Mathematics: The Twain Shall Meet	3
1.2	Inference Methods for Logic Models	6
1.3	Logic Modeling Meets Mathematical Modeling	7
1.4	The Difficulty of Inference	8
2	**Propositional Logic: Special Cases**	**11**
2.1	Basic Concepts of Propositional Logic	13
	2.1.1 Formulas	13
	2.1.2 Normal Forms	16
	2.1.3 Rules	22
2.2	Integer Programming Models	23
	2.2.1 Optimization and Inference	25
	2.2.2 The Linear Programming Relaxation	27
2.3	Horn Polytopes	31
	2.3.1 Horn Resolution	31
	2.3.2 The Integer Least Element of a Horn Polytope	34
	2.3.3 Dual Integrality of Horn Polytopes	37
2.4	Quadratic and Renamable Horn Systems	41
	2.4.1 Satisfiability of Quadratic Systems	42
	2.4.2 The Median Characteristic of Quadratic Systems	44
	2.4.3 Recognizing Renamable Horn Systems	45
	2.4.4 Q-Horn Propositions	53
2.5	Nested Clause Systems	55
	2.5.1 Nested Propositions: Definition and Recognition	55
	2.5.2 Maximum Satisfiability of Nested Clause Systems	58
	2.5.3 Extended Nested Clause Systems	61

2.6		Extended Horn Systems	63
	2.6.1	The Rounding Theorem	65
	2.6.2	Satisfiability of Extended Horn Systems	69
	2.6.3	Verifying Renamable Extended Horn Systems	70
	2.6.4	The Unit Resolution Property	71
	2.6.5	Extended Horn Rule Bases	73
2.7		Problems with Integral Polytopes	76
	2.7.1	Balanced Problems	78
	2.7.2	Integrality and Resolution	81
2.8		Limited Backtracking	87
	2.8.1	Maximum Embedded Renamable Horn Systems	89
	2.8.2	Hierarchies of Satisfiability Problems	91
	2.8.3	Generalized and Split Horn Systems	94

3 Propositional Logic: The General Case 97

3.1		Two Classic Inference Methods	98
	3.1.1	Resolution for Propositional Logic	99
	3.1.2	A Simple Branching Procedure	101
	3.1.3	Branching Rules	103
	3.1.4	Implementation of a Branching Algorithm	104
	3.1.5	Incremental Satisfiability	106
3.2		Generating Hard Problems	109
	3.2.1	Pigeonhole Problems	110
	3.2.2	Problems Based on Graphs	111
	3.2.3	Random Problems Hard for Resolution	113
	3.2.4	Random Problems Hard for Branching	114
3.3		Branching Methods	116
	3.3.1	Branch and Bound	117
	3.3.2	Jeroslow-Wang Method	121
	3.3.3	Horn Relaxation Method	122
	3.3.4	Bounded Resolution Method	123
3.4		Tableau Methods	123
	3.4.1	The Simplex Method in Tableau Form	124
	3.4.2	Pivot and Complement	128
	3.4.3	Column Subtraction	131
3.5		Cutting Plane Methods	133
	3.5.1	Resolvents as Cutting Planes	134
	3.5.2	Unit Resolution and Rank 1 Cuts	136
	3.5.3	A Separation Algorithm for Rank 1 Cuts	143
	3.5.4	A Branch-and-Cut Algorithm	144
	3.5.5	Extended Resolution and Cutting Planes	148
3.6		Resolution for 0-1 Inequalities	150

		3.6.1	Inequalities as Logical Formulas	153
		3.6.2	A Generalized Resolution Algorithm	154
		3.6.3	Some Examples	157
	3.7	A Set-Covering Formulation with Facet Cuts	159
		3.7.1	The Set-Covering Formulation	160
		3.7.2	Elementary Facets of Satisfiability	164
		3.7.3	Resolvent Facets Are Prime Implications	166
		3.7.4	A Lifting Technique for General Facets	171
	3.8	A Nonlinear Programming Approach	173
		3.8.1	Formulation as a Nonlinear Programming Problem	174
		3.8.2	An Interior Point Algorithm	175
		3.8.3	A Satisfiability Heuristic	176
	3.9	Tautology Checking in Logic Circuits	177
		3.9.1	The Tautology Checking Problem	177
		3.9.2	Solution by Benders Decomposition	179
		3.9.3	Logical Interpretation of the Benders Algorithm . .	182
		3.9.4	A Nonnumeric Algorithm	185
		3.9.5	Implementation Issues	188
	3.10	Inference as Projection	189
		3.10.1	The Logical Projection Problem	191
		3.10.2	Computing Logical Projections by Resolution	191
		3.10.3	Projecting Horn Clauses	193
		3.10.4	The Polyhedral Projection Problem	193
		3.10.5	Inference by Polyhedral Projection	195
		3.10.6	Resolution and Fourier-Motzkin Elimination	196
		3.10.7	Unit Resolution and Polyhedral Projection	197
		3.10.8	Complexity of Inference by Polyhedral Projection	198
	3.11	Other Approaches	199

4 Probabilistic and Related Logics 203

	4.1	Probabilistic Logic	205
		4.1.1	A Linear Programming Model	206
		4.1.2	Sensitivity Analysis	209
		4.1.3	Column Generation Techniques	211
		4.1.4	Point Values Versus Intervals	216
	4.2	Bayesian Logic	218
		4.2.1	Possible World Semantics for Bayesian Networks . .	220
		4.2.2	Using Column Generation with Benders Decomposition	224

		4.2.3	Limiting the Number of Independence Constraints	227
	4.3	Dempster-Shafer Theory	235
		4.3.1	Basic Ideas of Dempster-Shafer Theory	236
		4.3.2	A Linear Model of Belief Functions	239
		4.3.3	A Set-Covering Model for Dempster's Combination Rule	240
		4.3.4	Incomplete Belief Functions	244
		4.3.5	Dempster-Shafer Theory vs. Probabilistic Logic	246
		4.3.6	A Modification of Dempster's Combination Rule	248
	4.4	Confidence Factors in Rule Systems	250
		4.4.1	Confidence Factors	251
		4.4.2	A Graphical Model of Confidence Factor Calculation	254
		4.4.3	Jeroslow's Representability Theorem	259
		4.4.4	A Mixed Integer Model for Confidence Factors	262

5 Predicate Logic — 267

5.1	Basic Concepts of Predicate Logic	269
	5.1.1 Formulas	269
	5.1.2 Interpretations	270
	5.1.3 Skolem Normal Form	271
	5.1.4 Herbrand's Theorem	274
5.2	Partial Instantiation Methods	275
	5.2.1 Partial Instantiation	276
	5.2.2 A Primal Approach to Avoiding Blockage	278
	5.2.3 A Dual Approach to Avoiding Blockage	284
5.3	A Method Based on Hypergraphs	287
	5.3.1 A Hypergraph Model for Propositional Logic	288
	5.3.2 Shortest Paths in B-Graphs	290
	5.3.3 Extension to Universally Quantified Logic	290
	5.3.4 Answering Queries	295
5.4	An Infinite 0-1 Programming Model	295
	5.4.1 Infinite Dimensional 0-1 Programming	296
	5.4.2 A Compactness Theorem	297
	5.4.3 Herbrand Theory and Infinite 0-1 Programs	298
	5.4.4 Minimum Solutions	299
5.5	The Logic of Arithmetic	301
	5.5.1 Decision Methods for Arithmetic	301
	5.5.2 Quantitative Methods for Presburger Real Arithmetic	302
	5.5.3 Quantitative Methods for Presburger (Integer) Arithmetic	304

CONTENTS

6 Nonclassical and Many-Valued Logics   **307**
  6.1 Nonmonotonic Logics 308
  6.2 Many-Valued Logics 311
  6.3 Modal Logics 314
  6.4 Constraint Logic Programming 316
      6.4.1 Some Definitions 318
      6.4.2 The Embedding 319
      6.4.3 Infinite Linear Programs 321
      6.4.4 Infinite 0-1 Mixed Integer Programs 323

Appendix. Linear Programming   **325**
  Polyhedral Cones 325
  Convex Polyhedra 326
  Optimization and Dual Linear Programs 330
  Complexity of Linear Programming 331

Bibliography   **333**

Index   **358**

Preface

This book owes much to others. To a large degree it rests upon intellectual foundations laid by Robert E. Jeroslow and H. Paul Williams, both of whom explored the deep connections between logic and mathematical programming.

As authors, we are indebted to a number of people who read parts of the manuscript or just helped us to think about logic and optimization. They include Kim Allan Andersen, Raphael Araque, Egon Balas, Peter Barth, Alexander Bockmayr, Vivek Borkar, Endre Boros, Collette Coullard, Gérard Cornuéjols, Giorgio Gallo, Harvey Greenberg, Ignacio Grossmann, Peter Hammer, Jean-Louis Lassez, Sanjoy Mitter, Arun Pujari, Ron Rardin, Gerald Thompson, Klaus Truemper, M. Vidyasagar, V. Vinay, and John Wilson.

We have also benefited from collaboration with our present and former students: Srinivas Bollapragada, Geon Cho, Milind Dawande, Chawki Fedjki, Farid Harche, V. S. Jayachandran, Hak-Jin Kim, Miguel Montañez, Srinath Naidu, N. R. Natraj, Maria Auxillio Osorio, Greger Ottosson, Arnab Paul, Gabriella Rago, Ramesh Raman, Suman Roy, Anjul Shrivastava, and Hong Yan.

We acknowledge Neal Glassman and Abraham Waksman of the U.S. Air Force Office of Scientific Research, and particularly Donald Wagner of the U.S. Office of Naval Research, who foresaw the potential of combining logic and optimization and provided continuing support in the form of research grants.

Finally, we thank our editors at Wiley for their patience and support during the lengthy gestation of this book.

VIJAY CHANDRU
JOHN HOOKER

Bangalore, India
Pittsburgh, Pennsylvania, USA
January, 1999

Optimization Methods
For Logical Inference

Chapter 1

Introduction

Solving logical inference problems with optimization methods may seem a bit like eating sauerkraut with chopsticks, because the two come from vastly different worlds. Logical inference comes from a "left brain" world of formal languages and symbolic manipulation. Optimization comes from a "right brain" world of spatial models and numerical calculation. Logical inference is associated with artificial intelligence and computer science, and optimization with operations research and engineering.

But on second thought it should not be so surprising that optimization methods are useful in logic. It is the mathematical structure of a problem that determines whether an optimization model can help solve it, not the context in which the problem occurs. Tasks as different as routing soft drink trucks and manufacturing circuit boards, for instance, can pose the same optimization problem (in this case, the traveling salesman problem). We should therefore expect that logical inference might pose some familiar optimization problems.

The thesis of this book is that many deductive inference problems do in fact have the sort of mathematical structure that optimization methods, or methods suggested by optimization, can exploit. These problems arise in logics that have important applications in artificial intelligence, computer science, decision support, and manufacturing:

propositional logic
first-order predicate logic
probabilistic and related logics
logics that combine evidence (e.g., Dempster-Shafer theory)
rule systems with confidence factors
constraint logic programming systems

The last several years have already seen orders-of-magnitude improvements in inference algorithms for propositional and probabilistic logic, due in part to the techniques described in this book.

A mathematical analysis of inference problems also reveals some interesting parallels between logic and mathematics. These can lead to better algorithms, both numeric and symbolic. For instance, the inference problem in propositional logic can be solved as an integer programming problem. Furthermore, a well-known inference procedure in logic (unit resolution) generates cutting planes that help solve the integer programming problem. The result is a kind of symbiosis: logic helps integer programming to solve logic problems. Similar insights can allow one to identify special cases of inference problems that are easy to solve. This is important to do because the general inference problem in most logics of interest can be very hard computationally. For example, inference is easy for the "Horn" propositions used in most expert systems because they are associated with polyhedra that have a least element property. Since the same property is shared by other polyhedra, this leads to the extension of Horn propositions to a much larger class for which inference is equally easy.

Most of the pioneering work in the logic/optimization interface was done within the last three decades, notably by H. P. Williams [290, 291, 296], who showed some connections between inference in propositional logic and the projection of polyhedra, and R. Jeroslow [23, 165], who used integer programming to solve inference problems and introduced a number of other seminal ideas. In addition, T. Hailperin found in the work of George Boole the elements of a linear programming model for probabilistic logic [27, 28, 124, 126]. Most of the research in the area is even more recent, having taken place within the last dozen years (that is, after [141]).

It is therefore possible for a single book to describe a large fraction of what has been done, and this we undertake to do. We also present a number of new results of our own.

Although the book is addressed to a technical audience, we are aware that our readers represent a large variety of backgrounds: operations research, computer science, logic, artificial intelligence, and engineering. We therefore presuppose as little as possible. The entire book should be readily intelligible to an otherwise qualified reader with absolutely no background in logic. It is more difficult to make the same promise with respect to optimization, on which we draw more deeply. Some exposure to linear or integer programming would no doubt be helpful. But we try to explain everything we do and provide an appendix to cover basic concepts that we do not explain elsewhere.

The book is organized by the types of logic to which optimization methods have been applied. *Propositional logic*, the most basic sort of logic, has been the most thoroughly investigated from the point of view of optimization. It is allotted two long chapters. Chapter 2 deals with specialized methods for particular classes of propositional inference problems, and Chapter 3 with general methods. *Probabilistic logics* assign propositions probabilities rather than regarding them simply as true or false. The quantitative nature of probability naturally invites numerical methods, as witnessed by Boole's early contribution. Chapter 4 covers these as well as some applications to Dempster-Shafer theory (an approach to combining evidence) and rule systems with confidence factors.

First-order predicate logic, which contains such expressions as "for all" and "there exists," is a powerful system and consequently much harder to completely solve inference in. Chapter 5 contains a description of partial instantiation, a new methodology for inference in predicate logic, that was motivated by paradigms in large-scale optimization. Compactness and other structural properties of infinite-dimensional mathematical programming are used to analyze inference in first-order logic in this penultimate chapter. Chapter 6 completes our description of how optimization methods can be used to address logical inference by presenting brief applications of mathematical programming in many valued logics and nonclassical logics such as nonmonotonic logics, modal logics, and constraint logic programming. We have limited our discussion in this book to the use of optimization methods in deductive inference.

The rest of this introductory chapter puts the research described here into context. The application of quantitative methods to logic is part of a larger movement toward the merger of the two worlds of logic and mathematics. Moreover, logic has recently become a basic modeling tool alongside mathematics, and the two styles of modeling are beginning to combine. Thus the need for logical inference methods, particularly those that involve quantitative methods, is growing. The need becomes particularly urgent as logic models become larger, because the difficulty of inference increases very rapidly with the size of the model.

1.1 Logic and Mathematics: The Twain Shall Meet

For more than two millennia, methods of formal reasoning have followed two largely separate lines of development. On the one hand is formal logic, which stretches from the Jains of ancient India, to Aristotle's systematic investigations, through the medieval logicians, to such mathematical logicians as Gottlob Frege and Kurt Gödel, and finally to the large community of researchers now applying logic to problems in computer science and artificial intelligence. On the other hand is mathematical computation, which began even earlier in Egypt and Mesopotamia and developed into today's highly advanced methods.

Although largely parallel, these two paths have crossed on occasion, sometimes with spectacular results. This book aims to participate in what we believe is the latest crossing. But a proper appreciation of this phenomenon requires some more background.

The seventeenth century philosopher Gottfried Wilhelm von Leibniz brought about the first important synthesis of logical deduction and mathematical computation. His contribution was to point out that they are fundamentally the same. To follow the steps of a numeric algorithm is to perform a series of deductions. Conversely, a series of deductions in formal logic can be viewed as computation and can be automated just as long division can. Leibniz went so far as to envision a calculus of reasoning (*calculus ratiocanator*) in which all truths can in principle be obtained from self-evident premises by calculation.

Leibniz did not have the technical wherewithal even to begin to realize his vision, but George Boole did. He showed how to compute inferences in propositional logic with his famous algebra, and he did the same for probabilistic logic using an idea that anticipates linear programming. He began a series of developments, too long to recount here, that culminated in the field of artificial intelligence, which aims in part to carry out Leibniz's project of automating the reasoning process. More relevant for our purposes, however, is what was *not* accomplished by this and subsequent encounters of logic and mathematics, until very recently. Although Boole saw that mathematical and logical calculation are of a piece, he did not bring them together into a single calculus. On the contrary, a major point of Boolean algebra was to show that computation can be purely nonnumeric, or "symbolic" as we now sometimes say. Boole used arithmetical symbols to denote logical operations, but this was only a notational convenience [185].

Other encounters grew out of Leibniz's work but likewise fell short of true fusion. Leibniz realized that logical computation would be possible only within a formal language (*characteristica universalis*). This idea eventually led to the development of such formal languages as predicate logic and set theory, within which Bertrand Russell, Alfred North Whitehead, and their successors attempted to formalize mathematics in a "logicist" manner. (It is no accident that Russell wrote his dissertation on Leibniz.) Alfred Tarski and others devised decision procedures for logics that contain arithmetic. In fact, the whole formalist paradigm of doing mathematics, which in its starkest form (due to D. Hilbert) regards mathematics as the study of how uninterpreted symbols may be manipulated in a formal lan-

1.1 LOGIC AND MATHEMATICS: THE TWAIN SHALL MEET

guage, is largely inspired by the Leibnizian legacy. But through all these developments, it is hard to find an occasion on which someone actually introduced numerical methods into the logical deduction process, or logical methods into numerical computation.

Perhaps the first marriage of logical and mathematical computation (with the interesting exception of Boole's probabilistic logic) occurred with the arrival of digital electronic computers, which use logic gates to implement arithmetic operations in a circuit. But it was a marriage without consummation, because the logical operations remain at the micro level and leave the higher-level numeric algorithms unaffected.

The relationship was brought a bit closer when work in artificial intelligence revived interest in decision procedures for logics, including logics of arithmetic. Some algorithms were devised that solve numeric problems by replacing key numeric procedures with symbolic ones. There is, for instance, a partially symbolic algorithm for linear programming (the "SUP-INF method" [261]). These algorithms had limited impact on methods for numeric problems, however, because symbolic procedures are often unable to exploit the special structure of these problems to the extent that purely numeric algorithms do. The SUP-INF method, for instance, is extremely inefficient (it has doubly exponential complexity).

The latest mathematical/logical encounter is bidirectional. This book focuses on the application of numeric methods to logical deduction. The reverse influence is also underway. Logic programming and its successors are now used to solve problems that have long been attacked by the numerical methods of operations research. The successors include constraint logic programming and constraint satisfaction techniques [207, 286], which not only apply discrete and logical methods but integrate linear programming as well, adding another layer of interaction.

The work described in this book can therefore be seen as an episode in the cross-fertilization of logic and mathematics initiated by Leibniz. The word "optimization" appears in the title because most of the mathematical ideas applied to logical inference in the last few years have related to optimization methods, which turn out to be particularly suited to computing inferences. It is perhaps pure coincidence that the concept of optimization played a key role in Leibniz's philosophy.

Not all methods discussed herein are optimization methods, and not all are numeric. Yet all are inspired by such methods in one way or another, even if only in the sense that they result from a similar style of thinking. In fact, the main benefit of focusing on optimization may be simply that it encourages a research community rooted in the mathematical tradition to try its hand at inference problems.

1.2 Inference Methods for Logic Models

As inference problems grow in size and importance, it is more important than ever to solve them efficiently. Some of these problems come from specific domains that pose problems of a logical nature. Electronic circuit design and testing, for instance, have presented hard logic problems for some decades, and these problems get harder as circuits get more complex. Research in automated theorem proving for mathematics yielded a number of new inference methods. But there is a more general phenomenon that has made inference problems a standard feature of the technological landscape. This is the rise of *logic models* and, more recently, logico-mathematical models.

Just as logical and mathematical computation are essentially the same, a logic model is essentially the same kind of structure as a mathematical model. Either kind of model describes a problem in a formal language that allows one to deduce facts about its solution. In a mathematical model, the formalism is a mathematical theory, and the deduction takes the form of algebraic manipulation or numeric calculation.

The practice of logic modeling grew partly out of attempts to create artificial intelligence. An intelligent computer should not only be able to solve well-structured problems traditionally attacked with mathematical models, but it should be able to solve such "messier" problems as scheduling operations in a factory, designing a building layout, guiding a robot over unfamiliar terrain, diagnosing an illness, or interpreting natural language. The first two problems and perhaps the third may be formulable in mathematical terms but tend to be very hard to solve with mathematical methods. The remaining problems do not even admit a mathematical formulation.

One response to this dilemma in the artificial intelligence community has been to describe problems in a general-purpose formalism that permits deductions but presupposes no mathematical structure. Thus we see the development of such "knowledge-based systems" as expert systems. An expert system for diagnosing failures in a piece of machinery, for instance, might contain a few hundred "rules" that look something like, "If indicator light A is red and noise B is audible, then circuit C is defective." The user adds his observations to the set of rules, whereupon an "inference engine" deduces what may be wrong with the machine.

Another response has been the development of logic programs, normally written in PROLOG after a fashion initially advocated by Kowalski [188]. PROLOG accepts statements written in a restricted form of predicate logic and computes some of their implications. Inference is incomplete because of the lack of practical methods to solve the problem completely.

More recently, logic programming has been largely supplanted by *constraint programming* [207], which offers a number of specialized predicates that are useful for real-world problems, along with specialized algorithms to deal with them. They include both discrete predicates, such as "all-different," and arithmetical predicates.

Expert systems and other decision support systems are often designed to reason under uncertainty or incomplete information, the former most often captured by probabilities. The classical framework for reasoning with probabilities is Bayesian inference, which is the basis for decision trees. More recent variations include influence diagrams, which are based on Bayesian networks rather than trees.

Several logics are designed to account for uncertainty and incomplete information. Boole's original probabilistic logic does both, as do Dempster-Shafer theory and other logics of belief. These form the basis for "belief nets," of which Bayesian networks are only one example. Their nodes represent propositions and their arcs dependencies among propositions that permit inferences.

Data bases are important components of many decision support systems and can likewise be regarded as logic models in which predicate logic plays an important role. "Default logic" and "nonmonotonic logic" were developed partly to deal with logical problems they posed. Default logic allows one to make generalizations that are not strictly supported by data, and nonmonotonic logic allows one to retract an inference when additional information becomes available. Both types of reasoning (they are typically combined) now play a role in modeling systems other than data bases. Modal logics, including temporal logic, are also used to model knowledge acquisition as well as to verify computer programs, for example.

In addition to all this is a vast literature on fuzzy logic systems, which were originally intended to account for vagueness but now seem to serve other purposes as well.

1.3 Logic Modeling Meets Mathematical Modeling

Not only are logic models proliferating, but a newer trend is underway: their merger with more traditional mathematical models. Inference procedures that combine numeric and nonnumeric elements may be particularly appropriate for models that do the same.

One impetus for the merger lies in the fact that, in many situations, neither type of modeling alone is true to reality. At one point in history, an elegant mathematical reality seemed to underlie the world, and mathematical models sufficed. But the "modern" age of Newtonian simplicity has given way to a "postmodern" age of bewildering complexity that calls for a mixture of modeling styles. We want to understand and manage such complex systems as the economy and the atmosphere and large manufacturing plants, but there are no elegant mathematical models from which we can deduce predictions, or prescriptions for action, with only second-order error. To get even a first approximation we must work with a system description that may be nearly as complex as the system itself. One attraction of chaos theory was that it seemed to bring complex, baffling phenomena once again under the purview of simple mathematical models. But if such writers as Freeman Dyson [90] are right, it may only be a momentary respite from an overwhelming trend toward messiness.

We should therefore expect a postmodern model to be a mosaic of mathematical and nonmathematical elements. Some aspects of a problem may display mathematical structure, and a model would be remiss to neglect this—not only for the sake of verisimilitude, but to make deduction (i.e., computation) easier. A factory subsystem that admits a linear programming submodel, for instance, should be so modeled, if only to get a correct solution quickly for that part of the system. But other aspects of the problem will likely not submit easily to mathematical modeling and may call for logic-based modeling. It has become commonplace to mix the two in constraint programming, and a similar mixture seems on the horizon in mathematical programming.

It is hard to predict where logico-mathematical modeling will lead. Already, new formalisms are evolving that are classifiable as neither mathematical nor logical but have some structural similarities with both. In the meantime, the research described in this book can play a role in its development. One obvious contribution derives from its use of mathematical programming models to compute the implications of a logic model. But more generally, a persistent theme of the research is the discovery of structural and algorithmic commonality between logic and mathematics. One can expect this sort of discovery to hasten the fusion of logical models and methods with those of mathematics.

1.4 The Difficulty of Inference

Logical inference can be a very hard combinatorial problem. It is perhaps curious that one of the fundamental tasks of information science—deducing the implications of what we already know—could be so hard. It is hard in the sense that the amount of computation required to check whether a

1.4 THE DIFFICULTY OF INFERENCE

knowledge base implies a given proposition grows very rapidly with the size of the knowledge base. Thus as logic models grow larger, we face a need for continuing breakthroughs in the speed of inference algorithms, or else new ways to structure models so as to make inference easier.

The difficulty of inference in several logics can be stated quite precisely. In propositional logic, the most basic we consider, no known algorithm always solves the inference problem in better than exponential time (i.e., the solution time increases exponentially with the size of the knowledge base in the worst case). Technically, the problem is "NP-complete" [66, 108]). Robinson's well-known resolution algorithm [244], when specialized to propositional logic, requires exponential time in the worst case when applied to propositional logic [128]. Computational experience [139] indicates that the running time rapidly explodes in the typical case as well. We prove in Chapter 3 that a tree search algorithm (the Davis-Putnam-Loveland algorithm) is also exponential in the worst case. Franco and Paul [97] showed that it requires exponential time with probability approaching one when large random problems are chosen from a reasonable distribution.

The situation is even worse in first-order predicate logic. In this case the problem is not only hard but insoluble in general. There are procedures, such as the resolution procedure, that verify that any given implication in fact follows from the premises. But Alonzo Church proved in 1936 (and Alan Turing shortly thereafter) that there is no procedure that can always verify in finite time that a given nonimplication does not follow [57, 279]. To have a finite decision procedure one must restrict oneself to a fragment of predicate logic, and even then inference is very hard. One such fragment consists of formulas in Schönfinkel-Bernays form, in which all existential quantifiers ("there exists") precede all universal quantifiers ("for all"). Checking the satisfiability of Horn formulas in Schönfinkel-Bernays form, which comprise a very limited fragment, requires exponential time in the worst case [235]. Doing the same for an arbitrary formula in Schönfinkel-Bernays form requires "nondeterministic exponential" time and is therefore even harder [195].

Because propositional logic is a subset of probabilistic logic, inference in the latter is likewise hard and is easily shown to be an NP-complete problem [113].

The proper reaction to these observations is not to despair of solving large inference problems. Clever algorithms can overcome their difficulty in many cases, and analysis can identify subsets of logics for which inference is easier.

Chapter 2

Propositional Logic: Special Cases

We begin with propositional logic, the simplest sort of logic. Propositional logic, sometimes called sentential logic, may be viewed as a grammar for exploring the construction of complex sentences (propositions) from "atomic statements" using the logical connectives such as "and," "or," and "not." The fundamental problem of inference in logic is to ascertain if the truth of one formula (proposition) implies the truth of another. Even for elementary propositional logic, the inference problem is not entirely well-solved.

The problem of testing the satisfiability of a propositional formula is the core problem of inference. This *satisfiability problem* is simply stated as the problem of finding a set of truth assignments for the atomic propositions that renders the formula true. An obvious algorithm for this problem is to enumerate all possible truth assignments and evaluate the formula until it is satisfied or proved unsatisfiable. This would entail an unacceptable amount of work for all but small and uninteresting formulas. Our focus in this chapter will be on special formulas for which satisfiability can be tested extremely rapidly by algorithms whose run times are, typically, within a constant factor of the time it takes a computer to read the formula.

Quadratic and Horn formulas represent two classical examples of structured propositions that admit highly efficient (linear-time) inference algorithms. We shall explore these and related structures in this chapter using a quantitative or optimization-based perspective. This approach has two advantages. The first is that it leads to an understanding of the mathematical structure of these formulas which the purely syntactic approach of symbolic computation misses altogether. We will also demonstrate that this perspective has been very useful in identifying rich new classes of formulas which, like the Quadratic and Horn classes, admit fast inference algorithms. These developments, many of which are quite recent, have the potential for substantively increasing the expressive power of highly tractable fragments of propositional logic.

The main optimization-based perspective of satisfiability is developed on an (0-1) integer programming formulation. This formulation is a manifestation of a well-known reduction of satisfiability to testing the solubility of a system of linear inequalities defined on variables that are restricted to binary values of 0 or 1. Since solvability and optimization in integer programming are closely related we can view satisfiability as an optimization problem. A central idea in integer programming is to relax the integrality of the variables by treating the 0-1 variables as continuous variables taking values in the range [0,1] to obtain the *linear programming relaxation*. Linear programming problems are far easier to handle, both mathematically and computationally, than integer programming problems. The linear programming relaxations of satisfiability problems are particularly easy to solve because unit resolution can be adapted to design a complete solution method for these special linear programs.

The linear programming relaxation of a Horn formula turns out to have a remarkable property. The linear program has an integral least element that corresponds to the unique minimal model of the Horn formula. We will see that this least element property can be generalized in a very natural way to realize a class of Extended Horn Formulas [44]. The least element property of Horn formulas also permits a rich theory of proof signatures for these formulas based on the duality theorem of linear programming.

We will also study a class of "Q-Horn" formulas [34] that simultaneously generalizes Quadratic and Horn formulas. The class is characterized by a special property of a related linear programming problem. Another paradigm from optimization that we will find useful in studying special propositional formulas is that of forbidden minor characterizations of special structures. This paradigm will help in the study of Q-Horn formulas as well as some recursive structures built on Horn formulas that still yield tractable formulas that are called Generalized Horn Formulas [5, 106, 302].

We will start with a quick review of some of the basic concepts of propositional logic. This will lead us to the (0-1) integer programming formulation of satisfiability and on to the study of special structures in propositional logic.

2.1 Basic Concepts of Propositional Logic

In propositional logic we consider formulas (sentences, propositions) that are built up from atomic propositions (x_1, x_2, ..., x_n) that are unanalyzed. In a specific application, the meaning of these atomic propositions will be known from the context. The traditional (symbolic) approach to propositional logic is based on a clear separation of the syntactical and semantical functions. The syntactics deals with the laws that govern the construction of logical formulas from the atomic propositions and with the structure of proofs. Semantics, on the other hand, is concerned with the interpretation and meaning associated with the syntactical objects. Propositional calculus is based on purely syntactic and mechanical transformations of formulas leading to inference. The "principle of resolution" is the most important transformation rule for this purpose. We will maintain this traditional approach to propositional logic in this section. However, it should be noted that the quantitative or optimization perspective that will be introduced in the next section is not beholden to this doctrine of separating syntactics and semantics. In fact, much of its power derives from the integration of the two functions.

We will first review the syntactics of forming propositional logic formulas. The assignment of truth values (true/false) to atomic propositions and the evaluation of truth/falsehood of formulas is the essence of the semantics of this logic. The central problem of inference in propositional logic is the satisfiability problem. This is the problem of determining whether a given formula is true (is satisfied) for some assignment of truth values to the atomic propositions. Two formulas are (semantically) equivalent if their satisfying truth assignments are equivalent under some mapping. We will discuss rewriting "well-formed formulas" in equivalent canonical representations called "normal forms." A particular normal form, the "conjunctive normal form" (CNF), is the canonical representation of propositions that we use in the ensuing discussion of propositional logic. The artificial intelligence (AI) literature often describes the calculus of propositional logic in the syntax of "if-then" rules, which are sometimes called "inference rules." We shall see that this is just an alternate schema for CNF formulas.

2.1.1 Formulas

Propositional logic formulas are built up from atomic propositions using various logical connectives. The *primary connectives* are $\land, \lor, \neg$ which are understood to represent the semantics of *and, or, not*, respectively. An inductive definition of well-formed formulas (wffs) using these connectives is given by:

(a) Atomic propositions are wffs.

(b) If S is a wff so is $\neg S$.

(c) If S_1 and S_2 are wffs, so are $S_1 \vee S_2$ and $S_1 \wedge S_2$.

It is customary to use parentheses "(" and ")" to distinguish the start and end of the field of a connective operation. Consider the formula

$$((\neg x_1 \vee (x_1 \wedge x_3)) \wedge ((\neg(x_2 \wedge \neg x_1)) \vee x_3)) \tag{2.1}$$

We notice that since $\neg$ is a unary operator, the field on which it operates follows immediately. The subformula $\neg x_1$ has $\neg$ applied to the atomic proposition x_1 and so no parentheses are required. Whereas the subformula $(\neg(x_2 \wedge \neg x_1))$ has $\neg$ applied to the nonatomic proposition $x_2 \wedge \neg x_1$ and parentheses are required to delineate the field. The binary operators $\vee$ and $\wedge$ have their field to the left and right of them and the parentheses delineate them. For example, the first open parenthesis "(" and the last close parenthesis ")" delineate the field of the $\wedge$ connective that appears in the middle of the formula. A formula is a wff if and only if there is no conflict in the definition of the fields of the connectives. Thus a string of atomic propositions and primitive connectives, punctuated with parentheses, can be recognized as a well-formed formula by a simple linear-time algorithm. We scan the string from left to right while checking to ensure that the parentheses are nested and that each field is associated with a single connective. Incidentally, in order to avoid the use of the awkward abbreviation "wffs," we will henceforth just call them propositions or formulas and assume they are well formed unless otherwise noted.

The calculus of propositional logic can be developed using only the three primary connectives $\{\neg, \vee, \wedge\}$. However, it is often convenient to permit the use of certain additional connectives. Three such connectives that we will find occasion to use are $\rightarrow$, $\bigvee$, and $\bigwedge$. They are, respectively, the connective "implies," "p-ary disjunction," and the "p-ary conjunction." They are essentially abbreviations that have equivalent formulas using only the primary connectives. The equivalences are detailed below.

$$(S_1 \rightarrow S_2) \quad \text{is equivalent to} \quad (\neg S_1 \vee S_2)$$

$$(\bigvee_{i=1}^{p} S_i) \quad \text{is equivalent to} \quad (\cdots (S_1 \vee S_2) \vee S_3) \cdots S_p)$$

$$(\bigwedge_{i=1}^{p} S_i) \quad \text{is equivalent to} \quad (\cdots (S_1 \wedge S_2) \wedge S_3) \cdots S_p)$$

2.1 BASIC CONCEPTS OF PROPOSITIONAL LOGIC

The truth or falsehood of a formula is a semantic interpretation that depends on the values of the atomic propositions and the structure of the formula. The elements of the set {T,F} are called truth values with "T" denoting "true" and "F" denoting "false." A truth assignment is simply the assignment of values T or F to all the atomic propositions. To evaluate a formula we interpret the connectives $\neg$, $\vee$, and $\wedge$ with the appropriate meaning of "not," "or," and "and." As an illustration, consider the formula (2.1). Let us start with an assignment of true (T) for all three atomic propositions x_1, x_2, and x_3. At the next level, of subformulas, we have $\neg x_1$ evaluates to false (F), $(x_1 \wedge x_3)$ evaluates to T, $(x_2 \wedge \neg x_1)$ evaluates to F, and x_3 is T. The third level has $(\neg x_1 \vee (x_1 \wedge x_3))$ evaluating to T since it is the "or" of two subformulas one of which is false and the other true. Eventually $((\neg(x_2 \wedge \neg x_1)) \vee x_3))$ also evaluates to T. The entire formula is the "and" of two propositions both of which are true, leading to the conclusion that the formula evaluates to T. This process is simply the inductive application of the rules:

(a) S is T if and only if $\neg S$ is F.

(b) $(S_1 \vee S_2)$ is F if and only if both S_1 and S_2 are F.

(c) $(S_1 \wedge S_2)$ is T if and only if both S_1 and S_2 are T.

We now introduce a variety of inference questions related to the truth or falsehood of propositions. The *satisfiability problem* is the problem of determining whether a given formula is "satisfied" (evaluates to T) for some assignment of truth values to the atomic propositions. We showed that the formula (2.1) is indeed satisfiable, since it is satisfied by the truth assignment $(x_1, x_2, x_3) = $ (T,T,T). A satisfying truth assignment is called a *model*. A formula with no model is called unsatisfiable. A formula for which every truth assignment is a model is called a *tautology*. The formula $(x_1 \vee \neg x_1)$ is a tautology. A formula S_1 is said to imply another formula S_2, defined on the same set of atomic propositions as S_1, if every model of the former is also a model of the latter. Two formulas are said to be equivalent if they share the same set of models. We now have three basic inference problems in propositional logic.

- Is a given formula satisfiable?

- Is a given formula a tautology?

- Does one formula imply another?

As an illustration of Procedure 1, let us consider the formula (2.1). The step by step transformations on this sample are depicted below.

$$\begin{array}{ll}(\neg x_1 \vee (x_1 \wedge x_3)) \wedge & ((\neg(x_2 \wedge \neg x_1)) \vee x_3) \\ (\neg x_1 \vee (x_1 \wedge x_3)) \wedge & ((\neg x_2 \vee x_1) \vee x_3) \\ (\neg x_1 \vee x_1) \wedge (\neg x_1 \vee x_3) \wedge & (x_1 \vee \neg x_2 \vee x_3) \\ (\neg x_1 \vee x_3) \wedge & (x_1 \vee \neg x_2 \vee x_3)\end{array} \quad (2.6)$$

In the last step we removed the clause $(x_1 \wedge \neg x_1)$ since this is a tautology whose truth is invariant. In general, however, this simple procedure for reducing formulas to CNF runs into trouble. The main difficulty is that there may be an explosive growth in the length of the formula. The length of a formula is measured by the total number of literals in the description of the formula. Consider the action of Procedure 1 on the family of DNF formulas

$$(x_1 \wedge x_2) \vee (x_3 \wedge x_4) \vee \cdots \vee (x_{2n-1} \wedge x_{2n}) \quad (2.7)$$

It is not difficult to see that the CNF formula produced, by Procedure 1, is made up of the 2^n clauses

$$(x_{p(1)}, x_{p(2)}, \ldots, x_{p(n)}) \text{ where } x_{p(j)} \in \{x_{2j-1}, x_{2j}\} \ (j = 1, 2, \ldots, n) \quad (2.8)$$

These arguments prove that, in the worst case, Procedure 1 is an exponential-time algorithm.

Theorem 1 ([23]) *In the worst case, Procedure 1 may generate a CNF formula whose length is exponentially related to the length of the input formula.*

One may the tempted to argue that the exponential growth in the formula is caused by "redundant" clauses. Let us pursue this suspicion and convince ourselves that it is unfounded. A clause, in a CNF formula, is called redundant if it is implied by the remaining clauses. A single clause C *absorbs* clause D if every literal in C appears in D. Further, a clause D is logically implied by a single clause C if and only if C absorbs D or if D is a tautology. In a CNF formula therefore a clause will be redundant if it is absorbed by another. What makes inference, in propositional logic, so complex is the fact that redundancy can be deeply embedded. Fortunately, there is a simple principle that, on repeated application, reveals the embedded structure. This is the *principle of resolution*.

2.1 BASIC CONCEPTS OF PROPOSITIONAL LOGIC

In any CNF formula, suppose there are two clauses C and D with exactly one atomic proposition x_j appearing negated in C and posited in D. It is then possible to resolve C and D on x_j. We call C and D the *parents* of the *resolvent clause*, which is formed by the disjunction of literals in C and D, excluding x_j and $\neg x_j$. The resolvent is an implication of the parent clauses.

As an illustration of this principle consider the last line of formula (2.6).

$$\begin{aligned} C &= (\neg x_1 \vee x_3) \\ D &= (x_1 \vee \neg x_2 \vee x_3) \\ \hline & (\neg x_2 \vee x_3) \end{aligned}$$

The resolvent of C and D is the clause below the line. The resolution principle is a manifestation of case analysis, since if C and D have to be satisfied and (case 1) x_1 is true, then x_3 must be true; or if (case 2) x_2 is false then $(\neg x_2 \vee x_3)$ must be true. Hence every model for C and D is also a model for the resolvent. A *prime implication* of a CNF formula S, is a clause that is an implication of S that is not absorbed by any clause that is an implication of S, except by itself. In the above example, the prime implications are

$$\{(\neg x_1 \vee x_3), (\neg x_2 \vee x_3)\}$$

A simple procedure for deriving all prime implications of S is to repeatedly apply the resolution principle while deleting all absorbed clauses. If we are left with an empty clause, the formula S has no model. Otherwise, we will be left with all the prime implications. The correctness of this procedure will be established in Section 3.1.1, as will its computational complexity. This elegant principle of resolution therefore provides a complete inference mechanism for propositional logic based on purely syntactic techniques. However, akin to "truth tables", the principle of resolution is not a computationally viable inference method for all but small examples, in the general case. In the section, and indeed for much of this chapter, we will work with a weakened form of resolution, called "unit resolution", in which one of the parent clauses is always taken to be a clause with exactly one literal. This weaker principle is complete only on special classes of propositions that we will define.

Returning to the class of formulas (2.8) that proved Theorem 1, we see that no further resolution is possible and none of the clauses absorb each other. Therefore they are prime implications. This debunks the notion that "succinct" CNF representations of any formula can be guaranteed by the "distributive law" approach of Procedure 1. In order to obtain such "succinct" representations, we will have to resort to a "rewriting" technique that introduces additional atomic propositions corresponding to subformulas of the given formula. The two formulas will be equivalent in that every model for the original formula can be extended to a model for the rewritten one (with appropriate truth values for the new atomic propositions). Conversely, every model of the rewritten formula, when restricted to the original atomic propositions (projected), corresponds to a model of the original formula. Extended formulations have been the subject of study in integer programming for at least two decades now. The central theme of the monograph by Jeroslow [168] is on this interplay between reformulations in logic and their counterparts in integer programming.

That these rewriting techniques lead to CNF representations that are polynomially bounded in length was first noted by Tseitin [278] in a remarkable paper written in 1968. This rewriting technique was improved in later papers by several authors [23, 66, 297]. We will use the most recently published, and most efficient technique due to Wilson [297]. We will first illustrate the technique on an example and then state the general idea. Consider a small instance of the formulas that were the ruin of Procedure 1.

$$(x_1 \wedge x_2) \vee (x_3 \wedge x_4) \vee (x_5 \wedge x_6) \vee (x_7 \wedge x_8) \qquad (2.9)$$

Let us introduce new propositions $x_{12}, x_{34}, x_{56}, x_{78}$ that represent the four parenthetic conjunct subformulas. For each $x_{2j-1,2j}$ ($j = 1, 2, 3, 4$) we write the clauses

$$(\neg x_{2j-1,2j} \vee x_{2j-1}) \text{ and } (\neg x_{2j-1,2j} \vee x_{2j}) \qquad (2.10)$$

We also need a clause to knit the four subformulas together.

$$(x_{12} \vee x_{34} \vee x_{56} \vee x_{78}) \qquad (2.11)$$

Putting these clauses together we have rewritten the formula (2.9) in CNF using 12 atomic propositions and 9 constraints. Procedure 1 would have produced a CNF using only the original 8 atomic propositions but 16 clauses. The numbers are more dramatic for the general case (2.7). Procedure 1 obtained a CNF formula with $2n$ atomic propositions and 2^n clauses. The obvious extension of the rewriting technique introduced above would result in a CNF formula with $3n$ atomic propositions and $2n + 1$ clauses. Thus rewriting has brought the size down from exponential to linear for this class of examples. This result can be generalized for all formulas in propositional logic with a general rewriting technique described below.

Procedure 2: *Rewriting to CNF*

Input. A well-formed formula F with subformulas $\{F_i\}$.

2.1 BASIC CONCEPTS OF PROPOSITIONAL LOGIC

Output. A CNF formula S equivalent to F with literals F'_i and clause sets $C(F_i)$ corresponding to subformulas of F. $S = \bigwedge C(F)$.

Step 1. If F_i is a literal, $C(F_i) \leftarrow \emptyset$ and set $F'_i = F_i$.

Step 2. If $F_i = \bigwedge_{j=1}^{k} F_{i(j)}$, with all the $\{F'_{i(j)}\}$ and $C(F_{i(j)})$ constructed, define a new atomic proposition F'_i and the corresponding clause set $C(F_i)$ given by $\bigcup_{j=1}^{k} C(F_{i(j)})$ along with the new clauses,

$$\bigvee_{j=1}^{j=k} \neg F'_{i(j)} \vee F'_i$$

$$F'_{i(j)} \vee \neg F'_i, \quad j = 1, \ldots, k$$

Stop if $F = F_i$.

Step 3. If $F_i = \bigvee_{j=1}^{k} F_{i(j)}$, with all the $\{F'_{i(j)}\}$ and $C(F_{i(j)})$ constructed, define the clause set $C(F_i)$ by $\bigcup_{j=1}^{k} C(F_{i(j)})$ plus these clauses:

$$\bigvee_{j=1}^{k} F'_{i(j)} \vee \neg F'_i$$

$$\neg F'_{i(j)} \vee F'_i, \quad j = 1, \ldots, k$$

Stop if $F = F_i$.

Step 4. If $F_i = \neg F_k$ with F'_k and $C(F_k)$ constructed, set $C(F_i) = C(F_k)$ and $F'_i = \neg F'_k$. Stop if $F_i = F$.

Clearly, Procedure 2 constructs a CNF formula that is satisfiable if and only if the input formula is. Further, it is easy to prove by induction that the length of the clause sets constructed in steps 2,3 and 4 is, in each case, bounded by a constant factor of the length of the corresponding subformula. Hence we have the following remarkable "rewriting theorem".

Theorem 2 ([23, 297]) *Any formula in propositional logic can be rewritten to an equivalent CNF formula whose length is linearly related to the length of the original formula.*

Later in Section 2.2, we will encounter a spatial embedding of propositions in CNF. In that embedding it is possible to show that the CNF formula produced by Procedure 1 is the geometric projection of the formula produced by Procedure 2 (cf. [23]). Before we begin developing these remarkable connections between logic and the geometry of optimization, we need to set out the fundamentals of an alternate syntax for propositional logic that has been popularized in recent years by the Artificial Intelligence community. This is the syntax of rule-based systems.

2.1.3 Rules

A *rule system*, or *rule-based system*, is a type of propositional logic in which all the formulas are rules. A *rule* is an if-then statement of the form

$$(x_1 \wedge \cdots \wedge x_m) \to (y_1 \vee \cdots \vee y_n) \tag{2.12}$$

where $\to$ means "implies," and the *antecedents* $x_1, \ldots, x_m$ and the disjuncts $y_1, \ldots, y_n$ in the *consequent* are atomic propositions. The rule (2.12) is equivalent to the clause

$$\neg x_1 \vee \cdots \vee \neg x_m \vee y_1 \vee \cdots \vee y_n$$

Thus if there are no antecedents ($m = 0$), the rule simply asserts the consequent. If there is no consequent ($n = 0$), the rule denies the conjunction of the antecedents. Also, it is clear that any clause can be written as a rule, just by putting the positive literals in the consequent and treating the negative literals (stripped of their negations) as antecedents. If there is at most one disjunct in the consequent, the rule is called *Horn* and the corresponding clause is called a Horn clause. We will study this special structure in rules (and clauses) in great depth in this chapter.

Rules are often called "inference rules" in the AI literature, but they are not inference rules in the classical sense. They are simply formulas in a logical system, to which an inference rule may be *applied* to derive other formulas. In rule systems the most common inference rule is *modus ponens*, also known as *forward chaining*:

$$\begin{array}{c} (x_1 \wedge \cdots \wedge x_m) \to (y_1 \vee \cdots \vee y_n) \\ \to x_1 \\ \vdots \\ \to x_m \\ \hline \to (y_1 \vee \cdots \vee y_n) \end{array} \tag{2.13}$$

This means that one can infer the conclusion below the line from the premises above the line. An application of (2.12) is called *firing* the "rule" that is its first premise, perhaps due to an analogy with the "firing" of a synapse in the brain. Another popular inference rule is *modus tollens*, also

2.2 INTEGER PROGRAMMING MODELS

called *backward chaining*:

$$(x_1 \wedge \cdots \wedge x_m) \rightarrow (y_1 \vee \cdots \vee y_n)$$
$$y_1 \rightarrow$$
$$\vdots$$
$$\underline{y_n \rightarrow }$$
$$(x_1 \vee \cdots \vee x_m) \rightarrow \quad (2.14)$$

Modus ponens is obviously the same as a series of unit resolutions in which the unit clauses are positive. For example, the inference

$$(x_1 \vee x_2) \rightarrow y$$
$$\rightarrow x_1$$
$$\underline{\rightarrow x_2}$$
$$y$$

can be accomplished by resolving $\neg x_1 \vee \neg x_2 \vee y$ with x_1 to obtain $\neg x_2 \vee y$, and resolving the latter with x_2 to obtain y. Similarly, *modus tollens* is the same as a series of unit resolutions in which the unit clauses are negated.

Since unit resolution is not a complete inference method for propositional logic, forward and backward chaining are not complete for a rule system. The following valid inference, for example, cannot be obtained using these two inference rules:

$$x \rightarrow (y_1 \vee y_2)$$
$$y_1 \rightarrow z$$
$$\underline{y_2 \rightarrow z}$$
$$x \rightarrow z$$

But we will see that forward and backward chaining are a complete refutation method for a Horn rule system, since unit refutation is complete for Horn clauses.

2.2 Integer Programming Models

We now begin the true enterprise of this book, which is to apply the paradigms of mathematical programming to inference in logic. In order to take the first steps, we have to bridge the different worlds to which these subjects belong. In the previous section, we formulated the basic inference problems of propositional logic using symbolic valuations of propositions as either *True* or *False*. Mathematical programming, however, works with numerical valuations. Further, in order to usefully apply the methodology of mathematical programming to these inference problems, we will need to embed them in familiar forms. These "formulations" of inference in propositional logic are what we explore in this section. A reader, completely unfamiliar with the subject of mathematical programming, would do well to take a little time to browse in the Appendix as it contains a primer on some of the basic concepts of both the algebraic aspects of linear inequalities and corresponding geometry of polyhedra.

The solubility of systems of linear inequalities over the real (rational) numbers is closely related to the problem of *linear programming*. A standard approach is to deal with solubility as a special case of optimization over linear inequalities. *Integer programming* deals with the linear programming problems with the added restriction that some (possibly all) variables may take values restricted to the integers. By introducing suitable bounds on the variables (which are expressed by linear inequalities), it is possible to restrict variables to only take values in the nonnegative integers or even just values of 0 and 1. It is the latter "boolean" restriction that captures the semantics of propositional logic since the values of 0 and 1 may be naturally associated with *False* and *True*.

In integer programming, all the inequality constraints have to be satisfied simultaneously (in conjunction) by any feasible solution. It is natural therefore to attempt to formulate satisfiability of CNF formulas as integer programming problems with clauses represented by constraints and atomic propositions represented by 0-1 variables. Consider, for example, the single clause

$$x_2 \vee \neg x_3 \vee x_4$$

The satisfiability of this clause is easily embedded as solubility of an inequality over (0,1) variables as follows:

$$x_2 + (1 - x_3) + x_4 \geq 1$$

The satisfiability of the formula

$$(x_1) \wedge (x_2 \neg x_3 \vee x_4) \wedge (\neg x_1 \vee \neg x_4) \wedge (\neg x_2 \vee x_3 \vee x_5) \quad (2.15)$$

is equivalent to the existence of a solution to the system:

$$\begin{aligned} x_1 &\geq 1 \\ x_2 + (1 - x_3) + x_4 &\geq 1 \\ (1 - x_1) + (1 - x_4) &\geq 1 \\ (1 - x_2) + x_3 + x_5 &\geq 1 \\ x_1, \ldots, x_5 &= 0 \text{ or } 1 \end{aligned} \quad (2.16)$$

It is conventional in mathematical programming to clear all the constants to the right-hand side of a constraint. Thus a clause C_i is represented by

2.2 INTEGER PROGRAMMING MODELS

$a_i x \geq b_i$, where for each j, a_{ij} is $+1$ if x_j is a positive literal in C_i, is -1 if $\neg x_j$ is a negative literal in C_i, and is 0 otherwise. Also, b_i equals $1 - n(C_i)$, where $n(C_i)$ is the number of negative literals in C_i. We shall refer to such inequalities as *clausal*. So the linear inequalities (2.16) converted to clausal form are given by

$$x_1 \geq 1$$
$$x_2 - x_3 + x_4 \geq 0$$
$$-x_1 - x_4 \geq -1$$
$$-x_2 + x_3 + x_5 \geq 0$$
$$x_1, \cdots, x_5 = 0 \text{ or } 1$$

In general, satisfiability in propositional logic is equivalent to solubility of

$$Ax \geq b, \ x \in \{0,1\}^n \qquad (2.17)$$

where the inequalities of $Ax \geq b$ are clausal. Notice that A is a matrix of 0's and ± 1's, and each b_i equals 1 minus the number of -1's in row i of the matrix A. We are therefore looking for an extreme point of the unit hypercube in $\Re^n$ that is contained in all the half-spaces defined by the clausal inequalities. This is a *spatial* or geometric embedding of inference.

2.2.1 Optimization and Inference

The intersection of clausal half-spaces defines a convex polyhedron. If the box constraints $0 \leq x_j \leq 1$ ($j = 1, 2, ..., n$) are added, we obtain a bounded polyhedron or a polytope. Thus satisfiability is a special case of checking for a integer lattice (Z^n) point in a polytope defined implicitly by linear inequalities. The latter is exactly the feasibility problem of integer (linear) programming.

In optimizing a linear objective function over constraints, defined by linear inequalities and integer restrictions on variables, one must certainly also contend with checking the feasibility of the constraints. Thus it is natural to consider the optimization version of integer programming as a more general model than the feasibility version. This is also borne out by a simple formulation trick that introduces an extraneous variable called an *artificial variable* to pose the feasibility version of integer programming as a special case of optimization. Consider, for example, the CNF formula (2.15) whose satisfiability was posed as a feasibility question in integer programming (2.16). Satisfiability of (2.15) can also be tested by solving the optimization problem

$$\begin{aligned}
\min \quad & x_0 \\
\text{s.t.} \quad & x_0 + x_1 && \geq 1 \\
& x_0 \phantom{{}+x_1} + x_2 - x_3 + x_4 && \geq 0 \\
& x_0 - x_1 \phantom{{}+x_2 - x_3} - x_4 && \geq -1 \\
& x_0 \phantom{{}+x_1} - x_2 + x_3 \phantom{{}+x_4} + x_5 && \geq 0 \\
& x_j \in \{0,1\}, \quad j = 0, 1, \ldots, 5
\end{aligned}$$

The original formula (2.15) is satisfiable if and only if the optimization problem above is solved with x_0 at 0. The variable x_0 is called the *artificial* variable and the optimization problem a *phase I construction* in integer programming terminology. The optimization formulation is useful since the objective function provides a mechanism for directed search of truth assignments (0-1 values for $x_1, \ldots, x_n$) that satisfy the formula. In general, the phase I construction is of the form

$$\begin{aligned}
\min \quad & x_0 \\
\text{s.t.} \quad & x_0 e + Ax \geq b \\
& x_j \in \{0,1\}, \quad j = 0, 1, \ldots, n
\end{aligned}$$

where $Ax \geq b$ represents the original clausal inequalities and e is a column of 1's.

Inference in the form of implications can also be modeled as integer programming problems. An indirect route would be to transform the implication question to a satisfiability problem and then follow the Phase I construction discussed above. However, direct formulations are also possible. Let us consider an implication problem of the form: Does a formula S_1 imply another formula S_2? For simplicity let us first assume that S_1 is in CNF and S_2 is given by a single clause C. The corresponding optimization model for testing the implication is

$$\begin{aligned}
\min \quad & cx \\
\text{s.t.} \quad & Ax \geq b \\
& x \in \{0,1\}^n
\end{aligned}$$

where c is the incidence vector of clause C, and $Ax \geq b$ are the clausal inequalities representing S_1. If this optimization yields a minimum value $1 - n(C)$ or larger, we have established the valid implication of S_2 by S_1. Otherwise the implication does not hold.

Now, let us consider a situation where S_1 represents a consistent knowledge base in CNF with clausal inequalities $Ax \geq b$ and $\neg S_2$ a general formula in CNF with clausal inequalities $Bx \geq d$. To test if S_2 is logically

2.2 INTEGER PROGRAMMING MODELS

implied by S_1 we solve

$$\begin{aligned} \min \quad & x_0 \\ \text{s.t.} \quad & x_0 e + Bx \geq d \\ & Ax \geq b \\ & x \in \{0,1\}^n \end{aligned}$$

S_1 implies S_2 if and only if the minimum value of x_0 is 1. Notice that the artificial x_0 is not associated with $Ax \geq b$ since we have assumed that S_1 is consistent (satisfiable).

Rewriting inference problems in propositional logic as integer programming problems is easily accomplished. Since integer programming problems are reputed to be hard to solve in general, an initial reaction could be that we have only made a hard problem even harder. The proper response is that there are many integer programs that are easy to solve once special mathematical structure is detected in the model. As we shall see in this chapter and the next one, this appears to be true of the integer programming models of inference in propositional logic.

2.2.2 The Linear Programming Relaxation

We just saw that inference in propositional logic can be posed as a 0-1 integer programming problem. A fundamental technique in integer programming is to examine its continuous relaxation as a linear program. Properties of this linear programming relaxation often suggest computational strategies for solving the integer model. The linear programming relaxation of a 0-1 integer program is obtained by relaxing the condition that each variable takes value 0 or 1 to the weaker condition that each variable takes value in the interval [0,1]. Thus the linear programming relaxation of the 0-1 formulation (2.17) is given by

$$\begin{aligned} x_1 & & & & & \geq 1 \\ & x_2 & -x_3 & +x_4 & & \geq 0 \\ -x_1 & & & -x_4 & & \geq -1 \\ & -x_2 & -x_3 & +x_4 & & \geq 0 \\ & & 0 \leq x_1, ..., x_5 & & & \leq 1 \end{aligned} \qquad (2.18)$$

Clearly every 0-1 incidence vector of satisfying truth assignments for the underlying CNF formula is a solution to (2.18). But $(1, \frac{1}{2}, \frac{1}{2}, -1, \frac{1}{2})$ is also a solution to (2.18) and it is not clear what information such a noninteger solution provides vis-à-vis the inference question with which we started. Therefore a pertinent question is whether this linear programming relaxation retains sufficient structure to be a useful representation of the inference problem. We will, of course, argue that the answer to this question is in the affirmative.

First we will see that the linear programming relaxation encodes the power of syntactic reductions of propositions achieved by forward and backward chaining (cf. Section 2.1.3). In the next few sections we will also see that the relaxation provides a template for understanding special structures (syntactic restrictions) in propositions that permit efficient computation of inference. And further in the next chapter we will use the linear programming relaxation as the structure to build powerful inference algorithms for full propositional logic.

In Section 2.1.3, where we discussed forward and backward chaining for the rule systems, we also noted that forward chaining may be interpreted as unit resolution on unit clauses. Likewise backward chaining is unit resolution on negated unit clauses. Let us denote by *unit resolution* the combination of both forward and backward chaining.

Procedure 3: *Unit Resolution*

Input. A CNF formula $S = \bigwedge_{i=1}^{m} C_i$.

Output. Reduced CNF formula equivalent to S.

Step 1. If S contains no unit clause then stop and return S. Else if S contains two conflicting unit clauses $C_k = l_j$ and $C_l = \neg l_j$, stop and return $S = \{\}$.

Step 2. Let $C_s = l_j$ be a unit clause on literal l_j.

Step 3. Delete all clauses in S that contain l_j.

Step 4. Delete all appearances of $\neg l_j$ in the remaining clauses, and go to step 1.

If the procedure returns an empty formula S then unit resolution has succeeded in finding a satisfying truth assignment for S. On the other hand if it returns $S = \{\}$ (i.e., S equals the empty clause) then unit resolution has proved that S is unsatisfiable. However, in most cases, the procedure will simply return a nonempty reduced formula S. On our example CNF formula (2.15)

$$(x_1) \wedge (x_2 \vee \neg x_3 \vee x_4) \wedge (\neg x_1 \vee \neg x_4) \wedge (\neg x_2 \vee x_3 \vee x_5)$$

the procedure returns the reduced form

$$(x_2 \vee \neg x_3) \wedge (\neg x_2 \vee x_3 \vee x_5) \tag{2.19}$$

2.2 INTEGER PROGRAMMING MODELS

The correctness of the unit resolution is evident. The literal of any unit clause has to evaluate to *True* in any satisfying truth assignment. Unit resolution enforces this and simplifies the formula by deleting clauses that have consequently been satisfied and erases a literal evaluated *False* from the remaining clauses. Now let us see that these actions are implicitly encoded in the linear inequalities of the linear programming relaxation. In (2.18) the inequality corresponding to the unit clause (x_1) is given by $x_1 \geq 1$. However, there is an explicit upper bound $x_1 \leq 1$ on x_1. Thus x_1 is forced to have value exactly 1. Substituting this value for x_1 makes the first inequality redundant and the third inequality reduces to $-x_4 \geq 0$. Now combining $-x_4 \geq 0$ and $x_4 \geq 0$ the variable x_4 is forced to have value 0 and the linear inequalities are reduced to

$$\begin{aligned} x_2 - x_3 &\geq 0 \\ -x_2 + x_3 + x_5 &\geq 0 \\ 0 \leq x_2, x_3, x_5 &\leq 1 \end{aligned} \quad (2.20)$$

which is exactly the linear programming relaxation of the reduced formula (2.19). Also, if at any stage of unit resolution, two conflicting unit clauses (x_j) and $(\neg x_j)$ are obtained, the corresponding implied inequalities are $x_j \geq 1$ and $-x_j \geq 0$, which are in conflict. Consequently, in such a situation the linear program is inconsistent (infeasible). We have therefore justified our claim that the linear programming relaxation encodes the power of unit resolution.

A partial converse statement is also possible. This is because the structure of clausal inequalities permits a trivial fractional solution of the linear programming relaxation of a CNF formula devoid of unit clauses.

Lemma 1 *The linear programming relaxation of a unit clause free CNF formula is always feasible.*

Proof: The trivial functional solution $x_j = \frac{1}{2}$ for all j is feasible for such a linear program. To see this let C be an arbitrary clause and let $P(C)$ denote the positive literals in C and $N(C)$ the negative literals. Our assumption ensures that $|P(C) \cup N(C)| \geq 2$. The clausal inequality corresponding to C is

$$\sum_{x_j \in P(C)} x_j - \sum_{x_j \in N(C)} x_j \geq 1 - |N(C)| \quad (2.21)$$

Setting $x_j = \frac{1}{2}$ for all j yields a left-hand side with value

$$\begin{aligned} \tfrac{1}{2}(|P(C)| - |N(C)|) &= \tfrac{1}{2}(|P(C) \cup N(C)| - 2|N(C)|) \\ &\geq 1 - |N(C)| \end{aligned}$$

as required. □

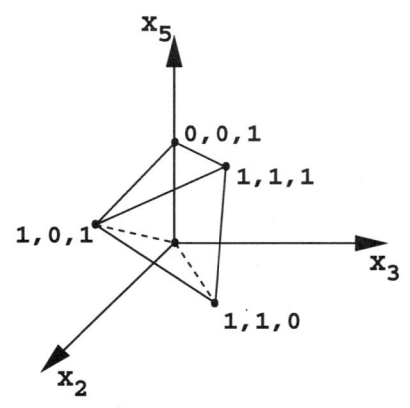

30   CHAPTER 2   PROPOSITIONAL LOGIC: SPECIAL CASES

Figure 2.1: The relaxation polytope.

So we have shown that

Theorem 3 ([23]) *A proposition S has a unit refutation (unsatisfiability of S proved by unit resolution) if and only if the linear programming relaxation of S is infeasible.*

For satisfiable propositions, however, the power of the linear programming relaxation exceeds that of unit resolution. The feasibility of the linear program may sometimes imply satisfiability even when unit resolution is unable to do so. To illustrate this point let us reconsider our example formula (2.15) in its reduced form (2.19) and the corresponding linear programming relaxation (2.20). We can actually map the feasible region as indicated in Figure 2.1.

Notice that all the corner (extreme) points of the feasible polytope are integral (0-1 valued). All we need to see is one integral point in the polytope to be convinced of the satisfiability of (2.15). And if we use a simplex method to solve the relaxation we are guaranteed a corner solution, which would prove the formula satisfiable. On the other hand, $(x_2, x_3, x_5) = (\frac{1}{2}, \frac{1}{2}, \frac{1}{2})$ is also a perfectly legitimate solution to the linear program from which we glean little information about the inference question.

To summarize, we have seen that the linear programming relaxation encodes the power of unit resolution. In fact on refutable propositions, the power of unit resolution and the relaxation are identical. For satisfiable propositions we may get lucky while solving the linear programming relaxation and hit on an integer solution and thus prove satisfiability.

Incidentally, the fact that the feasible region for the example above has all integer extreme points may seem to be a fluke. We happened to

2.3 HORN POLYTOPES

choose an example from a very special class of propositions called *balanced propositions* whose relaxations are integral polytopes. We will study balanced propositions in Section 2.7. In the next section we study Horn clause systems. This is a different special structure for which the linear programming relaxation is a complete representation of the inference (satisfiability) question even though the associated polytope may not be integral.

2.3 Horn Polytopes

A Horn rule was introduced in Section 2.1.3 as an implication rule with the consequent (head) either empty or a single atom. The equivalent clausal definition is that the clause must contain at most one positive literal. A Horn (or *definite*) clause system is a CNF proposition in which all clauses are Horn. Horn clause systems have been popularized by their use in expert systems and logic programming languages. They represent highly structured propositions for which satisfiability can be solved in linear time by a restricted form of unit resolution. And as we shall see, their linear programming relaxations reveal useful mathematical structure.

2.3.1 Horn Resolution

Consider the Horn clause system

$$(x_3) \wedge (x_1 \vee \neg x_2) \wedge (\neg x_1 \vee \neg x_3) \tag{2.22}$$

Unit resolution would begin with the unit positive clause (x_3), set x_3 to true, and simplify the formula to

$$(x_1 \vee \neg x_2) \wedge (\neg x_1) \tag{2.23}$$

Notice that the reduced formula is still Horn.

Unit resolution would continue with the negative unit clause. However, that is unnecessary since a Horn clause system with no unit positive clause is trivially satisfiable. Each clause must contain at least one negated literal (either a unit negative or a clause with two or more literals of which at most one is positive). Hence setting all remaining atoms to false satisfies the formula. In the example $x_1 = x_2 = False$ works. Thus we have formulated a complete inference procedure for Horn clause systems. We run unit resolution on only unit positive clauses. If unsatisfiability has not been detected we declare the formula satisfiable and set the remaining atoms to false. A simple property of Horn clause systems that we have exploited is that they are closed under deletion of literals and clauses.

The restricted form of unit resolution that we have described can be implemented to run in time that is a constant multiple of the length of the proposition (total number of occurrences of literal in the proposition). The linear-time implementation of Horn resolution due to Dowling and Gallier [87] (see also [217]) uses a multigraph representation of the proposition.

Given a Horn proposition $S = \wedge_{i=1}^{m} C_i$ we construct the graph representation G_S as follows.

- The nodes of G_S are $\{x_1, x_2, ..., x_n, t, f\}$, that is, the atomic propositions in S plus two special nodes t and f.

- The arcs of the directed graph G_S are defined by the clauses of G_S. Each arc a carries a label $l(a)$, which indicates the clause that defined it.

    – If $C = x_j$ is a unit positive clause then we introduce an arc (t, x_j) in G_S with label $l(t, x_j) = C$.

    – If $C = \neg x_{j1} \vee \cdots \vee \neg x_{jl}$ is a negative clause then we introduce arcs (x_{jt}, f) for $t = 1, 2, \ldots, k$ with corresponding labels $l(x_{jt}, f) = C$.

    – If $C = x_l \vee \neg x_{j1} \vee, \cdots, \vee \neg x_{jk}$ is a definite clause ($k \geq 1$) then we introduce the arcs (x_{jt}, x_l) for $t = 1, 2, ..., k$ with corresponding labels $l(x_{jt}, f) = C$.

The algorithm also maintains a clause counter $\mu(i)$ for each clause C_i ($i = 1, 2, ..., m$), which represents the number of negative literals remaining in the clause as resolution proceeds. When $\mu(i)$ reaches 0 a unit positive clause is revealed and the corresponding atom is set to true. We will place a node of G_S in the set $\mathcal{T}$ when the corresponding atom is forced to take value *True*. Initially only node $t \in \mathcal{T}$. The algorithm stops if at any stage node f is placed into $\mathcal{T}$, because we will have produced an inconsistency (empty clause) and hence will have proved the proposition unsatisfiable. Otherwise the algorithm terminates with a satisfying solution in which the nodes in $\mathcal{T}$ are set to *True* and the rest to *False*.

Procedure 4: *Horn Resolution*

Step 0. Construct the labeled multigraph G_S. Initialize clause counters $\mu_i \leftarrow$ number of negative literals in C_i. Let $\mathcal{T} \leftarrow \{t\}$.

Step 1. Pick a node $\alpha \in \mathcal{T}$ that has not been scanned. If all nodes in $\mathcal{T}$ have been scanned, stop and report $\mathcal{T}$.

2.3 HORN POLYTOPES

Step 2. (Scan α.) For each arc (α, β) of G_S, do the following. Let $\mu(i) \leftarrow \mu(i) - 1$, where $C_i = l(\alpha, \beta)$. If $\mu(i) = 0$ and $\beta = f$, stop and report "S unsatisfiable"; otherwise let $\mathcal{T} \leftarrow \mathcal{T} \cup \{\beta\}$.

Step 3. Repeat Step 1.

As the procedure examines each arc of G_S at most once, the complexity is linear in the number of arcs. But the number of arcs is no larger than the total number of occurrences of literals in S. Hence this is a linear-time algorithm.

It should be evident that for a satisfiable S, the satisfying truth assignment found by Horn resolution sets only those atoms to *True* that must be *True* in all models of S. Therefore, under the order relation $False < True$ we have proved constructively the following theorem.

Theorem 4 *A satisfiable Horn proposition has a unique minimal model.*

This theorem also follows from a very strong closure property satisfied by the set of models of a Horn proposition.

Lemma 2 *If T_1 and T_2 are models for a Horn proposition S, then so is $T_1 \wedge T_2$ (an atom is set to true in $T_1 \wedge T_2$ if and only if it is true in both T_1 and T_2)*

Proof: Suppose T_1 and T_2 are models and $T_1 \wedge T_2$ falsifies a clause C of the proposition. If C is a negative clause then $T_1 \wedge T_2$ must evaluate all atoms in C to true. Consequently, both T_1 and T_2 falsify C, which is a contradiction. If C contains a positive literal x_k then one of T_1 and T_2 (say, T_1) must set x_k to false, for otherwise $T_1 \wedge T_2$ would satisfy C. But T_1 satisfies C, and so C must contain a negative literal $\neg x_j$ such that T_1 falsifies x_j. But then $T_1 \wedge T_2$ falsifies x_j, which means $T_1 \wedge T_2$ satisfies C, again a contradiction. $\square$

The closure of satisfying truth assignments of propositions under the "$\wedge$" operation is a characteristic of Horn propositions. Define S_1 and S_2 to be equivalent propositions if they are built on the same ground set of atoms and if they have the same set of models. A proposition S is called *H-equivalent* (or Horn equivalent) if it is equivalent to some Horn proposition.

If a proposition S is H-equivalent there must be a Horn proposition H-equivalent to S. From Lemma 2 it follows that the set of models of H (and hence of S) is closed under the $\wedge$ operation. Now suppose that S is a proposition whose models are closed under the $\wedge$ operation and that S is not Horn. S must contain a clause C of the form $x_k \vee x_l \vee D$, where D is a disjunction (possibly empty) of literals. If every model of S satisfies $x_k \vee D$ we can delete x_l from C and obtain an equivalent proposition. Likewise if $x_l \vee D$ is satisfied by all models, we delete x_k from C. We continue this process, gradually whittling the excess positive literals from clauses in S, until we obtain a Horn proposition equivalent to S or we are unable to apply the deletion criteria. If we are unable to apply the deletion criteria it is because we have two models T_1 and T_2 and a clause $x_k \vee x_l \vee D$ such that

$$T_1(x_k) = True, \quad T_1(x_l) = T_1(D) = False$$
$$T_2(x_l) = True, \quad T_2(x_k) = T_2(D) = False$$

But then $T_1 \wedge T_2$ cannot satisfy C, which contradicts our assumption that the models of S are closed under the $\wedge$ operation. Therefore the deletion process does not get stuck and S is reduced to an equivalent Horn formula. We have proved the following.

Theorem 5 *The set of models of a proposition is closed under the $\wedge$ operation if and only if the proposition is H-equivalent.*

We have proved a slightly stronger result. Let $\mathcal{L}$ be a collection of 0-1 n-vectors. We say that $\mathcal{L}$ is *lower comprehensive* if given any x and y in $\mathcal{L}$ we are guaranteed that $z = \min\{x, y\}$ is also in $\mathcal{L}$. Now, interpreting a 0 as *False* and 1 as *True*, we observe that the proof of the theorem actually says the following.

Corollary 1 *A collection $\mathcal{L}$ of 0-1 n-vectors is lower comprehensive if and only if $\mathcal{L}$ is the collection of models of some Horn proposition.*

The rich syntactic and semantic structure of Horn propositions that we have just examined has a natural manifestation in the spatial embedding of these propositions. The structure is revealed as special integrality properties and shape characteristics of the linear programming relaxation polytopes of Horn propositions, that is, Horn polytopes.

2.3.2 The Integer Least Element of a Horn Polytope

We have seen that a satisfiable Horn proposition S has a unique minimal model found by a restricted form of unit resolution. We have also seen that unit resolution is encoded in the linear programming relaxation of the 0-1 integer programming formulation of satisfiability of S. These facts collude to prove a fundamental structural property of Horn polytopes, the linear programming relaxations of Horn propositions.

A least element of a polyhedron P is a point $x_{\min} \in P$, all of whose components are no larger than the corresponding components of any x

2.3 HORN POLYTOPES

in P. Clearly, not every convex polyhedron has a least element. As we shall see, however, every nonempty Horn polytope has a least element. Further, this least element is integer valued and corresponds precisely to the unique minimal model of the Horn proposition. This remarkable confluence of syntactics and geometry will pay rich dividends in understanding and generalizing the special structure in the class of Horn propositions.

Suppose we have a nonempty polyhedron P defined by a system of linear inequalities
$$P = \{x \in \Re^n \mid A'x \geq b', x \geq 0\}$$
with the qualification that each row of A' has at most one positive component. Now if $b' \leq 0$ it is evident that P contains a least element $x_{\min} = 0$. If not, choose b'_i, the largest positive component of b'. Since P is non-empty we are assured that the row A'_i of A' has exactly one positive entry, say, $a'_{ij} > 0$. The ith inequality of the system $Ax' \geq b'$ has the form

$$x_j \geq \frac{1}{a'_{ij}} \left(b'_i - \sum_{k \neq j} a'_{ik} x_k \right) \tag{2.24}$$

The nonpositivity of a'_{ik} ($k \neq j$) and nonnegativity of the x_k's imply the positive lower bound
$$x_j \geq b'_i / a'_{ij} \tag{2.25}$$
Let x^* denote a lower bound on all x in P. So we can take $x^*_j = b'_i/a'_{ij}$ and $x'_k = 0$ ($k \neq j$). Now we rewrite the inequalities of P using the lower bound substitution $\tilde{x} = x - x^*$. Thus
$$\tilde{P} = \{\tilde{x} \in \Re^n \mid A'\tilde{x} \geq \tilde{b} = (b' - A'x^*), \tilde{x} \geq 0\}$$
is a translation of P. We repeat the process and escalate the lower bound x^* until $\tilde{b}$ becomes nonpositive. When that happens, and it must since the lower bound x^* cannot increase indefinitely, x^* will be the least element of P. Thus we have constructively shown the following.

Theorem 6 ([70]) *A non-empty convex polyhedron of the form*
$$P = \{x \mid A'x \geq b', x \geq 0\} \tag{2.26}$$
has a least element if each row of A' has at most one positive component.

Suppose now that the positive entries of A' are all $+1$s and that the initial right-hand-side vector b' has all integer components. It is easy to verify that the lower bound escalation scheme used to prove Theorem 6 actually proves the existence of an integral least element of P.

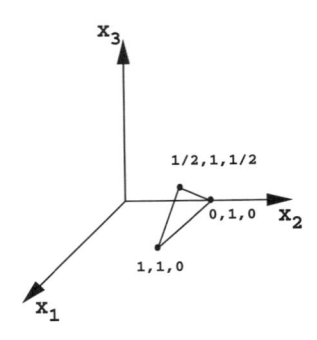

Figure 2.2: A Horn polytope.

Corollary 2 ([41]) *A convex polyhedron of the form (2.26) has an integral least element for any integral b' for which P is non-empty, provided each row of A' has at most one positive component and all positive components of A' are equal to 1.*

This corollary is all we need for establishing the existence of an integral least element of a Horn polytope. Consider the Horn proposition

$$(x_1 \vee \neg x_2 \vee \neg x_3) \wedge (x_2) \wedge (\neg x_1 \vee \neg x_3)$$

The corresponding Horn polytope is defined by the linear inequalities

$$\begin{aligned} x_1 - x_2 - x_3 &\geq -1 \\ x_2 &\geq 1 \\ -x_1 \phantom{{}-x_2} - x_3 &\geq -1 \\ -x_j &\geq -1, \quad j = 1,2,3 \\ x_j &\geq 0, \quad j = 1,2,3 \end{aligned}$$

Clearly each inequality has at most one positive coefficient on the left-hand side (this follows from the syntactic definition of Horn formulas) and further all positive left-hand-side coefficients are equal to 1. Hence the conditions of Corollary 2 are met and we are assured the existence of an integral least element. Figure 2.2 shows the structure of the Horn polytope for this example. Notice that although the polytope has an integral least element in the extreme point $(0,1,0)$, the polytope is itself not integral since it also has the fractional extreme point $(\frac{1}{2}, 1, \frac{1}{2})$.

The argument on the example above serves for all Horn polytopes since the only fact we used was that each clause has at most one positive literal,

2.3 HORN POLYTOPES

a property belonging to all Horn propositions. Combining this observation with Corollary 2 we have the following theorem.

Theorem 7 ([44, 171]) *A nonempty Horn polytope (the linear programming relaxation of satisfiability of a Horn proposition) has an integral least element. Further, this least element is the incidence vector of the unique minimal model of the proposition found by Horn resolution.*

Of course, one could now find the integral least element by an optimization technique. In fact for any vector $c \in \Re^n$, all of whose components are positive, we know that the linear program

$$\min\{c^T x \mid x \in \text{Horn polytope}\} \qquad (2.27)$$

is optimized uniquely by the integral least element. This optimization model is unlikely to suggest an inference algorithm anywhere near as efficient as Horn resolution. However, duality theory for linear programming applied to optimization over a Horn polytope reveals remarkable structural properties of unit resolution proofs derived from the corresponding Horn proposition.

2.3.3 Dual Integrality of Horn Polytopes

The clausal satisfiability problem of becomes a linear programming problem when all the clauses are Horn clauses. In this case the linear programming dual has an interesting interpretation, discovered by Jeroslow and Wang [171]. When the clauses are unsatisfiable, the values of the dual variables are proportional to the number of times the corresponding clauses serve as premises in a refutation.

A Horn clause is a clause with at most one positive literal. Recall that in the unit resolution algorithm, at least one parent of every resolvent must be a unit clause. In positive unit resolution, the unit clause must be a positive literal. When unit resolution is applied to a Horn set, a Horn set results. We saw that positive unit resolution and hence linear programming can check Horn sets for satisfiability. This is basically because the resolvent of a clause with a unit clause is simply the sum of the two clauses, when the clauses are written as 0-1 inequalities.

Now for the interpretation of the dual. First, the satisfiability problem can be written as a 0-1 problem,

$$\begin{aligned} \min \quad & x_0 \\ \text{s.t.} \quad & x_0 e + Ax \geq a \\ & x_j \in \{0, 1\}, \text{ all } j \end{aligned} \qquad (2.28)$$

38 CHAPTER 2 PROPOSITIONAL LOGIC: SPECIAL CASES

The linear relaxation of this problem is

$$\begin{array}{ll} \min & x_0 \\ \text{s.t.} & x_0 e + Ax \geq a \quad (u) \\ & -x \geq -e \quad\quad\quad (v) \\ & x \geq 0 \end{array} \qquad (2.29)$$

and its dual is

$$\begin{array}{ll} \max & ua - v \\ \text{s.t.} & ue \leq 1 \\ & uA - v \leq 0 \\ & -u \leq 0 \\ & -v \leq 0 \end{array} \qquad (2.30)$$

The dual solution indicates how many times each clause is "used" in a resolution proof of unsatisfiability, in the following sense. Each of the original clauses is used once to obtain itself. Furthermore, if an original clause C is used u_1 and u_2 times, respectively, to obtain the parents of some resolvent in the proof, it is used $u_1 + u_2$ times to obtain the resolvent.

Theorem 8 ([171]) *Let $Ax \geq a$ in (2.28) represent an unsatisfiable set of Horn clauses. Then if (u, v) is any optimal extreme point solution of the dual (2.30), there is an integer N and a refutation proof of unsatisfiability such that Nu_i is the number of times that each clause i is used to obtain the empty clause in the proof.*

Jeroslow and Wang [171] actually state a slightly different result in which some x_j is minimized subject to a satisfiable Horn set $Ax \geq a$. If x_j follows from $Ax \geq a$, the dual solution is integral and indicates (without multiplication by N) the number of times each clause is used in a resolution proof of x_j. This result, however, does not apply to all Horn satisfiability problems.

Example. Consider the following unsatisfiable Horn clauses:

$$\begin{array}{c} x_1 \\ \neg x_1 \vee x_2 \\ \neg x_2 \end{array} \qquad (2.31)$$

To check for satisfiability by minimizing a single variable, one can add an artificial variable x_0:

$$\begin{array}{ll} \min & x_0 \\ \text{s.t.} & x_0 + x_1 \geq 1 \quad (u_1) \\ & x_0 - x_1 + x_2 \geq 0 \quad (u_2) \\ & x_0 \quad\quad - x_2 \geq 0 \quad (u_3) \end{array} \qquad (2.32)$$

2.3 HORN POLYTOPES

But the clauses are no longer Horn, and the original Jeroslow-Wang result cannot be applied. In fact, the optimal solution is $(x_0, x_1, x_2) = (\frac{1}{3}, \frac{2}{3}, \frac{1}{3})$, and the optimal dual solution $u = (u_1, u_2, u_3) = (\frac{1}{3}, \frac{1}{3}, \frac{1}{3})$ is nonintegral. Theorem 8 can be applied, however, because (2.31) is Horn. It states that for some N, Nu gives the number of times each clause is used to obtain the empty clause. The refutation is obvious: resolve the first two clauses to obtain x_2, and resolve x_2 with the third clause to obtain the empty clause. So each clause is used once and $N = 3$. □

Example. For a slightly more interesting illustration, suppose that:

1. A real estate development firm wishes to build a shopping center.
2. Marketing studies suggest that the center must be conjoined with a supermarket if it is to be successful.
3. The traffic generated by both a shopping center and supermarket will require a new access road.
4. The firm can raise enough investment capital to build at most two of the three projects (shopping center, supermarket, access road).

The undertaking is obviously infeasible, but it is instructive to apply Theorem 8. If x_j indicates that project j will be funded (with $j = 1, 2, 3$ for shopping center, supermarket, and access road, respectively), the desiderata can be written in rule form and Horn clausal form as follows:

1. $\rightarrow x_1$ $\qquad\qquad$ x_1
2. $x_1 \rightarrow x_2$ $\qquad\qquad$ $\neg x_1 \vee x_2$
3. $(x_1 \wedge x_2) \rightarrow x_3$ $\qquad\qquad$ $\neg x_1 \vee \neg x_2 \vee x_3$
4. $(x_1 \wedge x_2 \wedge x_3) \rightarrow$ $\qquad\qquad$ $\neg x_1 \vee \neg x_2 \vee \neg x_3$

A resolution refutation appears in Figure 2.3. Note that clauses 1, 2, 3, and 4 are used 4, 2, 1, and 1 times, respectively, to obtain the empty clause. Theorem 8 is applied to the minimization problem,

$$\begin{aligned}
\min \quad & x_0 \\
\text{s.t.} \quad & x_0 + x_1 \geq 1 & (u_1) \\
& x_0 - x_1 + x_2 \geq 0 & (u_2) \\
& x_0 - x_1 - x_2 + x_3 \geq -1 & (u_3) \\
& x_0 - x_1 - x_2 - x_3 \geq -2 & (u_4)
\end{aligned}$$

An optimal dual solution is $u = (u_1, u_2, u_3, u_4) = (\frac{1}{2}, \frac{1}{4}, \frac{1}{8}, \frac{1}{8})$. So for $N = 8$, Nu gives the desired result. □

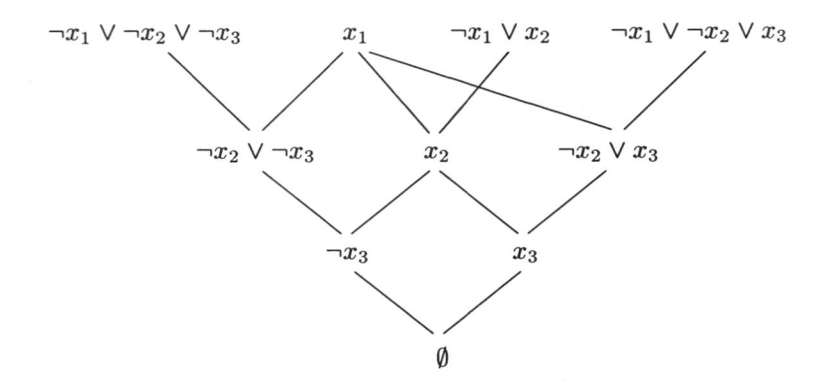

40   CHAPTER 2   PROPOSITIONAL LOGIC: SPECIAL CASES

```
¬x₁ ∨ ¬x₂ ∨ ¬x₃      x₁        ¬x₁ ∨ x₂        ¬x₁ ∨ ¬x₂ ∨ x₃

        ¬x₂ ∨ ¬x₃         x₂          ¬x₂ ∨ x₃

                    ¬x₃          x₃

                         ∅
```

Figure 2.3: A unit resolution proof of unsatisfiability.

In the way of sensitivity analysis, the fact that each $u_i > 0$ indicates that, in at least one proof of infeasibility, every clause is essential. The clauses i corresponding to larger u_i's might be said to be more important, in the sense that they are used more often in the proof.

Unfortunately, the dual multipliers u_i do not in general encode the structure of a refutation proof and therefore do not represent a complete dual solution. For instance, the dual multipliers $u = (\frac{1}{3}, \frac{1}{3}, \frac{1}{3})$ obtained for (2.32) are consistent with two refutations: one in which the first two clauses of (2.31) are resolved first, and one in which the last two clauses are resolved first.

Proof of Theorem 8. Let (u, v) be an optimal extreme point solution of the dual (2.30). Then (u, v) satisfies $n + m$ linearly independent constraints of (2.30) as equations. Furthermore, $ue = 1$ must be one of these $n + m$ equations. For if it were not, the equations would all have a right-hand side of zero, which implies $(u, v) = (0, 0)$, due to their independence. But this is impossible, because the objective function has value 1.

So one can write (2.30) as

$$\begin{array}{ll} \max & ua - v \\ \text{s.t.} & ue = 1 \\ & [u\ v]B = 0 \\ & [u\ v]N \geq 0 \end{array} \qquad (2.33)$$

where $A = [B\ N]$ after rearrangement of columns, and

$$\left[\begin{array}{c|c} e & B \\ 0 & \end{array} \right] \qquad (2.34)$$

is an $(n+m) \times (n+m)$ nonsingular matrix.

One can now show that the solution (u, v) has the interpretation stated above. First, since (u, v) is feasible and optimal in the dual, $uA - v \leq 0$ and $ua - ve \geq 1$. It follows that the constraints $Ax \geq a, -x \geq -1$ are infeasible, which means that the positive unit resolution algorithm yields the empty clause when applied to the clauses in this constraint set. Assign to each clausal inequality i in $Ax \geq a$ a weight w_i equal to number of times that clause is used to obtain the empty clause. Then since each resolvent is the sum of its parents, the weighted sum of the clausal inequalities in $Ax \geq a$, using the weights w_i, yields $0 \geq 1$. In particular, $[w\ 0]B = 0$. Now if $u' = w/\Sigma_i w_i$ and $v' = 0$, then

$$[u'\ v'] \left[\begin{array}{c} e \\ 0 \end{array} \middle| B \right] = [1\ 0]$$

Because the matrix (2.34) is nonsingular, the above has a unique solution, which means that $(u', v') = (u, v)$. The theorem follows. □

2.4 Quadratic and Renamable Horn Systems

The simple syntactic restriction of allowing no more than two literals per clause defines the class of *quadratic propositions*. Like Horn propositions, quadratic propositions admit a fast linear-time algorithm for checking satisfiability. This algorithm is also symbolic and uses an underlying graph representation of the quadratic formula to reveal either a satisfying truth assignment, if one exists, or the source of unsatisfiability.

The linear-time solubility of inference on quadratic propositions is a fundamental result that finds application as a subroutine in many contexts. We will illustrate one such application in this section. A proposition is called *disguised* or *renamable Horn* if it can be made Horn by complementing some of the atomic propositions. We will see that the problem of recognizing a renamable Horn proposition reduces to the satisfiability of a quadratic formula.

The linear relaxation polytype of a quadratic formula has no special structure that sheds light on the inference problem. However, there is a connection to linear programming. We will see that Q-Horn systems, a class of propositions defined via a linear programming property, properly generalize quadratic, Horn, and renamable Horn systems simultaneously. Satisfiability and recognition of Q-Horn systems can both be achieved in linear time [30, 34].

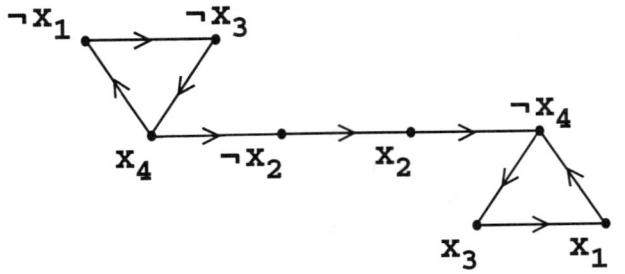

Figure 2.4: Implication graph for a quadratic system.

2.4.1 Satisfiability of Quadratic Systems

We begin with a description of the well-known linear-time satisfiability algorithm for quadratic systems due to Aspvall, Plass, and Tarjan [7] (see also [43]). A quadratic system is a CNF formula with no more than two literals per clause. As an example, consider

$$Q = (x_1 \vee \neg x_3) \wedge (\neg x_1 \vee \neg x_4) \wedge (x_2) \wedge (\neg x_2 \vee \neg x_4) \wedge (x_3 \vee x_4) \quad (2.35)$$

The implication graph $G(Q)$ of a quadratic system Q is defined as follows. The vertices of the directed graph are the literals. For each clause of the form $(\ell_i \vee \ell_j)$ we create two directed edges $(\neg \ell_i, \ell_j)$ and $(\neg \ell_j, \ell_i)$ in the graph. And for each unit clause of the form (ℓ_k) we create one directed edge $(\neg \ell_k, \ell_k)$. So for the example (2.35) the corresponding $G(Q)$ appears in Figure 2.4.

Clearly an edge of the form (ℓ_i, ℓ_j) has the interpretation that, in any satisfying truth assignment, if ℓ_i is set to T then so is ℓ_j. Therefore, if for any atomic proposition x_j we have a directed path from x_j to $\neg x_j$ and one from $\neg x_j$ to x_j in the graph $G(Q)$, then we may conclude that Q is not satisfiable. In graph theoretic terms this is equivalent to saying that a sufficient condition for unsatisfiability of Q is the membership of a literal and its complement in a strong component of $G(Q)$. We will see that it is also a necessary condition.

The definition of $G(Q)$ implies that these graphs of quadratic systems satisfy a duality property. An implication graph $G(Q)$ is isomorphic to the graph obtained by reversing the directions of all the edges of $G(Q)$ and by complementing the names of all the vertices (recall that the literals of Q are the names of the vertices).

A fundamental result in algorithmic graph theory is that for any directed graph, the depth first search technique may be used to identify, in topological order, all the strong components in linear time [1, 268]. The topological order simply means that for two strong components S_1 and S_2,

2.4 QUADRATIC AND RENAMABLE HORN SYSTEMS

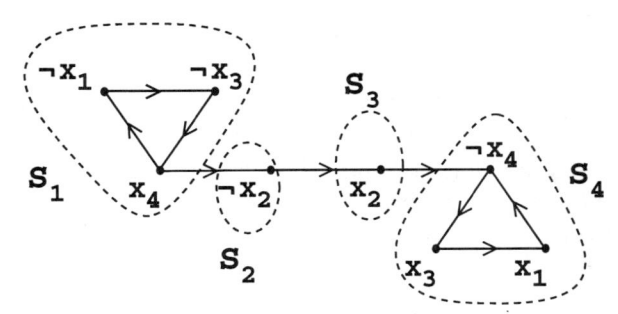

Figure 2.5: Strong components of an implication graph.

if there is an edge from a vertex in S_1 to a vertex in S_2, then S_2 is a successor of S_1. Note that the duality property of implication graphs ensures that if S is a strong component of G then so is its dual $\bar{S}$. The strong components, in topological order, of our example implication graph are depicted in Figure 2.5. Notice that $S_1 = \bar{S}_4$ and $S_2 = \bar{S}_3$.

Observe also that if a strong component S of any implication graph satisfies $S = \bar{S}$ then it must contain only complementary pairs of literals. Conversely, any literal x_j and its complement $\neg x_j$ are both in the same strong component S only if $S = \bar{S}$. Now we are ready to describe the linear-time satisfiability algorithm for quadratic systems.

Procedure 5: *Quadratic Satisfiability*

Input. Quadratic CNF formula Q.

Step 1. Construct the implication graph $G(Q)$.

Step 2. In reverse topological order, process the strong components S of $G(Q)$ as follows. If S is marked true or false do nothing. If S is unmarked, then if $S = \bar{S}$ stop (Q is unsatisfiable), and otherwise mark S true and $\bar{S}$ false.

Step 3. Assign *True* (*False*) to all literals in a component S marked true (false).

The procedure when applied to the example (2.35) marks S_1 and S_2 false, and S_3 and S_4 true. Thus x_1, x_2, and x_3 are true and x_4 is false in the resulting truth assignment that satisfies Q. The correctness of the procedure follows from the fact that when it ends with a true/false marking of the vertices of the graph $G(Q)$ (i.e., when it concludes that Q is satisfied) there is no directed path from any vertex marked true to a vertex marked false. Thus all implications are satisfied.

2.4.2 The Median Characteristic of Quadratic Systems

In Section 2.3 we saw that the set of models of a Horn formula is closed under the "∧" operation. Conversely, we argued that every set of vertices of the unit hypercube closed under the "∧" operation corresponds to the set of models of some Horn formula. In this section we will see that quadratic formulas have analogous properties with respect to the *median* operator.

Given three 0-1 n-vectors x, y, z, their median, written median$\{x, y, z\}$, is defined as follows:

$$m_i = \text{median}\{x_i, y_i, z_i\} = \begin{cases} 1 & \text{if two of } x_i, y_i, z_i \text{ are 1} \\ 0 & \text{otherwise} \end{cases} \quad (2.36)$$

A collection of 0-1 n-vectors M is called a *median set* if the median of every subset of three vectors is in M. Now let Q be any quadratic formula and let x, y, z denote three models of Q. We may, of course, assume that x, y, z are 0-1 n-vectors with 1 denoting *True* and 0 denoting *False*. Now let $\ell_i \vee \ell_j$ denote an arbitrary clause C of Q, and let $m = \text{median}\{x, y, z\}$. Since x, y, z are models of Q, each of them evaluates (ℓ_i, ℓ_j) to either $(1,0)$ or $(0,1)$ or $(1,1)$. Clearly it is impossible then for m to evaluate (ℓ_i, ℓ_j) to $(0,0)$. Therefore m satisfies C and we have proved the following.

Lemma 3 *The set of models of a quadratic proposition is a median set.*

The converse statement can also be shown to hold and hence we have the following theorem.

Theorem 9 ([56]) *A collection of 0-1 n-vectors is a median set if and only if it is the set of models of some quadratic proposition.*

Proof: We have already proved the "if" part of the theorem statement. To prove the "only if" part, let M be a median set and let Q denote the conjunction of all two-literal clauses satisfied by all the 0-1 vectors in M. Now let N denote the collection of models of Q. Clearly $M \subseteq N$. Since by construction we have ensured that every member of M is a model of Q, to show that $M = N$ and thus complete the proof of the theorem, we pick an arbitrary model $\eta \in N$ and show that η is in M.

We will actually prove that for every i, j with $1 \leq i \leq j \leq n$ there is some $m^{ij} \in M$ such that $m^{ij}_k = \eta_k$ for all $i \leq k \leq j$. Since $m^{1n} = \eta$ as a consequence we will have proved the theorem. We proceed by induction on the quantity $j - i$. When $i = j$, we first observe that the unit clause $x_i \neq \eta_i$ (i.e., the clause x_i if $\eta_i = 0$ and $\neg x_i$ if $\eta_i = 1$) cannot be a part of Q. For otherwise η would not be in N. Thus $x_i \neq \eta_i$ is not satisfied by all members of M, and hence there is some $m^{ii} \in M$ such that $m^{ii}_i = \eta_i$.

2.4 QUADRATIC AND RENAMABLE HORN SYSTEMS

Now for the inductive step, assume the existence of $m^{i+1,j}$ and $m^{i,j-1}$ in M with the requisite properties. Since η is in N, Q does not contain the two-literal clause $(x_i \neq \eta_i) \vee (x_j \neq \eta_j)$. Thus there exists some $\bar{m}$ in M with $\bar{m}_i = \eta_i$ and $\bar{m}_j = \eta_j$. Now let $m^{ij} = \text{median}\{\bar{m}, m^{i+1,j}, m^{i,j-1}\}$. Clearly m^{ij} meets the requirements since

$$m^{ij}_i = \bar{m}_i = m^{i,j-1}_i = \eta_i$$
$$m^{ij}_j = \bar{m}_j = m^{i+1,j}_j = \eta_j$$
$$m^{ij}_k = m^{i+1,j}_k = m^{i,j-1}_k = \eta_k \text{ for } i < k < j$$

and the induction is complete. □

2.4.3 Recognizing Renamable Horn Systems

The linear-time solubility of inference on Horn systems motivates the search for classes of propositions that generalize Horn systems and yet admit efficient inference algorithms. Renamable Horn systems represent the simplest such generalization. A CNF sentence is said to be renamable Horn if it can be made Horn by complementing some of its atomic propositions. In order for this class to be useful we need an efficient algorithm, preferably linear-time, that can identify a renamable Horn system and disclose which atomic propositions have to be complemented to reduce the system to a Horn formula.

In an elegant paper, Lewis [197] showed that the disclosure of a renamable Horn formula is reducible to solving the satisfiability of an auxiliary quadratic system. Aspvall [6] refined this approach to provide the first linear-time recognition algorithm for renamable Horn systems. Mannila and Mehlhorn [205] have also described a linear-time algorithm for recognizing renamable Horn formulas. We will first examine the work of Aspvall and then provide a more direct linear-time recognition algorithm for renamable Horn systems based on a graph labeling method that is described in Chandru et al. [43]. The reduction of renamable Horn disclosure to satisfiability of a quadratic proposition is best explained through an example. Consider the CNF formula

$$S = (x_1 \vee \neg x_2 \vee \neg x_3) \wedge (x_2 \vee x_3) \wedge (\neg x_1) \qquad (2.37)$$

Clearly, S is not Horn since the second clause contains two positive literals. Let $u_i \in \{T, F\}$ denote the decision to complement the atomic letter x_i (i.e., $u_i = T$) or to leave it alone ($u_i = F$). The conditions on the u_i to ensure that S is renamable Horn are given by the quadratic system

$$Q_S = (u_1 \vee \neg u_2) \wedge (u_1 \vee \neg u_3) \wedge (\neg u_2 \vee \neg u_3) \wedge (u_2 \vee u_3) \qquad (2.38)$$

The first clause $(u_1 \vee \neg u_2)$ simply states that if x_2 is complemented then so must x_1 for otherwise the first clause of S would contain two positives. Since Q_S represents the necessary and sufficient conditions for renaming S a Horn formula we conclude that S is renamable Horn if and only if Q_S is satisfiable. In this example, Q_S is satisfiable ($u_1 = u_2 = T$, $u_3 = F$) and S is renamed

$$S' = (\neg x_1 \vee x_2 \vee \neg x_3) \wedge (\neg x_2 \vee x_3) \wedge (x_1) \qquad (2.39)$$

a Horn clause system. Note that the unit clause of S places no restrictions on the $\{u_i\}$ and that Q_S contains one clause for every pair of literals in a non-unit clause of S. Thus the length of Q grows quadratically with the length of S.

To obtain a quadratic system linear in the size of S that captures the Horn renamability of S, Aspvall [6] invoked the simple but powerful technique of reformulation with auxiliary variables. (We have seen this power of auxiliary variables in Section 2.1.2, where we obtained a linear size CNF reformulation of any well-formed formula.) Aspvall's trick is as follows.

(i) Given any clause $C = (\ell_1 \vee \ell_2 \vee \cdots \vee \ell_k)$ of S with $k \geq 2$, we introduce $(k-1)$ auxiliary variables $w_1^C, w_2^C, \ldots, w_{k-1}^C$.

(ii) The quadratic system corresponding to C is now

$$\tilde{Q}_C = (\ell_1 \vee w_1^C) \wedge$$
$$\left(\bigwedge_{2 \leq t \leq k-1} (\ell_t \vee \neg w_{t-1}^C) \wedge (\neg w_{t-1}^C \vee w_t^C) \wedge (\ell_t \vee w_t^C) \right) \wedge$$
$$(\ell_k \vee \neg w_{k-1}^C)$$

(iii) $\tilde{Q}_S = \bigwedge_{\substack{C \in S \\ |C| \geq 2}} \tilde{Q}_C$

First let us measure the length of $\tilde{Q}_S$. From the definition of $\tilde{Q}_C$ in (ii) we see that the total length of $\tilde{Q}_C$ is $2(3 \cdot \max\{0, k-2\} + 2) \leq 6|C|$, that is, within a constant multiple of the length of C. Therefore $\tilde{Q}_S$ is linear in the size of S, as required. What is left is to convince ourselves that a solution to the satisfiability of $\tilde{Q}_S$ is a disclosure of the Horn renamability of S.

Theorem 10 ([6]) *S is renamable Horn if and only if $\tilde{Q}(S)$ is satisfiable. If $\tilde{Q}(S)$ is satisfiable by (x^*, w^*) then a Horn renaming of S is given by the rule for all i.*

$x_i^* = 1$	$\longrightarrow$	*complement x_i*
$x_i^* = 0$	$\longrightarrow$	*do nothing*

2.4 QUADRATIC AND RENAMABLE HORN SYSTEMS

Proof: Let ℓ_i and ℓ_j be any two literals that appear in a clause C of S. Then from the definition of $\tilde{Q}_C$ we know that the clauses

$$(\ell_i \vee w_i^C), (\neg w_i^C \vee w_{i+1}^C), \ldots, (\neg w_{j-2}^C \vee w_{j-1}^C), (\ell_j \vee \neg w_{j-1}^C) \qquad (2.40)$$

are all contained in $\tilde{Q}_C$. Hence the clause $(\ell_i \vee \ell_j)$ is implied by the clauses in $\tilde{Q}_C$. Therefore a satisfying truth assignment (x^*, w^*) for $\tilde{Q}_S$ provides a Horn renaming for S as stated in the theorem.

Conversely, let us assume that S is renamable Horn. Let $\bar{x}_i = T$ if x_i is renamed and $\bar{x}_i = F$ otherwise. For an arbitrary clause $C = (\ell_1 \vee \ell_2 \vee \cdots \vee \ell_k)$ of S, the truth assignment $\bar{x}$ corresponding to the Horn renaming of S must make every $\ell_i (1 \leq i \leq k)$ evaluate to true with the possible exception of one literal, say, ℓ_t. Now extend the truth assignment $\bar{x}$ to $(\bar{x}, \bar{w})$ by assigning $\bar{w}_i^C = F$ for all $1 \leq i < t$ and $\bar{w}_i^C = T$ for $t \leq i \leq k-1$. It follows from the definition of $\tilde{Q}_C$ that this extended truth assignment $(\bar{x}, \bar{w})$ is in fact a model for $\tilde{Q}_C$. Since C was chosen arbitrarily we conclude that $\tilde{Q}_S$ is satisfiable. □

In what follows we describe an alternate, more direct linear-time algorithm for recognizing disguised Horn sentences [43]. Our approach is based on a graph labeling technique. We first note that any sentence S in CNF may be represented as a $(0, \pm 1)$ matrix $D = (d_{rc})$ as follows. The row indices R of D range over the clauses of S and the column indices C range over the atomic propositions of S. Entry d_{rc} is $+1$ if the cth proposition appears as a *positive* literal in the rth clause, and is -1 if the cth proposition appears as a *negated* literal in the rth clause. And finally d_{rc} is 0 if the cth proposition does not appear in the rth clause. The matrix D will be called the *clause-proposition incidence matrix of S*.

Using this terminology, the problem of recognizing whether a sentence in CNF is renamable Horn is stated as follows.

(i) Given a $(0, \pm 1)$ matrix D, determine whether the columns of D may be scaled by ± 1's to obtain a matrix $\tilde{D}$ in which every row has at most on $+1$ entry.

If D is a $(0, \pm 1)$ matrix, with row index set R and column index set C, there is a natural representation of D as a bipartite digraph $G = (R \cup C, A)$, where

$$A = \{(r, c) \mid d_{rc} = +1\} \cup \{(c, r) \mid d_{rc} = -1\}. \qquad (2.41)$$

We say that G can be labeled if there exists a subset C' of C, such that the reversal of all arcs incident to members of C' results in a digraph $\tilde{G}$ in which every vertex in R has out-degree at most 1. Translated to G, the recognition problem (1) is stated as follows.

(ii) Given a bipartite directed graph $G = (R \cup C, A)$, determine whether or not G can be labeled.

Our graph algorithm that solves this problem assigns labels $+1$ and -1 to the vertices in C. A label of -1 indicates membership in C', the set of vertices whose incident arcs will be reversed and a label of $+1$ indicates membership in $C \setminus C'$. (Note: No arcs are actually reversed during the algorithm. The reversals are performed after the entire set C' has been determined.)

Suppose some vertex $c_0 \in C$ has been assigned a label. Then in order for C' to satisfy the condition in (ii), some of the vertices of C that are distance two from c_0 have a forced label. Specifically, if vertex c is distance two from c_0, then the labels on c_0 and c must be such that the common vertex $r \in R$ adjacent to c_0 and c has out-degree at most one after the reversals are performed. The table below enumerates the possible ways in which a vertex c can be distance two from c_0 and, for each possibility, indicates which label on c_0 forces a label on c.

Label	c_0 r c	Forced label
-1	• ← • → •	no
$+1$		-1
-1	• → • ← •	$+1$
$+1$		no
-1	• ← • ← •	no
$+1$		$+1$
-1	• → • → •	-1
$+1$		no

The following analysis allows us to identify the types of paths from labeled vertex c_0 that imply forced labels on vertices of arbitrary distance from c_0. let P be the length-two path from c_0 to c. Notice that if vertex c_0 has label -1, then vertex c has a forced label exactly when c_0 is the tail of the incident arc of P. Similarly, if c_0 has label $+1$, then c has a forced label exactly when c_0 is the head of the incident arc of P. Next, notice that provided vertex c has a forced label, that label depends only on the direction of the arc of P incident with c; that is, if c is the head of that arc, then its forced label is -1, and if c is the tail of that arc, then its forced label is $+1$.

Let P be a path starting at a labeled vertex $c_0 \in C$. Path P is called a *labeling path* if the following conditions hold.

(a) If c_0 is labeled $+1$, then it is the head of the incident arc of P, and if c_0 is labeled -1, then it is the tail of the incident arc of P.

2.4 QUADRATIC AND RENAMABLE HORN SYSTEMS

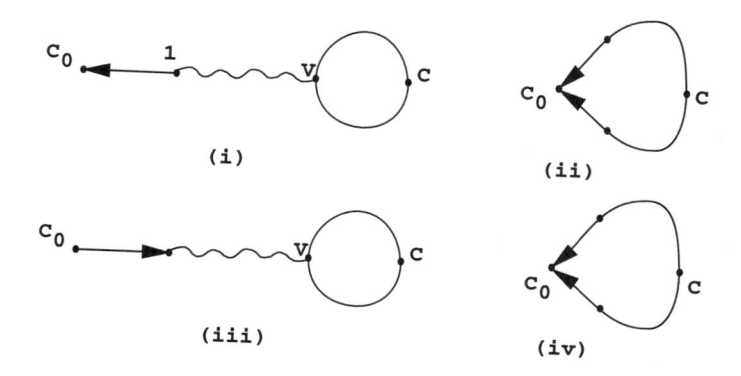

Figure 2.6: Critical subgraphs for determining whether a graph can be labeled.

(b) Every internal vertex $c \in C$ of P has in-degree and out-degree one.

Now assume P is a labeling path starting at vertex c_0. For vertices $c \in C$ of P, let $p(c)$ denote the arc of the (c_0, c)-subpath of P that is incident with vertex c. By the above analysis (applied inductively on the length of P), if c is the head of arc $p(c)$, then c has forced label -1, and if c is the tail of arc $p(c)$, then c has forced label $+1$.

A subgraph H of G is called *type J* if it is homeomorphic to (i) or (ii) in Figure 2.6 and is called *type K* if it is homeomorphic to (iii) or (iv) in the figure, where $c_0 \in C$ and is called the root, each vertex $c \in C$ in H has in-degree and out-degree in H at least one, and if $v \in C$, then both arcs of the cycle incident with v are directed toward or away from v.

Lemma 4 *If $G = (R \cup C, A)$ has a subgraph of type J rooted at vertex $c_0 \in C$ and a subgraph of type K rooted at c_0, then G cannot be labeled.*

Proof: Suppose G can be labeled. First suppose that vertex c_0 has label $+1$. Let H be a subgraph of type J rooted at c_0, and let $c \in C$ be a vertex of the cycle of H. Now H contains two distinct labeling paths starting at c_0 and ending at vertex c. One of these labeling paths forces vertex c to have label $+1$, and the other forces c to have label -1. This contradiction implies vertex c_0 has label -1. Similarly, considering the subgraph of type K rooted at c_0 leads to a contradiction. Thus, G cannot be labeled. □

Lemma 4 gives a condition sufficient to conclude that graph G cannot be labeled. It turns out that this condition is also necessary. The algorithm presented later in the section either finds a labeling G or finds a vertex

50   CHAPTER 2   PROPOSITIONAL LOGIC: SPECIAL CASES

$c_0 \in C$, together with subgraphs of type H and K rooted at c_0. This provides a "forbidden subgraph" characterization of the class of bipartite graphs that can be labeled.

Let $\bar{R} \subseteq R$ and let $\bar{C} \subseteq C$. Let $\bar{G}$ be the subgraph of G induced by $\bar{R} \cup \bar{C}$. Assume the vertices of $\bar{C}$ have each been labeled $+1$ or -1. Then $\bar{G}$ is said to be a *properly labeled subgraph* if the following two conditions hold.

(a) Every vertex $r \in \bar{R}$ is adjacent only to vertices in $\bar{C}$ and has out-degree at most one in the digraph G' obtained from G by reversing all arcs incident with some vertex in $\bar{C}$ that has label -1.

(b) There are no arcs of the form (c,r), where $r \in R \setminus \bar{R}$ and $c \in \bar{C}$.

Lemma 5 *If $\bar{G} = (\bar{R} \cup \bar{C}, \bar{A})$ is a properly labeled subgraph of G, and if $G^* = G \setminus \bar{R} \cup \bar{C}$, then G can be labeled if and only if G^* can be labeled.*

Proof: Clearly if G can be labeled, then G^* can be labeled. Assume G^* can be labeled. Then this labeling, together with the labeling of $\bar{G}$, provides a labeling of G. This follows from the definition of properly labeled subgraph. □

The main step of our recognition algorithm is the following layer/label procedure, in which a particular vertex c_0 is assigned a label, and a search is initiated at c_0. This search labels all vertices in C that are in some labeling path starting at c_0 and terminates with either a properly labeled subgraph of type J or K rooted at c_0.

Procedure 6: *Layer/Label*

Input. Bipartite digraph $G = (R \cup C, A)$, vertex $c_0 \in C$ with label $+1$ or -1.

Output. Either a properly labeled subgraph $\bar{G} = (\bar{R} \cup \bar{C}, \bar{A})$, where $c_0 \in \bar{C}$, or a subgraph H, where H is type J if c_0 has label $+1$ and H is type K if c_0 has label -1.

Step 0. $C_0 \leftarrow \{c_0\}$. If c_0 has label $+1$, then $C_0^+ \leftarrow \{c_0\}$. If c_0 has label -1, then $C_0^- \leftarrow \{c_0\}$.

Step 1. For $i = 1, 2, \cdots$, do the following.

(a) Let $A_i^R \leftarrow A' \cup A''$, where

$$A' = \{(c,r) \in A \mid c \in C_{i-1}^-, r \in R \setminus \bigcup_{j<i} R_j\}$$
$$A'' = \{(r,c) \in A \mid c \in C_{i-1}^+, r \in R \setminus \bigcup_{j<i} R_j\}$$

2.4 QUADRATIC AND RENAMABLE HORN SYSTEMS

If $A_i^R = \emptyset$, then go to step 4.

(b) Let

$$R_i \leftarrow \{r \in R \setminus \bigcup_{j<i} R_j \mid r \text{ is incident to an arc in } A_i^R\}$$

If some $r \in R_i$ is incident to 2 arcs in A_i^R, then go to step 3.

(c) Let $A_i^C \leftarrow A' \cup A''$, where

$$A' = \{(c,r) \in A \mid r \in R_i \text{ and } c \in C \setminus \bigcup_{j<i} C_j\}$$
$$A'' = \{(r,c) \in A \mid r \in R_i \text{ and } c \in C \setminus \bigcup_{j<i} C_j\}$$

If $A_i^C = \emptyset$, then go to step 4.

(d) Let

$$C_i^+ \leftarrow \{c \in C \setminus \bigcup_{j<i} C_j \mid (c,r) \in A \text{ for some } r \in R_i\}$$
$$C_i^- \leftarrow \{c \in C \setminus \bigcup_{j<i} C_j \mid (r,c) \in A \text{ for some } r \in R_i\}$$

If $C_i^+ \cap C_i^- \neq \emptyset$, then go to step 2.

(e) Let $C_i \leftarrow C_i^+ \cup C_i^-$. Label vertices in C_i^+ with $+1$. Label vertices in C_i^- with -1.

Step 2. Let $c \in C_i^+ \cup C_i^-$. Let $r_1 \in R_i$ such that $(r_1, c) \in A$ and $r_2 \in R_i$ such that $(c, r_2) \in A$. Let

$$\bar{G} \leftarrow G[\bar{R} \cup \bar{C}, \bar{A}], \text{ where} \qquad (2.42)$$
$$\bar{R} = \bigcup_i R_i, \quad \bar{C} = \bigcup_i C_i, \quad \bar{A} = \bigcup_i A_i^C \cup \bigcup_i A_i^R$$

Let P_1 be a (c, c_0)-path in $\bar{G}$ containing r_1, and let P_2 be a (c, c_0)-path in $\bar{G}$ containing r_2. Let H be a subgraph of $P_1 \cup P_2$ of type J, if c_0 has label $+1$, and type K, if c_0 has label -1. Return H and stop.

Step 3. Let $r \in R_i$ be adjacent to 2 arcs in A_i^R. Let $\bar{G}$ be given by (2.42). Let P_1, P_2 be two distinct (r, c_0)-paths in $\bar{G}$. Let H be a subgraph of $P_1 \cup P_2$ of type J, if c_0 has label $+1$ and of type K, if c_0 has label -1. Return H and stop.

Step 4. Let $\bar{G}$ be given by (2.42). Return $\bar{G}$ and the labels on $\bar{C}$ and stop.

Lemma 6 *If Procedure Layer/Label stops and returns a subgraph $\bar{G} = (\bar{R} \cup \bar{C}, \bar{A})$, where vertices in $\bar{C}$ are each labeled $+1$ or -1, then $\bar{G}$ is a properly labeled subgraph.*

Proof: Let $r \in \bar{R}$ be arbitrary, and let $c \in C$ be adjacent to r. Since $r \in \bar{R}$, $r \in R_i$, for some i. By construction, it follows that $c \in C_j$, for some $0 \leq j \leq i$.

Let $G' = (R \cup C, A')$ be the digraph obtained from G by reversing all arcs incident with some vertex in $\bar{C}$ that has label -1. We wish first to show each vertex $r \in \bar{R}$ has out-degree at most one. Let $r \in R$ be arbitrary. Then $r \in R_i$ for some i. Every arc of G adjacent to r is either in A_i^R or A_i^C. Of those in A_i^C, exactly those whose tail is r are reversed in obtaining G'. By step 1, there is at most one arc of A_i^R incident to r. Thus, vertex r has out-degree at most one in G'.

Next we show there are no arcs of the form $(r, c) \in A'$, where $r \in R \setminus \bar{R}$ and $c \in \bar{C}$. Suppose (r, c) is an arc. Since $c \in \bar{C}$, $c \in C_i$, for some i. Moreover either $c \in C_i^+$ and $(r, c) \in A$ or $c \in C_i^-$ and $(c, r) \in A$. In either case, the vertex r is either in R_{i+1} or r is in $\bigcup_{j<i} R_j$ contradicting the fact $r \in R \setminus \bar{R}$. □

This establishes the correctness of Procedure Layer/Label. The three lemmata together suggest a simple recursive scheme for solving problem (ii).

Procedure 7: *Scaling*

Input. The bipartite digraph $G(R \cup C, A)$.

Output. A complete labeling of G, i.e., every $c \in C$ is assigned a label of ± 1 or a proof that G cannot be labeled.

Step 1. Select some $c_0 \in C$, stop if $C = \emptyset$. Call Procedure Layer/Label with c_0 labeled $+1$. If output is a properly labeled subgraph $\bar{G} = (\bar{R} \cup \bar{C}, \bar{A})$, then let $G \leftarrow G \setminus (\bar{R} \cup \bar{C})$ and go to step 1. Otherwise, output is a type J subgraph H_1 of G containing c_0.

Step 2. Call Procedure Layer/Label with c_0 labeled -1. If output is a properly labeled subgraph $\bar{G} = (\bar{R} \cup \bar{C}, \bar{A})$, then let $G \leftarrow G \setminus (\bar{R} \cup \bar{C})$ and go to step 1. Otherwise, output is a type K subgraph H_2 of G containing c_0.

Step 3. Return H_1 and H_2 and stop.

Remarks

(a) If the procedure stops in step 1, then we have a complete labeling of G.

2.4 QUADRATIC AND RENAMABLE HORN SYSTEMS

(b) If the procedure stops in step 3, it follows from Lemma 4 that G cannot be labeled. The subgraphs H_1 and H_2 provide a short proof of this conclusion.

(c) Since Procedure Layer/Label is essentially a modification of breadth-first-search, it can be implemented to run in time that is linear in the number of arcs it encounters. (See Ahu et al. [1].) Procedure Scaling calls Layer/Label at most twice for a given vertex c_0, after which the algorithm terminates, or q arcs get deleted from G. This number q is the number of arcs encountered in the application of Layer/Label that terminated with a completely labeled subgraph. If the two applications of Layer/Label are run in parallel, then the number q is guaranteed to be linear in the total number of arcs encountered in both. It follows that Procedure Scaling runs in time proportional to the number of arcs in G.

Thus we have the following theorem.

Theorem 11 *Procedure Scaling is correct and runs in time $O(L)$, where L is the number of arcs of G.*

2.4.4 Q-Horn Propositions

We have encountered three special types of propositions: Horn, renamable Horn, and quadratic, for which efficient satisfiability algorithms are possible. Boros, Crama, and Hammer [30] observed that these three structures could be unified into a new structure they called Q-Horn propositions. A Q-Horn proposition is a CNF formula S for which the following linear inequality system is soluble:

$$\sum_{\substack{j \\ x_j \in P(C_i)}} \alpha_j + \sum_{\substack{j \\ x_j \in N(C_i)}} (1 - \alpha_j) \leq 1, \text{ for each clause } C_i$$

$$0 \leq \alpha_j \leq 1, \text{ for all } j$$

Note that we can always assume that if the linear inequality system is soluble then it has a solution with all $\alpha_j \in \{0, \frac{1}{2}, 1\}$. This is because any feasible solution can be rounded to a $\{0, \frac{1}{2}, 1\}$ solution (if $\alpha_j < \frac{1}{2}$ set it to 0, and if $\alpha_j > \frac{1}{2}$ set it to 1).

It is also easy to verify that Horn, renamable Horn, and quadratic propositions are Q-Horn.

- *Horn:* Set all $\alpha_j = 1$.

- *Renamable Horn:* Set α_j at 0 or 1, depending on the renaming.
- *Quadratic:* Set all $\alpha_j = \frac{1}{2}$.

We might ask if Q-Horn is much more than the union of these three classes of special structures. The answer is that it is more, but only slightly. Let us assume we have an arbitrary formula that is Q-Horn and further assume that we have renamed variables so that all the α_j corresponding to this formula are in $\{\frac{1}{2}, 1\}$. Now let X_1 denote the variables that have $\alpha_j = 1$ and X_2 denote the remaining variables. It is not hard to visualize that the clauses of the Q-Horn formula must have the following incidence structure.

$$\begin{pmatrix} H & 0 \\ \Theta & Q \end{pmatrix}$$

The columns of the incidence matrix are partitioned into X_1 and X_2 columns. The notation H stands for Horn, Θ for entries that are either 0 or -1, and Q for quadratic. A linear-time satisfiability algorithm for a Q-Horn formula given to us with the above structural decomposition is evident. Run Horn resolution on the H clauses. Simplify the remaining clauses using the least model of H. Set all the remaining X_1 variables to false. What is left now is a quadratic formula whose satisfiability can also be solved in linear time.

Recognizing if a given formula is Q-Horn can be accomplished in polynomial time by solving the corresponding linear inequality system. However, there is a far more efficient (linear-time) combinatorial recognition algorithm described in [34].

The linear inequality description of Q-Horn propositions leads us to consider an index for any CNF proposition. Let $z(S)$ denote the minimum value attained by the linear program

$$\begin{aligned} \min \quad & z \\ \text{s.t.} \quad & \sum_{\substack{j \\ x_j \in P(C_i)}} \alpha_j + \sum_{\substack{j \\ x_j \in N(C_i)}} (1 - \alpha_j) \leq z, \text{ for all } i \\ & 0 \leq \alpha_j \leq 1, \text{ for all } j \end{aligned}$$

In a remarkable paper, Boros et al. [31] show that this index sharply delineates between classes of easy and hard satisfiability problems. They show that any class of propositions in which each proposition S satisfies $z(S) \leq 1 + (c \log n)/n$, where n is the number of atoms in S, admits polynomial time satisfiability algorithms for any $c \in \Re$. Of course, Q-Horn formulas S are exactly those satisfying $z(S) = 1$. They also show that any class of propositions in which every proposition S satisfies $z(S) \leq 1 + n^{-\beta}$ is NP-complete for any $\beta < 1$.

2.5 NESTED CLAUSE SYSTEMS

Thus it may seem that with Q-Horn formulas we may be close to the limit of special structure in propositions that admit polynomial-time satisfiability algorithms. The dichotomy theorems of Schaefer [247] and of Jeavons et al. [163] also support this thesis. However, fresh perspectives can lead to pleasant surprises, and we still have many ahead.

2.5 Nested Clause Systems

This chapter deals with a number of generalizations of Horn clause systems. Renamable Horn (Section 2.4), extended Horn (Section 2.6), Q-Horn (Section 2.4.4), and generalized Horn (Section 2.8), are such generalizations which have been and are to be discussed. Quadratic clause systems, the other cornerstone of specially structured propositions, seem to have enjoyed not nearly as much attention.

In a recent paper, Knuth [186] introduces a new class of propositions called *nested propositions* for which he provides a linear-time satisfiability algorithm. The algorithm works by essentially replacing each clause by a quadratic clause and by solving a dynamic satisfiability problem on quadratic clauses. Even more recently, Hansen and Jaumard [131] have shown that nested and quadratic propositions share a common parent (generalization) which they called extended nested propositions, a class of propositions on which satisfiability can be solved in linear time as well.

2.5.1 Nested Propositions: Definition and Recognition

Let X be the set of n atomic propositions totally ordered by $<$ as follows.

$$x_1 < x_2 < \cdots < x_n$$

Let $\bar{X} = \{\neg x_1, \neg x_2, \ldots, \neg x_n\}$ be the set of the negations (or complements) of elements of X. $X \cup \bar{X}$ is the set of $2n$ *literals* corresponding to these variables; literals that belong to X are called *positive* and the rest are *negative*. The total order on X can be extended to an ordering on the literals in the following (natural) way by disregarding the signs. In other words, for the literals, we have

$$x_1 \equiv \neg x_1 < x_2 \equiv \neg x_2 < \ldots < x_n \equiv \neg x_n$$

A clause C that contains a single literal is a *unit* clause; if C contains two literals, then it is a quadratic clause; if C contains more than two literals, then we call it a *large* clause.

We now make some definitions and observations that will be useful later.

1. The literals of a clause can be written in *increasing* order (this is because a clause is just a set of literals on *distinct* variables).

2. Clauses other than unit clauses have a *least* literal σ or a *greatest* literal τ such that $\sigma \neq \tau$; all variables *strictly* between σ and τ are said to be *interior* to that clause.

3. A variable that has not yet appeared interior to any clause is called a *partition variable*. To start with, the set of partition variables is the same as the set of variables (X). As the algorithm processes a new clause, we update the set of partition variables by removing variables that are interior to the new clause. The set of partition variables plays a key role in the algorithms for nested clauses.

A clause C_i *straddles* C_j if there exists literals σ and τ in C_i and α in C_j such that α lies strictly between α and τ. In other words,

$$\text{there exist } \sigma, \tau \in C_i \text{ and } \alpha \in C_j \text{ such that } \sigma < \alpha < \tau$$

Two clauses *overlap* if and only if they straddle each other. A set of clauses is called *nested* if there is *some* ordering of the variables such that no two clauses overlap.

Let $X = \{x_1, x_2, x_3, x_4\}$ and consider the proposition F defined by

$$F = (C_1 \wedge C_2 \wedge C_3) \\ C_1 = (\neg x_1 \vee x_3), \ C_2 = (x_2 \vee \neg x_4), \text{ and } C_3 = (x_3 \vee x_4) \quad (2.43)$$

Let $<_1$ and $<_2$ be total orders on X defined by

$$x_1 <_1 x_2 <_1 x_3 <_1 x_4 \\ x_1 <_2 x_3 <_2 x_2 <_2 x_4$$

We first discuss Example (2.43) with the order on X being $<_1$. Here, C_1 straddles C_2 because $x_2 \in C_2$, $x_1, \neg x_3 \in C_1$, and $x_1 <_1 x_2 <_1 \neg x_3$ (i.e., we have a literal of C_2 that lies *strictly* between two literals of C_1). It is easy to check that, with respect to $<_1$, (C_1, C_2), (C_2, C_1) and (C_2, C_3) are the pairs of *straddling* clauses. (*Note:* (C_i, C_j) indicates that C_i straddles C_j.) C_1 and C_2 overlap as they straddle each other and so S is not nested with respect to this ordering. However, we *cannot* conclude that F is not a nested formula, as there could be some other total order on X that makes S nested. For instance, let the total order on X now be $<_2$. With respect to this ordering C_1 and C_2 do not straddle any clause, whereas C_3 straddles C_2. As we do not have a pair of overlapping clauses, S is nested with respect to this ordering and hence S belongs to the class of nested propositions.

2.5 NESTED CLAUSE SYSTEMS

Consider the relation $>$ defined as follows:

$$C_i > C_j \text{ if } C_i \text{ straddles } C_j \text{ but } C_j \text{ does not straddle } C_i$$

It is easy to check that $>$ is a transitive relation. For nested clauses, clearly $C_i > C_j$ if and only if C_i straddles C_j. As $>$ is transitive, we can topologically sort (in linear time) any set of nested clauses into a *linear* arrangement in which each clause appears *after* every clause it straddles. We assume that the clauses are presented to us in this order; we also assume that the literals of each clause are presented to us in increasing order. We call this a *valid* ordering of the clauses of S. Observe that if a variable appears *interior* to a clause C, then it cannot appear in *any* of the subsequent clauses (this is because of the assumption on the order in which the clauses are presented). As a result, it is enough to remember information regarding the *partition variables* (because any future clause *must* have all its variables from this set of partition variables).

In Example (2.43), with $<_2$ as the order on X, $>$ relates only one pair, (C_3, C_2). Thus any ordering of the clauses in which C_3 appears after C_2 is valid.

The problem of recognizing when a given proposition S is nested has a striking relationship to testing graph planarity. First we define a bipartite graph G_S as follows. The vertex sets of G_S are the atomic propositions X on one side and the clauses $\{c_1, c_2, \ldots, c_m\}$ on the other. An edge (x_i, c_j) of G_S represents the membership of x_i or $\neg x_i$ in c_j. Now we extend G_S to G'_S by adding a new vertex y and edges (y, x_i) for all i. Note that G'_S is still bipartite.

Consider the proposition

$$S = c_1 \wedge c_2 \wedge c_3 = (\neg x_1 \vee x_3) \wedge (x_2 \vee \neg x_4) \wedge (x_3 \vee x_4)$$

G_S is shown in part (a) of Figure 2.7. G'_S is shown in part (b) and alternate embedding of G'_S is shown in part (c). The embedding in Figure 2.7(c) is planar, proving that S is nested for the order $x_1 < x_3 < x_2 < x_4$ since the clauses do not overlap (cross).

Theorem 12 ([46]) *S is a nested proposition if and only if G'_S is planar.*

Proof: Suppose first that S is nested with respect to a total order $x_{i_1} < x_{i_2} < \cdots < x_{i_n}$ on the atoms. As in the example, we can construct a planar embedding of G_S by arranging the x_i in the total order (left to right) in a row above the placement of vertex y. Because the clauses only straddle and do not overlap, we can position the clause vertices c_i so that there are no edge crossings. Conversely, a planar embedding of G'_S reveals the total

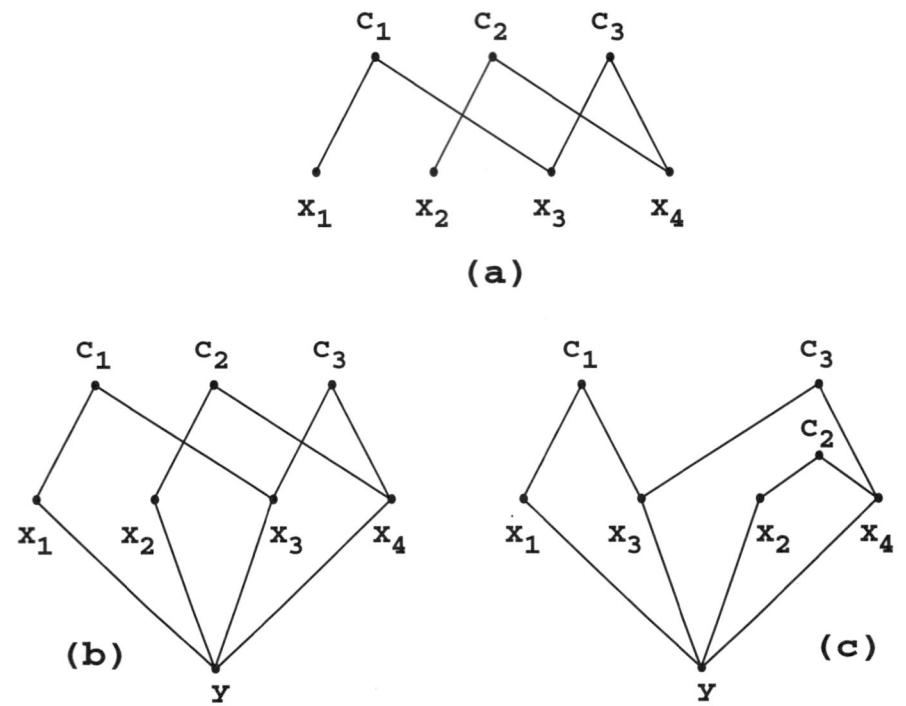

Figure 2.7: Graphs for checking whether a proposition is nested.

order of the x_i (from the order of the (y, x_i) in the star graph of edges incident on y) that makes S a nested proposition. □

Using well known linear-time algorithms for planarity testing of graphs [155], we realize a linear-time disclosure of nested propositions.

2.5.2 Maximum Satisfiability of Nested Clause Systems

As we mentioned earlier, Knuth [186] describes a satisfiability algorithm for nested propositions using an incremental approach that solves a dynamic version of the satisfiability problem on quadratic propositions. We will describe here an extension of Knuth's algorithm that solves a more general version of the satisfiability problem. The MAX-SAT problem (or maximum satisfiability problem) asks for a truth assignment that maximizes the number of clauses that are satisfied. Clearly, MAX-SAT subsumes the satisfiability problem and is often considered a much harder problem. It has

2.5 NESTED CLAUSE SYSTEMS

been shown, for example, that MAX-SAT on quadratic and Horn propositions is NP-hard [108, 162] (see Section 2.4 for a discussion of what this means).

We will see that MAX-SAT on nested propositions can be solved in linear time. Thus nested propositions and balanced propositions (see Section 2.7) seem to be the only non-trivial classes of propositions for which MAX-SAT can be handled effectively. In Chapter 4 we will see that the maximum satisfiability problem plays an important role in inference problems of probabilistic and related logics.

The MAX-SAT algorithm for nested propositions processes the formula clause by clause (incrementally). As noted before, we assume that the clauses are presented to us in a *valid* order. Observe that the total number of literals in a set of m nested clauses is at most $2m+n$. For convenience, we introduce *two* dummy variables x_0 and x_{n+1} and a *dummy* clause $C_{m+1} = x_0 \lor x_{n+1}$ (x_0 is the *least* element in X and x_{n+1} is the *greatest*). The clauses are specified in two arrays: **lit**$[j]$, $j = 1, \ldots, 2m + n$, and **start**$[i]$, $i = 1, \ldots, m + 1$, where the literals of clause i are

$$\text{lit}[j], \text{ for } \text{start}[i] \leq j < \text{start}[i+1]$$

in increasing order as j increases.

We noted that the interior variables of a clause are not present in any of the subsequent clauses; so it is enough to maintain information only about the partition variables. Suppose that at a certain stage of the algorithm, the set of partition variables is $x_0 < x_{i_1} < x_{i_2} < \ldots < x_{i_k} < x_{n+1}$ (note that x_0 and x_{n+1} will *always* belong to the set of partition variables). The set of clauses processed is conceptually split into intervals $[x_0, x_{i_1}]$, $[x_{i_1}, x_{i_2}], [x_{i_2}, x_{i_3}], \ldots, [x_{i_{k-1}}, x_{i_k}], [x_{i_k}, x_{n+1}]$, such that all literals of every processed clause lie in *exactly* one of those intervals. As a convention, for unit clauses, the left endpoint of an interval is included and the right endpoint is excluded. For example, if the current partition variable set is $x_0 < x_1 < x_2 < x_3 < x_4 < x_5$ and if x_3 is the next clause appearing in our list, then we assume that x_3 belongs to the interval $[x_3, x_4]$ and *not* to the interval $[x_2, x_3]$. This is just to ensure that each processed clause belongs to *exactly* one interval in the current partition. The current intervals are maintained in the array **next**, where **next**$[x_{i_j}] = x_{i_{j+1}}$ for $1 \leq j \leq k$.

The only other data structure that we need is the three-dimensional array **maxsat**, which has the following interpretation. If $[x_{i_j}, x_{i_{j+1}}]$ is an interval belonging to the current partition, then **maxsat**$[x_{i_j}, s, t]$ is the maximum number of already-processed clauses in the interval $[x_{i_j}, x_{i_{j+1}}]$ that can simultaneously be satisfied by setting x_{i_j} to s and $x_{i_{j+1}}$ to t. In the algorithm below, **maxsat**$[0, \textit{false}, \textit{false}]$ is the maximum number of clauses that can be satisfied.

Procedure 8: *Nested MAX-SAT*

For $i \leftarrow 0$ to n: Let **next**$[x_i] \leftarrow x_{i+1}$.
For $i \leftarrow 0$ to n:
    For $s \leftarrow$ *false* to *true*:
        For $l \leftarrow$ *false* to *true*:
            Let **maxsat**$[x_i, s, l] \leftarrow 0$.
For $i \leftarrow 1$ to $m + 1$:
    Let *least* $\leftarrow$ abs(**lit**[**start**$[i]$]).
    Let *greatest* $\leftarrow$ abs(**lit**[**start**$[i + 1]$] $- 1$).
    If (*least* = *greatest*) or (**next**[*least*] = *greatest*) then
        (Update the **maxsat** array directly)
    Else
        (Compute the New MAX-SAT value)
    Let **next**[*least*] $\leftarrow$ *greatest*.
    For $s \leftarrow$ *false* to *true*:
        For $t \leftarrow$ *false* to *true*:
            Let **maxsat**[*least*,s,t] $\leftarrow$ *newmax*$[s,t]$.

It turns out that the **maxsat** array has all the information to solve the MAX-SAT problem. The algorithm's main task is to maintain the **maxsat** array as it examines a new clause. Suppose we have processed the clauses $C_1, \ldots, C_{i-1}$, with the current partition variables $x_0, x_{i_1}, x_{i_2}, \ldots, x_{i_k}, x_{n+1}$. Suppose the next clause C_i is $y_1 \vee y_2 \vee \cdots \vee y_q$. The variables $y_1 < y_2 < \ldots < y_q$ will be a subset of the current partition variable set. All of the current partition variables between y_1 and y_q, whether they appear in the new clause or not, are interior to the clause and hence will be removed.

The updating is done as follows. If the new clause has no interior variables, then updating is trivial. If the new clause *has* interior variables, then by our assumption on the order, these variables cannot appear in any of the subsequent clauses. So, a truth assignment to such a variable will affect only the processed clauses. We choose truth-assignments to such variables which satisfy *as many processed clauses as possible*. Suppose there are I interior variables. Does this mean we have to check all 2^I truth assignments to pick the best? The answer is (obviously) in the negative because of the way partition variables interact with each other. Suppose z_1, z_2, z_3 are successive variables in the current partition. Then, clearly, z_1 and z_3 could not have occurred simultaneously in any of the previous clauses. This means that assigning T/F to z_1 does not affect any of the clauses in which z_3 occurs (although it may affect clauses that have z_2). As the variables in the current partition interact only with their neighbors, we can do this checking incrementally. Thus the updating is carried out by essentially doing an exhaustive check on the variables of the new clause.

2.5 NESTED CLAUSE SYSTEMS

We have to show that the algorithm preserves the meaning of the **maxsat** array after processing a new clause. This is quite obvious from the remarks made above and can be proved formally by induction on the number of variables interior to the new clause.

The complexity is $O(m + n)$ because each variable is either first or last in the current clause or it is being permanently removed from the partition. The former accounts for $2(m + 1)$ cases (we have added an extra clause) while the latter accounts for n cases.

2.5.3 Extended Nested Clause Systems

We begin with a definition of *hypergraphs*. A *hypergraph* G consists of a finite nonempty set $V = V(G)$ of *vertices* together with a prescribed set $E = E(G)$ of subsets of *distinct* vertices of V, called *hyperedges*. Vertices u and v are said to be *adjacent* if there is some hyperedge e containing both u and v. A hypergraph in which all the hyperedges are of size 2 is called a *graph*.

Hansen et al. [131] define the class of extended nested satisfiability problems in terms of two hypergraphs that are associated with any given formula F. The first hypergraph $H = (V(H), \mathcal{E}(H))$ has a vertex set $V(H) = \{x_1, x_2, \ldots, x_n\}$ associated with the variable set X of F; to each clause C_i of F corresponds an edge $E_i \in \mathcal{E}(H)$ containing the vertices associated with the variables appearing in that clause. An edge E is called *large* if it has more than two vertices. To define the second hypergraph $\hat{H}$, consider the set W of vertices associated with variables appearing only in unary or binary clauses of F and the subhypergraph $H_W = (W, E_W)$ induced by W. Clearly, H_W is a *graph*. Let W_i, for $i = 1, 2, \ldots, \bar{w}$, denote the connected components of H_W. For each W_i, let N_i denote its set of neighbors, that is, vertices x_k of $V \setminus W$ with x_ℓ in W_i such that H contains the edge $\{x_k, x_\ell\}$. Then construct $\hat{H} = (V(\hat{H}), \mathcal{E}(\hat{H}))$ from H as follows: (i) set $V(\hat{H})$ to $V(H)$ and $\mathcal{E}(\hat{H})$ to $\mathcal{E}(H)$; (ii) add to $\mathcal{E}(\hat{H})$ large edges $E_i = (W_i \cup N_i)$ for each connected component W_i of H_W, that is, for $i = 1, 2, \ldots, \bar{w}$; (iii) delete from $\mathcal{E}(\hat{H})$ all unary or binary edges contained in a large edge.

Next we look at the *erasure* operation defined on $\hat{H}$. This operation consists in either (a1) finding an edge E of $\hat{H}$ with at most one vertex belonging also to other edges and (a2) removing that edge from $\hat{H}$, or (b1) finding a large edge E of $\hat{H}$ with exactly two vertices, say, x_k and x_ℓ, belonging also to other edges, (b2) removing that edge E from $\hat{H}$, and (b3) adding the edge $\{x_k, x_\ell\}$ to $\hat{H}$ if no edge of $\hat{H}$ contains both x_k and x_ℓ. $\hat{H}$ is *erasable* if recursive application of the erasure operation leads to removal of all *large* edges.

Definition: *A formula F belongs to the class of extended nested formulas if and only if the corresponding hypergraph $\hat{H}$ is erasable.*

A crucial observation here is that erasability of an edge is independent of the previous erasure operations. This is because the set of vertices belonging to several edges of $\hat{H}$ never increases. This observation makes the recognition question trivial. Another observation is that if we start with a quadratic proposition there are no large edges and therefore no erasure is necessary. Thus all quadratic propositions are automatically also extended nested propositions.

Let $X = \{x_1, x_2, \cdots, x_8\}$ and consider the proposition,

$$F = \{C_1 \wedge \cdots \wedge C_{14}\}$$

$C_1 = (x_1 \vee x_2)$	$C_7 = (x_5 \vee x_6)$
$C_2 = (x_1 \vee \neg x_2)$	$C_8 = (\neg x_5 \vee x_6)$
$C_3 = (\neg x_2 \vee \neg x_3)$	$C_9 = (\neg x_5 \vee x_8)$
$C_4 = (x_2 \vee \neg x_4)$	$C_{10} = (x_5 \vee \neg x_7)$
$C_5 = (x_2 \vee x_4)$	$C_{11} = (x_5 \vee x_7)$
$C_6 = (\neg x_2 \vee x_4)$	$C_{12} = (\neg x_5 \vee x_7)$
$C_{13} = (x_2 \vee x_3 \vee \neg x_7)$	$C_{14} = (\neg x_1 \vee x_2 \vee \neg x_7)$

The hypergraph H associated with the formula F in Example (2.5.3) is represented in Figure 2.8. The variables that appear in *only* unary or binary clauses are x_4, x_5, and x_6, and so $W = \{x_4, x_5, x_6\}$. The *graph* (W, E_W) is shown in Figure 2.9. The graph (W, E_W) has two connected components with $W_1 = \{x_4\}$ and $W_2 = \{x_5, x_6\}$. The corresponding neighbor sets are $N_1 = \{x_2\}$ and $N_2 = \{x_7, x_8\}$. The hypergraph $\hat{H}$ associated with F is

$$\{\{x_2, x_3, x_7\}, \{x_1, x_2, x_7\}, \{x_5, x_6, x_7, x_8\}, \{x_2, x_4\}\}$$

and is shown in Figure 2.10.

In Example (2.5.3), it is evident that all the *large* edges of $\hat{H}$ can be erased and so F belongs to the class of extended nested formulas.

Hansen et al. [131] present a linear algorithm to solve SAT for extended nested clauses. The key idea of their algorithm is to split the formula F into a pair of formulas F_R and $F \setminus F_R$ such that:

1. F_R and $F \setminus F_R$ share at most two variables. These shared variables are referred to as *parameter* variables.

2. F_R has *at most one* large clause C; all the variables present in the *binary* part of F_R are present in C. For F to be satisfiable, there must be truth values of the parameter variables such that both F_R and $F \setminus F_R$ are simultaneously satisfiable.

2.6 EXTENDED HORN SYSTEMS

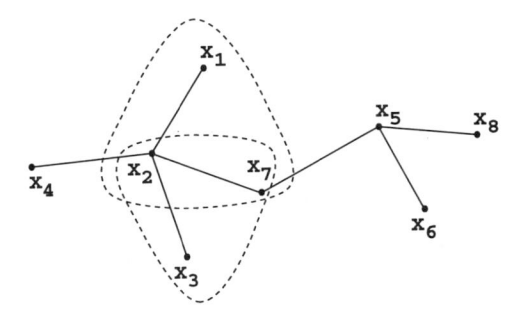

Figure 2.8: The hypergraph H corresponding to a formula.

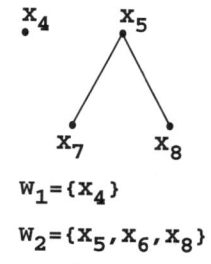

$W_1 = \{X_4\}$
$W_2 = \{X_5, X_6, X_8\}$

Figure 2.9: The graph (W, E_W) corresponding to hypergraph H.

The algorithm first checks for satisfiability of F_R using standard algorithms, stops if F_R is unsatisfiable or excludes those truth assignments to the parameter variables that make F_R unsatisfiable by adding appropriate clauses to $F \setminus F_R$. The algorithm now works on the new $F \setminus F_R$.

2.6 Extended Horn Systems

The attraction of Horn systems is that unit resolution alone checks them for satisfiability. In fact, it suffices to use a restricted form of unit resolution that resolves only on positive unit clauses. Unit resolution not only runs in linear time but is easy to understand and to implement. It would therefore be very useful to identify larger classes of satisfiability problems that can be solved by unit resolution.

One way to do this is to exploit a connection between logic and linear programming already stated in Section 2.2. Namely, unit resolution detects unsatisfiability if and only if the linear relaxation of the problem is infeasible. This means one can find satisfiability problems soluble by unit

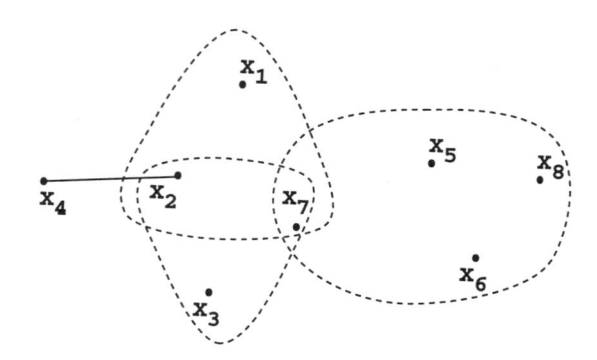

Figure 2.10: The hypergraph $\hat{H}$.

resolution by finding 0-1 programming problems that are feasible exactly when their linear relaxations are.

The operations research community has already developed at least two approaches to identifying such 0-1 problems. One is to characterize problems whose linear relaxations have only integral extreme points. The next section will explore this avenue of attack.

Another approach stems from a result of Chandrasekaran [41]. He discovered a class of 0-1 problems that can be solved by rounding a solution of the linear relaxation in a prespecified way (not necessarily by rounding down). Remarkably, the clausal problems in this class form a generalization of Horn problems. We call them *extended Horn* problems.

Extended Horn problems have a purely combinatorial characterization. A set of clauses is extended Horn if its atomic propositions correspond to the arcs of some rooted arborescence (i.e., a rooted directed tree in which all arcs point away from the root) in such a way that each clause in the set describes an "extended star-chain" pattern on the arborescence. That is, for each positive literal in the clause one marks the corresponding arc with a forward flow of one unit and for each negative literal a reverse flow of one unit. The overall flow pattern of a clause must consist of arc-disjoint chains, each of which carries a unidirectional flow of one unit into the root, plus at most one chain that is free to exist anywhere on the arborescence.

Ordinary Horn clause systems correspond to arborescences that are stars (i.e., all arcs are incident to a central root). This shows that extended Horn systems indeed represent a substantial generalization of Horn systems.

Since an extended Horn problem can be checked for satisfiability by testing the linear relaxation for feasibility, it can also be checked by unit resolution. But suppose one desires not only a satisfiability check, but a

2.6 EXTENDED HORN SYSTEMS

satisfying solution if one exists. The linear programming approach yields one if the fractions are rounded in the proper way. It turns out that this same solution can be obtained subsequent to unit resolution by setting an unfixed variable to false if its corresponding arc lies an odd distance from the root of the arborescence and to true otherwise. As a special case, all unfixed variables in a classical Horn problem are set to false, because their corresponding arcs are adjacent to the root.

Finding a solution in this manner, however, requires not only that the problem be extended Horn, but that one know how it can be mapped to a rooted arborescence. Fortunately, this knowledge is unnecessary if one uses a slight modification of unit resolution introduced by Schlipf et al. [248]. The modified algorithm, which one might call *extended unit resolution*, runs in quadratic rather than linear time. It checks for satisfiability in any class that has the *unit resolution property*: that is, any problem in the class can be checked for satisfiability by unit resolution and remains in the class after any subset of variables are fixed. Extended Horn problems have the unit resolution property, as do the balanced problems discussed in the next section.

Schlipf et al. point out that, curiously, a slightly more general class than extended Horn has the unit resolution property and can therefore be solved in quadratic time by extended unit resolution. This chapter concludes with a characterization of this larger class that could prove useful in practice.

2.6.1 The Rounding Theorem

Chandrasekaran's rounding theorem [41] provides the basis for defining extended Horn propositions. It begins with a fact proved in Section 2.3.2: that a nonempty polyhedron of the form $P = \{x \mid Dx \geq f, x \geq 0\}$ has an integral least element if each row of D has no more than one positive component and all positive components of D are $+1$ (assuming D and f are integral). A completely analogous argument shows that a nonempty polyhedron $\bar{P} = \{y \mid Ey \geq g, y \leq a\}$ has an integral largest element provided that $a, E,$ and g are integral and each row of E has at most one negative entry, namely, -1. Note that polyhedra with integral least (largest) elements have an interesting "rounding" property. If $\bar{P}$ has an integral largest element and $y \in \bar{P}$ then $\lceil y \rceil$ also belongs to $\bar{P}$. As a consequence we have the following theorem.

Theorem 13 (Chandrasekaran [41]) : *Consider the linear system $Ax \geq b, x \geq 0$, where A is an $m \times n$ integral matrix and b is a commensurate integral vector. Let T be a nonsingular $n \times m$ matrix that satisfies the following conditions:*

66   CHAPTER 2   PROPOSITIONAL LOGIC: SPECIAL CASES

(i) T and T^{-1} are integral.

(ii) Each row of T^{-1} contains at most one negative entry, and all such entries are -1.

(iii) Each row of AT^{-1} contain at most one negative entry, and all such entries are -1.

Then if x is a solution to the linear system, so is $T^{-1}\lceil Tx \rceil$.

Proof: Let A, b, and T (T^{-1}) meet all the conditions of the theorem. Now suppose that $\bar{y}$ is a solution to the system $AT^{-1} \geq b$. Setting $\bar{a} = \lceil \bar{y} \rceil$ we note that the polyhedron $\bar{P} = \{y \ \ AT^{-1}y \geq b, y \leq \bar{a}\}$ has an integral largest element and therefore $\lceil \bar{y} \rceil$ is in $\bar{P}$. What we have shown is that

$$AT^{-1}y \geq b \text{ implies that } AT^{-1}\lceil y \rceil \geq b$$

Setting $x = T^{-1}y$ we have

$$Ax \geq b \text{ implies that } AT^{-1}\lceil Tx \rceil \geq b \tag{2.44}$$

Similarly, supposing $\tilde{y}$ satisfies $T^{-1}y \geq 0$ and setting $\tilde{a} = \lceil \tilde{y} \rceil$, we have that $\tilde{P} = \{y : T^{-1}y \geq 0, y \leq \tilde{a}\}$ has an integral largest element and therefore $\lceil \tilde{y} \rceil$ is in $\tilde{P}$. So,

$$T^{-1}y \geq 0 \text{ implies } T^{-1}\lceil y \rceil \geq 0$$

and using $T^{-1}y = x$ we have

$$x \geq 0 \text{ implies } T^{-1}\lceil Tx \rceil \geq 0 \tag{2.45}$$

The implications (2.44) and (2.45) together prove the theorem. □

We have seen that the integer programming formulation of satisfiability of a CNF proposition S may be expressed as

$$Hx \geq d$$
$$-x \geq -e$$
$$x \geq 0, \text{ integral}$$

where $Hx \geq d$ expresses the clausal inequalities, and $-x \geq -e$ (e is a column vector of 1's) expresses the upper bound of 1 on the x variables.

To apply Theorem 13 to propositional logic we need to identify propositions S for which the linear relaxation

$$\begin{array}{c} Ax \geq b \\ x \geq 0 \end{array} \quad \text{where } A = \begin{bmatrix} H \\ -I \end{bmatrix}, b = \begin{bmatrix} h \\ -e \end{bmatrix}$$

meets the conditions of the theorem. Thus we have to realize a matrix T that satisfies

2.6 EXTENDED HORN SYSTEMS

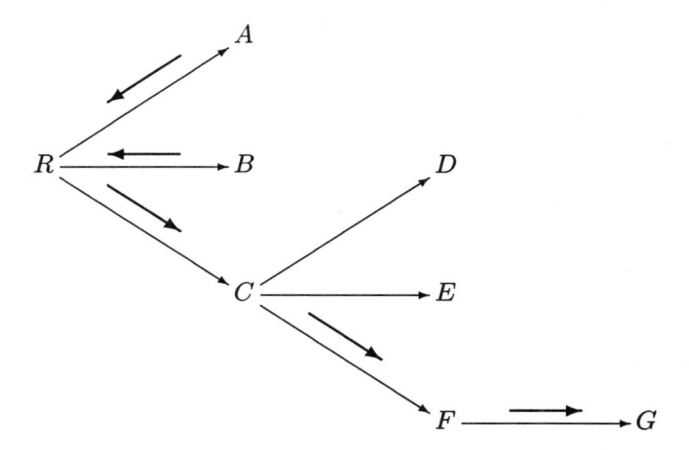

Figure 2.11: Rooted arborescence for an extended Horn proposition.

(i′) T and T^{-1} are non-singular and integral $n \times n$ matrices.

(ii′) Each row of T^{-1} contains at most one $+1$ and at most one -1 and no other nonzero entries.

(iii′) Each row of HT^{-1} contains at most one negative entry, which must be -1.

Condition (ii′) implies that T^{-1} is the edge-vertex incidence matrix of a directed graph. Combining this with the nonsingularity of T^{-1} (i′) we can conclude that T^{-1} is the edge-vertex incidence matrix of a directed tree on $n+1$ vertices. Consider, for example, the following matrix T^{-1} and its corresponding directed tree, shown in Figure 2.11.

$$T^{-1} = \begin{array}{c} \\ 1 \\ 2 \\ 3 \\ 4 \\ 5 \\ 6 \\ 7 \end{array} \begin{pmatrix} A & B & C & D & E & F & G \\ -1 & 0 & 0 & 0 & 0 & 0 & 0 \\ 0 & -1 & 0 & 0 & 0 & 0 & 0 \\ 0 & 0 & -1 & 0 & 0 & 0 & 0 \\ 0 & 0 & 1 & -1 & 0 & 0 & 0 \\ 0 & 0 & 1 & 0 & -1 & 0 & 0 \\ 0 & 0 & 1 & 0 & 0 & -1 & 0 \\ 0 & 0 & 0 & 0 & 0 & 1 & -1 \end{pmatrix} \begin{array}{c} R \\ 1 \\ 1 \\ 1 \\ 0 \\ 0 \\ 0 \\ 0 \end{array}$$

The column corresponding to vertex R is shown to the right of the T^{-1} matrix. There always is an extra column since a directed tree of n edges requires $n+1$ vertices. The entry of column R corresponding to the row of a vertex i is a $+1$ if row i of T^{-1} has no $+1$, is a -1 if row i of T^{-1} has no -1, and is a 0 otherwise. It is the convention in network flow theory to call R the *root* of the directed tree.

Now all that remains is to translate the condition (iii') to the clause structure of our proposition S. Let any row h of our clause matrix H be a vector of E unit flows on the corresponding edges of the directed tree. (Since we are trying to interpret conditions on HT^{-1} we assume that the column indices of H are identified with the row indices of T^{-1}, that is, the edges of the tree.) A $+1$ entry is a unit flow in the corresponding edge in the forward direction, and a -1 is flow in the reverse direction. In our example if $h = [-1\ -1\ 1\ 0\ 0\ 1\ 1]$ the flows on the directed tree are as marked with the heavy arrows in Figure 2.11.

In this example, the vector $hT^{-1} = [\ 1\ 1\ 0\ 0\ 0\ 0\ -1]$, which we interpret as an extraneous supply of one unit of flow each at vertices A and B, and a demand of one unit at vertex G. Flow conservation has to be satisfied at all vertices. At vertex R, flow can be added or removed extraneously to achieve flow balance. The example makes clear what the interpretation of condition (iii') has to be. There can be only one demand vertex other than R. Also, since flow conservation has to be maintained, a set of flows must be along directed chains that empty into the root vertex R, with possibly one exception, namely, a chain that ends at an arbitrary vertex.

Let an *extended star* denote a rooted tree consisting of one or more arc-disjoint chains. Given a rooted tree $\mathcal{T}$ and a clause matrix H, with columns of H in some correspondence with the arcs of $\mathcal{T}$, we say that H has the *extended star-chain property* with respect to $\mathcal{T}$ if each row of H represents a flow that can be partitioned into a set of unit flows into the root on some (possibly empty) extended star subtree of $\mathcal{T}$ and a unit flow on one (possibly empty) chain in $\mathcal{T}$. We can summarize this discussion now.

Lemma 7 *Conditions (i'), (ii'), and (iii') are equivalent to the requirement that there exist some tree $\mathcal{T}$ such that H has the extended star-chain property with respect to $\mathcal{T}$.*

We say that a CNF proposition $\mathcal{H}$ is *extended Horn* if there exists a rooted *arborescence* $\mathcal{A}$ such that the clause incidence matrix H of $\mathcal{H}$ has the extended star-chain property with respect to $\mathcal{A}$. $\mathcal{H}$ is renamable extended Horn if H has the extended star-chain property with respect to a rooted directed tree $\mathcal{T}$. The renaming of $\mathcal{H}$ corresponds to arc reversals in $\mathcal{T}$ that convert $\mathcal{T}$ into an arborescence. Notice also that a (renamable) Horn proposition is a very special case of a (renamable) extended Horn proposition whose (tree) arborescence structure is simply a star graph with the root at the center.

2.6.2 Satisfiability of Extended Horn Systems

We arrived at the class of extended Horn propositions by meeting the conditions of the Rounding Theorem (Theorem 13). So there is an obvious satisfiability algorithm; namely, solve the linear relaxation to obtain $\bar{x}$ and round $\bar{x}$ to $T^{-1}\lceil T\bar{x}\rceil$ to obtain a satisfying assignment.

Theorem 14 ([44]) *An extended Horn proposition $\mathcal{H}$ (explicit or renamable) is satisfiable if and only if the linear relaxation of $\mathcal{H}$ is feasible (i.e., has a solution).*

Since unit resolution solves the linear relaxation of satisfiability, we have the following.

Corollary 3 *An extended Horn proposition $\mathcal{H}$ (explicit or renamable) is unsatisfiable if and only if there is a unit resolution proof of refutation.*

In unit resolution we recursively fix the truth value of a unit clause and simplify the proposition. The simplification involves erasure of clauses and erasure of literals. The erasure of a literal corresponds to contracting the corresponding edge of the arborescence (tree). Note that the star-chain property of a clause is maintained under such an operation. Thus when unit resolution ends we have either established the refutation of satisfiability of $\mathcal{H}$ or reduced $\mathcal{H}$ to $\hat{\mathcal{H}}$ and the corresponding $\mathcal{T}$ to $\hat{\mathcal{T}}$. Obviously $\hat{\mathcal{H}}$ contains no unit clauses, and its linear relaxation has the trivial solution $(\frac{1}{2}, \frac{1}{2}, \ldots, \frac{1}{2})$. In order to construct the satisfying truth assignment, we can compute the rounding $T^{-1}\lceil T(\frac{1}{2}, \frac{1}{2}, \ldots, \frac{1}{2})\rceil$, where T^{-1} is defined by $\hat{\mathcal{T}}$. A simple parity rule solves this problem. Let P_i denote the unique path from the root to vertex i in the arborescence (tree), let arc k be denoted by $(i(k), j(k))$, and let $|P_i|$ be the number of arcs in P_i.

Theorem 15 ([44]) *Let T^{-1} be the $(n \times n)$ vertex edge incidence matrix of a rooted directed tree on $(n + 1)$ vertices (with the column of the root omitted). The kth component of $T^{-1}\lceil(\frac{1}{2}, \ldots, \frac{1}{2})\rceil$ is $|P_{i(k)}|$ (mod 2).*

Proof: Let ℓ_{ij} denote the entries of T^{-1} and t_{ij} those of T. For each edge $k = (i(k), j(k))$ we know that $\ell_{ki(k)} = 1$, $\ell_{kj(k)} = -1$, and $\ell_{kh} = 0$ for $h \neq i(k), j(k)$. It is a well-known result in network flow theory that by row permutations we can bring T^{-1} to triangular form. In such a form, the diagonal entries ℓ_{kk} indicate the orientation of the edge k with respect to the root: $\ell_{kk} = 1$ indicates that edge k is pointing to the root, and $\ell_{kk} = -1$ otherwise. For each vertex i let $F_i = \{k \in P_i | \ell_{kk} = 1\}$ and $B_i = \{k \in P_i | \ell_{kk} = -1\}$, where F stands for forward arc and B for backward arc.

70   CHAPTER 2   PROPOSITIONAL LOGIC: SPECIAL CASES

Another well-known fact about network matrices that is easily checked is that the inverse of a tree matrix is a path incidence matrix. Thus $t_{ik} = 1$ if $k \in F_i$, $t_{ik} = -1$ if $k \in B_i$, and $t_{ik} = 0$ otherwise. If follows that the ith entry of $\lceil T(\frac{1}{2}, \ldots, \frac{1}{2}) \rceil$ is given by $\lceil \frac{1}{2}(|F_i| - |B_i|) \rceil$. Therefore, the kth component of $T^{-1} \lceil T(\frac{1}{2}, \ldots, \frac{1}{2}) \rceil$ is given by

$$\lceil \tfrac{1}{2}(|F_{i(k)}| - |B_{i(k)}|) \rceil - \lceil \tfrac{1}{2}(|F_{j(k)}| - |B_{j(k)}|) \rceil \qquad (2.46)$$

Because $|P_{i(k)}| = |F_{i(k)}| + |B_{i(k)}|$, the quantity (2.46), which is the kth component of $T^{-1} \lceil T(\frac{1}{2}, \ldots, \frac{1}{2}) \rceil$, evaluates to $|P_{i(k)}|$ (mod 2). □

This parity rule simply states that, in an arborescence, edges at odd levels (from the root) are assigned the value false and even levels the value true. Notice that this assignment does indeed satisfy the formula. If the extended star component of a clause C is nonempty, there has to be at least one negative literal in C at level 1 in the arborescence (i.e., an edge incident at the root) that is set to true. And if the extended star of C is empty, there must be a chain of length at least two that contains an odd-level edge pointing into the root or an even-level edge pointing away from the root and again the clause will be satisfied by the parity rule.

It is not difficult to convince oneself that the unit resolution plus parity rule can be implemented to run in linear time on extended Horn systems.

2.6.3 Verifying Renamable Extended Horn Systems

We saw in Section 2.4 that the recognition of renamable Horn propositions can be reduced to a satisfiability question of a quadratic proposition. We also saw the use of auxiliary variables in obtaining a succinct reduction. These ideas carry over to the problem of verifying renamable extended Horn systems.

Suppose we are given a proposition R in CNF and we are also given a rooted arborescence $\mathcal{A}$ whose edges correspond exactly to the atomic letters of R. Now the verification problem of interest is whether there is a renaming of the letters that makes R an extended Horn system with respect to $\mathcal{A}$.

As usual we need a letter (atomic proposition) y_j for each letter x_j of R. If y_j is set to true, then x_j is to be renamed and if y_j is false, then x_j remains as is.

Each clause of R induces a flow structure on $\mathcal{A}$. If the flow structure has more than one component disjoint from the root, then R cannot be renamed to an extended Horn system. Similarly, if other than the root, there are two or more vertices of $\mathcal{A}$ with degree larger than 2 in the flow structure, then R cannot be renamed to an extended Horn system. Also, if a component containing the root has a nonroot vertex of degree greater than 2, then this has to be the only component of the flow structure. And finally, if the flow structure has a component disjoint from the root, the component must be a path of $\mathcal{A}$ (ignoring the edge directions). Having discarded these cases, let us now look at the nontrivial situations. The easiest case is when a clause induces two disjoint flow structures, with one that contains the root. The structure containing the root must be renamed to an extended star and this corresponds to a set of unit clauses on the $\{y_j\}$. The structure disjoint from the root must correspond to a flow chain. Thus, for every pair of adjacent edges in the path of the flow structure, we must ensure that the renaming makes their orientations consistent. This can be enforced by quadratic clauses on the corresponding $\{y_j\}$.

The case where there is a single component in the flow structure gets more complicated, since now there is a choice for the chain in the extended star-chain property being enforced. However, this can be handled by quadratic clauses by introducing variables corresponding to various paths of the flow structure. The details are somewhat tedious but straightforward and lead to a quadratic formula whose satisfiability determines the renamability of R to an extended Horn system on $\mathcal{A}$. The interested reader is directed to the paper [44] wherein a complete description of the verification procedures are given.

The more interesting problem is that of recognition of extended Horn systems: in this problem, we are given only the proposition S and are asked if there is any arborescence $\mathcal{A}$ that makes S extended Horn with respect to it. This is related to classical problems in graph realization. Recent work [267] on this recognition question has shown that it is indeed possible to recognize large subclasses of extended Horn systems with effort that is only slightly more than linear time. However, the question of whether there is even a polynomial-time algorithm for recognizing the entire class of extended Horn formulas remains unanswered as of this writing.

2.6.4 The Unit Resolution Property

When a clause set is renamable extended Horn but no underlying arborescence is known, satisfiability can still be checked in quadratic time by a simple extension of the unit resolution algorithm. This is because renamable extended Horn sets have the *unit resolution property*, defined as follows. A clause set S has the unit resolution property if:

(a) S is unsatisfiable if and only if unit resolution finds a contradiction.

(b) For any given literal L, if S' is the result of removing from S all clauses containing L and deleting all occurrences of literal $\neg L$, then S' has the unit resolution property.

A simple extension of unit resolution [248] not only checks for satisfiability but finds a solution if one exists, provided the clause set in question has the unit resolution property. It is essentially a search with one-step lookahead.

Procedure 9. *Extended Unit Resolution*

Step 1. Let S be a set of clauses; let S' be the result of applying the unit resolution algorithm to S. Stop if a contradiction is found; S is unsatisfiable.

Step 2. If S' is empty, then stop; the variables have been fixed to values that satisfy the clauses in S.

Step 3. Let $S'' = S' \cup \{x_j\}$ for some x_j occurring in S' and apply unit resolution to S''. If S'' is not found to be unsatisfiable, fix x_j to true, let $S' = S''$, and repeat this step.

Step 4. Let $S'' = S' \cup \{\neg x_j\}$ and apply unit resolution to S''. If S'' is not found to be unsatisfiable, fix x_j to false, let $S' = S''$, and return to step 3.

Step 5. Stop without determining whether S is satisfiable.

If n is the number of variables and ℓ the number of literals in S, this algorithm has complexity $O(n\ell)$.

Theorem 16 *If S has the unit resolution property, then the extended unit resolution algorithm correctly decides whether S is satisfiable.*

The algorithm is clearly correct. The only issue is whether it can terminate in step 5 when S has the unit resolution property. But clearly it cannot. Step 5 can be reached only if unit resolution detects a contradiction in both $S' \cup \{x_j\}$ and $S' \cup \{\neg x_j\}$, which means S' is unsatisfiable. This is impossible because (a) unit resolution did not detect a contradiction in S', and (b) S' has the unit resolution property because it is obtained from S by fixing variables.

It has already been noted that the extended star-chain property is preserved by the erasure of literals and clauses. Extended Horn sets therefore have the unit resolution property, and similarly for renamable extended Horn sets. A solution of such a problem (if it exists) can be found in quadratic time by the above algorithm.

In fact, Schlipf et al. [248] point out that a slight generalization of extended Horn sets has the unit resolution property. The chains that carry flows into the root need not be arc-disjoint. So rather than requiring that

2.6 EXTENDED HORN SYSTEMS

the flow induced by each clause have an extended star-chain pattern, one can require only that it have an arborescence-chain pattern.

To put this more precisely, let H be the coefficient matrix representing a set of clauses. H has the generalized extended Horn structure if there is an arborescence $\mathcal{A}$, and a one-to-one correspondence between the variables of H and the arcs of $\mathcal{A}$, with the following properties. Let each row of H induce a directed subgraph $\hat{\mathcal{H}}$ of $\mathcal{A}$ whose arcs correspond to the variables in that row, where the orientation of an arc is reversed if the corresponding variable is negated in the row. Then there must be a partition of the arcs of $\hat{\mathcal{H}}$ into those forming a (possibly empty) arborescence rooted at the root of $\mathcal{A}$ and those forming a (possibly empty) directed chain.

It is not hard to see that if unit resolution removes all unit clauses without finding a contradiction, the remaining clauses can be satisfied by the same alternating pattern of truth values that satisfies an extended Horn set. Furthermore, contraction of an arc does not destroy the arborescence-chain structure. Generalized extended Horn sets therefore have the unit resolution property, and a solution (if one exists) can be found in linear time by extended unit resolution. The same is true after renaming of variables.

2.6.5 Extended Horn Rule Bases

Because of the difficulty of checking whether a clause set is renamable extended Horn, it is useful in practice to build a knowledge base in such a way as to maintain extended Horn structure from the start. This is normally how conventional Horn knowledge bases are obtained. Rather than build a knowledge base and then check whether it is renamable Horn, one simply takes care to use only (explicit) Horn clauses when building it.

This approach is a little more complicated for extended Horn structure, because it is a property of the clause set as a whole rather than individual clauses. Nonetheless one can build an extended Horn set by specifying in advance what the underlying arborescence is to be, and by checking whether each new clause defines an acceptable flow pattern on the arborescence.

The burden of this section is to show that a predefined arborescence can have a natural interpretation in practice, so that it may make sense to build a knowledge base around it. The discussion will deal with rules rather than clauses, because they are more commonly used in applications.

Rules were introduced in Section 2.1.3, which also discussed Horn rules. Briefly again a rule is an implication of the form

$$(y_1 \wedge y_2 \wedge \cdots \wedge y_p) \rightarrow (z_1 \vee z_2 \vee \cdots \vee z_q)$$

The conjunction of the y_i forms the antecedent and the disjunction of the z_j the consequent. Horn rules are those in which the consequent can contain no more than one letter. In considering any generalization or extension of Horn systems, one would hope that this very restrictive syntactic specification can be suitably relaxed.

We now characterize sets of rules that have the generalized extended Horn structure discussed in the previous section. Any such set must correspond to an arborescence $\mathcal{A}$ in the following way. Since the atomic propositions in the consequent of an extended Horn rule are positive in the associated clause, they must form a chain of $\mathcal{A}$ in which every arc is directed away from the root; we can call it the *consequent chain*. To characterize the antecedents, let us also say that a conjunction of premises is a *complete premise* if the individual premises correspond to a chain of $\mathcal{A}$ with all arcs directed toward the root and one endpoint at the root. A conjunction of premises is a *partial premise* if it corresponds to a chain of $\mathcal{A}$ that is not incident to the root, and whose arcs are all directed toward the root. A partial premise is *coterminal* with the consequent chain if the two chains have an endpoint in common.

Thus a set of rules is extended Horn if and only if there is some arborescence $\mathcal{A}$ such that every consequent corresponds to a chain in $\mathcal{A}$, and the antecedent of every rule is a conjunction of zero or more complete premises and at most one partial premise, where the latter is coterminal with the consequent chain. Because the rule base is to be extended Horn in the generalized sense of the previous section, the complete premises can have letters in common (i.e., flows into the root need not be arc-disjoint).

The arborescence that corresponds to an extended Horn set can be viewed as defining paths of inquiry, corresponding to chains leading out from the root. Ascertaining the truth of each proposition leads one to raise certain other questions represented by the children (immediate successors) of the propositions in the tree. If a certain child is found to be true, one investigates *its* children, and so on. The premises of a typical rule are satisfied when one pursues one or more lines of inquiry and, in each case, finds the relevant propositions to be true. Thus one would expect the antecedent of a rule to consist of one or more complete premises, each representing a line of inquiry. For the consequent to be a chain, it must assert that at least one question in some partial line of inquiry has an affirmative answer.

Consider, for example, the arborescence in Fig. 2.12, in which lines of inquiry relate to determining whether a person should receive government benefits. Consider the extended Horn set of rules:

(i) If an applicant cannot work, then he/she is either retired or disabled.

(ii) If an applicant is poor and cannot work, and if he/she is retired, then he/she is entitled to benefits.

2.6 EXTENDED HORN SYSTEMS

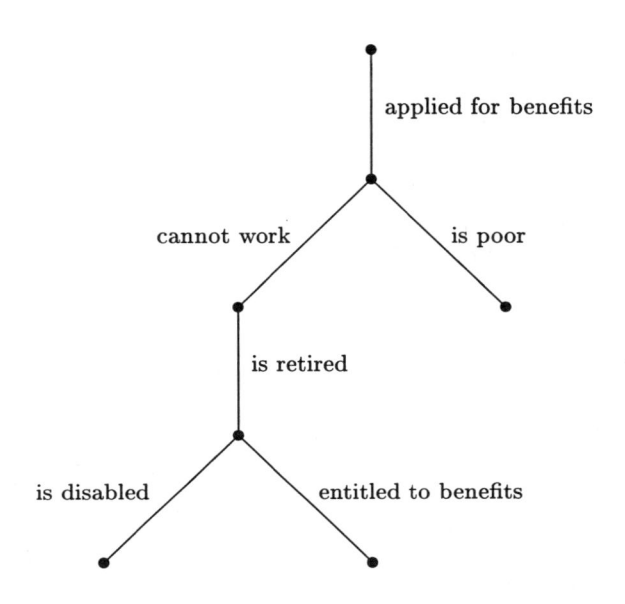

Figure 2.12: Arborescence for an extended Horn rule base.

(iii) If an applicant is disabled, he/she is entitled to benefits.

The first rule contains one complete premise, which is satisfied when one successfully follows a two-step line of inquiry. The rule is non-Horn because of the disjunctive consequent, but it is extended Horn with respect to the arborescence because the consequent arcs form a chain.

The second rule contains two complete premises. One establishes that one is an applicant, cannot work, and is retired. The other establishes that one is an applicant and poor. Note that the premises overlap, which is permitted by generalized extended Horn sets.

The third rule contains a partial premise that is coterminal with the consequent. Thus a partial premise represents a partial line of inquiry that need not start at the beginning. The only restriction is that it start at the same point as the conclusion drawn from it.

Rules (i)–(iii) permit a limited form of disjunctive reasoning to infer the following:

(iv) If an applicant is poor and cannot work, he is entitled to benefits.

If this rule is negated and added to rules (i)–(iii), the resulting clause set can be proved unsatisfiable by unit resolution.

This example, although non-Horn, happens to be renamable Horn. The following clause set, however, is extended Horn but not renamable Horn.

$$x_1 \to (x_2 \vee x_3)$$
$$(x_2 \wedge x_3) \to x_1$$

The underlying arborescence contains two arcs adjacent to the root (corresponding to x_1, x_3) and a third arc (corresponding to x_2) adjacent to the x_3 arc.

2.7 Problems with Integral Polytopes

We have already seen that a Horn or extended Horn satisfiability problem can be solved by linear programming. This is because the problem is satisfiable if and only if its linear relaxation is feasible.

Another class of satisfiability problems soluble by linear programming are those whose linear relaxations have only *integral* extreme point solutions. If the relaxation has an extreme point solution, we know right away that the satisfiability problem has a solution: just let the atomic proposition x_j be true when $x_j = 1$ in the extreme point solution, and false otherwise. Since the relaxation has an extreme point solution if it has a solution, the problem is satisfiable if and only if the relaxation is feasible.

Consider, for instance, the clause set

$$\begin{aligned} x_1 \vee x_2 \vee x_3 \\ \neg x_1 \vee \neg x_2 \vee \neg x_3 \end{aligned} \quad (2.47)$$

The linear relaxation is

$$\begin{aligned} x_1 + x_2 + x_3 &\geq 1 \\ -x_1 - x_2 - x_3 &\geq -2 \\ 0 \leq x_j \leq 1, \quad j &= 1, \ldots, 3 \end{aligned} \quad (2.48)$$

It has extreme point solutions $(x_1, x_2, x_3) = (0, 0, 1)$, $(0, 1, 0)$, $(0, 1, 1)$, $(1, 0, 0)$, $(1, 0, 1)$, $(1, 1, 0)$, all of which are integral. Therefore (2.47) has a solution because (2.48) does.

Another advantage of constraint sets with integral polytopes is that linear programming finds integral solutions for optimization problems. The typical linear programming algorithm finds an optimal extreme point. Thus when all extreme points are integral, it finds an integral optimum. For instance, one can find an integral point that maximizes $2x_1 - 3x_2 + x_3$ subject to (2.48) simply by using a linear programming algorithm to find

2.7 PROBLEMS WITH INTEGRAL POLYTOPES

the optimal extreme point $(1, 0, 1)$. This sort of optimization problem is important in probabilistic logic (Section 4.1.3).

Horn satisfiability problems, even though they are soluble by linear programming, do not necessarily have integral polytopes. For instance, the linear relaxation of the clause set

$$\begin{aligned} x_1 \vee \neg x_2 \\ \neg x_1 \vee \neg x_2 \end{aligned} \tag{2.49}$$

has the extreme point solution $(x_1, x_2) = (\frac{1}{2}, \frac{1}{2})$. If a linear programming algorithm finds this solution, we know that the clauses are satisfiable, but only because a satisfying solution can be obtained in Horn systems by rounding the $\frac{1}{2}$'s down to 0. Also, linear programming cannot in general find integral solutions for optimization problems over a Horn constraint set.

Linear constraint sets with integral polyhedra have been studied intensely, because they represent integer programming problems that can be solved by linear programming. The most basic result is that a linear system

$$\begin{aligned} Ax \geq a \\ 0 \leq x_j \leq 1, \text{ all } j \end{aligned} \tag{2.50}$$

has an integral polytope if the matrix A is *totally unimodular*, meaning that every square submatrix of A has a determinant of 0, 1, or -1. But this is a strict condition, and it is hard to check whether it is satisfied. See [249] for a survey of results in this area.

This section explores two additional integrality results for satisfiability problems. The first, obtained by Conforti and Cornuéjols [63], states that any satisfiability problem represented by a *balanced* matrix has an integral polytope. Problems with balanced matrices not only have the unit resolution property discussed in the previous section, but they have a stronger property that allows one to find a solution (when it exists) in linear time.

A second result [147] uncovers a connection between integrality and the well-known resolution algorithm for inference, to be discussed more fully in Chapter 3. The result relies on the fact that a satisfiability problem can be viewed as a union of overlapping set covering subproblems. It says that if one augments a clause set with all of the inferences that can be drawn by resolution, the resulting problem has an integral polytope if and only if its set-covering subproblems do. Resolution therefore reduces the integrality question for satisfiability problems to that for set-covering problems, which are essentially satisfiability problems with all positive literals.

It is unclear at this writing whether these two results have practical benefits. The former is limited by the fact that problems with balanced matrices are relatively few and hard to recognize. The second falls victim to two liabilities: the resolution algorithm tends to be very inefficient (it requires exponential time in the worst case), and a satisfiability problem has in general exponentially many set-covering subproblems. But both results afford intriguing glimpses into some of the deeper connections between logic and linear programming.

2.7.1 Balanced Problems

A matrix with components in $\{1, -1, 0\}$ is *balanced* if every square submatrix with exactly two nonzero entries in each row and column has the property that its entries sum to a multiple of four. It can be shown that if (2.50) is the linear relaxation of a satisfiability problem, and A is a balanced matrix, then (2.50)'s polytope is integral. For convenience, we say that a set of clauses is balanced if A is balanced.

Theorem 17 ([63]) *The satisfiability problem for a balanced set of clauses has an integral polytope.*

For instance, (2.47) is balanced and indeed has an integral polytope. A balanced problem need be neither Horn nor renamable Horn, since (2.47) is neither. Conversely, Horn problems need not be balanced, as (2.49) demonstrates. A proof of Theorem 17 appears at the end of this section.

It is unknown whether balancedness has a natural logical interpretation. But two things can be said. First, balanced problems have the unit resolution property discussed in the previous section and can therefore be solved by extended unit resolution. It is clear from the definition of balanced matrices that when variables of a balanced problem are fixed, the remaining problem is still balanced. Also, unit resolution checks balanced problems for satisfiability because the latter is equivalent to feasibility of the linear relaxation.

Second, balanced problems have a property that is even stronger than the unit resolution property: only unit clauses can fix variables.

Corollary 4 *If a balanced clause set contains no unit clauses, then for any atomic proposition x_j, the clause set has a solution in which x_j is true and one in which x_j is false.*

Proof: Since every clause has at least two literals, the linear relaxation is satisfied by setting $x_j = \frac{1}{2}$ for all j. This point must be the convex combination of vertices of the polytope described by the relaxation, which by Theorem 17 are integral. Clearly, for every j there is a vertex x in the convex combination at which $x_j = 0$ and a vertex at which $x_j = 1$. □

2.7 PROBLEMS WITH INTEGRAL POLYTOPES

This means that extended unit resolution solves a balanced problem even without the one-step lookahead. First apply unit resolution to eliminate all unit clauses. Then pick any remaining atom x_j and add the unit clause x_j or $\neg x_j$, arbitrarily. By Corollary 4, the system remains satisfiable if it was satisfiable to begin with. Repeat the procedure until unit resolution detects inconsistency, in which case the problem is unsatisfiable, or eliminates all clauses, in which case it is satisfiable. This algorithm obviously runs in linear time.

The problem of recognizing balances matrices with coefficients in $\{0,+1\}$ is discussed in [65]. The recognition algorithm persented there is based on decomposition results that have been extended to $0,+1,-1$ matrices in [64].

To illustrate the style of argument in this area, we will conclude with a demonstration of the main theorem.

Proof of Theorem 17. Suppose to the contrary that A in (2.50) is balanced but that (2.50) has an extreme point solution x^* with at least one fractional component. Also, let A be a minimal counterexample with respect to the number of rows plus the number of columns.

Since x^* is an extreme point solution, some subset of the constraints (2.50) (possibly including bounds $x_j \geq 0$, $-x_j \geq -1$) forms a system $Bx \geq b$ for which $Bx = b$ has the unique solution x^*. In fact, the minimality of A allows us to show that x^* is the unique solution of $Ax = a$. If $Bx \geq b$ contains a bound $x_j \geq 0$ or $-x_j \geq -1$, we can add the unit clause $\neg x_j$ or x_j (respectively) to the clause set corresponding to $Bx \geq b$ and perform the unit resolution procedure. This yields a smaller system without x_j and its bound. By eliminating all bounds in this way we obtain a system $B'y = b'$ whose unique solution y^* contains all the fractional components of x^*, and for which B' is a submatrix of A and therefore balanced. Since A is minimal, we have $A = B'$, and therefore $Ax^* = a$.

Our strategy from here out is to show that $Ax = a$ must have an integral solution, which contradicts the fact that the nonintegral vector x^* is the unique solution.

Let $\bar{A}x = \bar{a}$ be the result of removing the last constraint from $Ax = a$. By complementing variables, we can suppose that the last constraint contains no negative coefficients and write it $cx \geq 1$. $\bar{A}x = \bar{a}$ is feasible because x^* satisfies it. All of its extreme points are integral, because A is a minimal counterexample. Thus x^* is a convex combination of integral points satisfying $\bar{A}x = \bar{a}$. Since x^* satisfies $cx = 1$ but none of the integral points do, there must be integral points y, z for which $cy = 0$ and $cz > 1$.

Let U be the set of indices for which $y_j = 1$ and $z_j = 0$, and V the set for which $y_j = 0$ and $z_j = 1$. We will need the following fact.

Claim: In the submatrix of $\bar{A}$ consisting of columns $U \cup V$, every nonzero row contains exactly two nonzero entries, and they have the same sign if and only if one belongs to U and one to V.

Proof of claim: Let us say that a 0-1 point x "activates" a_{ij} if it makes the literal $a_{ij}x_j$ true. That is, x activates a_{ij} if $a_{ij} = x_j = 1$, or if $a_{ij} = -1$ and $x_j = 0$. Since $\bar{A}y = \bar{a}$, y activates exactly one entry in each row of $\bar{A}$, and similarly for z. But in any given row, y activates only 1's in U and -1's in V, whereas z does the reverse. Thus there is at most one 1 and at most one -1 in U, and similarly for V.

It remains to show that there are at least two nonzeros in $U \cup V$, if any. But if there is a 1 in U, y activates it. So z must activate some other entry in $U \cup V$. This is because z does not activate the 1 in U, and if z activated an entry outside $U \cup V$, y would activate it, too, and would therefore activate two entries in the row. So there is another nonzero in $U \cup V$. The same sort of argument applies if there is a -1 in U, or a 1 or -1 in V. □

We will now construct an integral solution of $Ax = a$ by modifying y. Since $cz > 1$, z must activate at least one entry (in fact, at least two) in c. But since y activates no entries in c and $c \geq 0$, we must have $c_k = 1$ for some $k \in V$. Consider the graph G that contains a node corresponding to each column j in $U \cup V$ and an arc (j, j') whenever a_{ij} and $a_{ij'}$ are both nonzero in some row i of $\bar{A}$. Let C be the connected component of G containing node k. Then we modify y by changing any $y_j = 1$ to 0 when $j \in C$, and any $y_j = 0$ to 1 when $j \in C$, to obtain a new point y^*.

It remains to show that the integral point y^* satisfies $Ay^* = a$. We first note that $\bar{A}y^* = \bar{a}$, due to the fact that $\bar{A}y = \bar{a}$. Suppose y activates an entry a_{ij} that y^* does not. Then either (a) $y_j = 1$ and $y_j^* = 0$ or (b) $y_j = 0$ and $y_j^* = 1$. In case (a), $a_{ij} = 1$ and $j \in U$, and the above Claim implies that row i contains a -1 in some column j' of U or a 1 in some column j' of V. Thus $j' \in C$, and y^* activates the 1 or -1. In case (b), $a_{ij} = 0$ and $j \in V$, and similarly y^* activates another 1 or -1. If y^* activates an entry that y does not, then by analogous reasoning y^* fails to activate an entry that y activates. Since y activates exactly one entry in every row of $\bar{A}$, it follows that y^* does the same, so that $\bar{A}y^* = \bar{a}$.

We must now show that $cy^* = 1$, from which it follows that $Ay^* = a$. Here at last we use the fact that A is balanced. Since $cy = 0$, y activates none of the 1's in c. It therefore suffices to show that y^* activates exactly one of the c_j for which $y_j = 0$ and $y_j^* = 1$; that is, for which $j \in C$. For this it is enough to show that $c_j = 1$ for exactly one $j \in C$.

We already know that $c_j = 1$ for at least one $j \in C$, because $c_k = 1$. To show that $c_j = 0$ for all other $j \in C$, suppose to the contrary that $c_{k'} = 1$ for some $k' \in C$, $k' \neq k$. Choose k' so that the length of a shortest path P

2.7 PROBLEMS WITH INTEGRAL POLYTOPES 81

from k to k' in G is as short as possible. Let B be the square submatrix of A with columns corresponding to the nodes of P and rows corresponding to the arcs of P, along with the row c. Since $k, k' \in V$, the path P must pass from a node in V to a node in U, or from a node in U to a node in V, an odd number of times. Then by the above Claim, B contains an odd number of rows with two 1's; the remaining rows contain a 1 and -1 or else all zeros. Thus B's entries do not sum to a multiple of 4, which contradicts the fact that A is balanced. This completes the proof of Theorem 17. $\square$

2.7.2 Integrality and Resolution

Although a balanced satisfiability problem has an integral polytope, the reverse need not hold. Consider, for example, the satisfiability problem,

$$\begin{aligned} x_1 + x_2 + x_3 &\geq 1 \\ x_1 - x_2 - x_3 &\geq -1 \end{aligned} \tag{2.51}$$

The problem is not balanced because the entries of the square submatrix of the first two columns, for example, sum to 2 (not a multiple of 4). But the problem defines an integral polytope.

No known structural property of satisfiability problems is both necessary and sufficient for integrality. There is a connection, however, between integrality and logical inference that leads to a necessary and sufficient condition for a certain class of problems. Consider the following satisfiability problem.

$$\begin{aligned} x_1 + x_2 &\geq 1 \quad \text{(a)} \\ x_2 + x_3 &\geq 1 \quad \text{(b)} \\ x_2 - x_3 &\geq 0 \quad \text{(c)} \end{aligned} \tag{2.52}$$

It does not define an integral polytope because the polytope has two nonintegral extreme points, $(x_1, x_2, x_3) = (\frac{1}{2}, \frac{1}{2}, \frac{1}{2})$ and $(1, \frac{1}{2}, \frac{1}{2})$. Note, however, that constraints (b) and (c) have an obvious implication: $x_2 \geq 1$. If this constraint is added to the set, the two noninteger extreme points are cut off; they are no longer feasible.

The implication $x_2 \geq 1$ is a special case of a *resolvent*. In general, two clauses can be *resolved* when exactly one variable x_j occurs posited in one and negated in the other. Their resolvent is a clause consisting of all the literals in either parent except x_j and $\neg x_j$. For instance, the third clause on the left below is the resolvent of the first two:

$$\begin{array}{ll} x_1 \vee x_2 \vee \neg x_3 & \quad x_1 + x_2 - x_3 \geq 0 \\ \neg x_1 \vee \neg x_3 \vee x_4 & \quad -x_1 - x_3 - x_4 \geq -2 \\ x_2 \vee \neg x_3 \vee x_4 & \quad x_2 - x_3 - x_4 \geq -1 \end{array}$$

The inequality form appears on the right. A resolvent is not in general the sum of its parents, but it is implied by its parents. The reasoning is by cases. In the example, the first parent says that if x_1 is false then x_2 or $\neg x_3$ is true, whereas the second parent says that if x_1 is true then $\neg x_3$ or x_4 is true; in either case the resolvent is true.

Example (2.52) suggests that if a constraint set satisfies a logical closure property—that is, it contains any resolvent that can be obtained from a pair of its clauses—then it defines an integral polytope. Actually this is not true in general, but it is true if and only if the *set-covering subproblems* that make up the satisfiability problem are all integral.

A set-covering constraint set has the form $Ax \geq e$, where A is a matrix of components in $\{0, 1\}$ and e is a vector of 1's. For example, constraints (a) and (b) in (2.52) define a set-covering problem that one may view as a set-covering subproblem of (2.52). Constraints (a) and (c) also define a set-covering subproblem after complementing x_3 (i.e., substituting $1 - x_3$ for x_3). In general, a set-covering subproblem of a satisfiability problem $Ax \geq a$ is a *monotone* subset of its constraints; that is, a subset in which no variable occurs both posited and negated. A subproblem is *maximal* if the addition of any inequality from $Ax \geq a$ destroys monotonicity. Both subproblems just mentioned are maximal.

A set of clauses may be described as *complete* when any resolvent that can be obtained from two of its clauses is absorbed by a clause already in the set. This leads to the following necessary and sufficient condition for integrality of complete problems.

Theorem 18 ([147]) *Let $Ax \geq a$ represent a complete set of clauses. The inequalities $Ax \geq a$, $0 \leq x \leq e$ describe an integral polytope if and only if $A'x \geq a'$, $0 \leq x \leq e$ describe an integral polytope for every maximal set-covering subproblem $A'x \geq a'$ of $Ax \geq a$.*

This does not provide a structural characterization of integrality, but it shows that a logical inference procedure, resolution, in some sense reduces the integrality question for satisfiability to that for set covering problems.

Before proving this result, it is useful to observe that integral satisfiability problems have the unit resolution property, because of the following elementary fact.

Lemma 8 *If any variable is fixed to 0 or 1 in any constraint set $Ax \geq a$, $0 \leq x \leq e$ that defines an integral polytope, the resulting constraint set likewise defines an integral polytope.*

Proof: Suppose without loss of generality that x_1 is fixed to δ, where δ is 0 or 1. Let $\bar{A}\bar{x} \geq \bar{a}$ be the resulting constraint set, where $\bar{x} = (x_2, \ldots, x_n)$.

2.7 PROBLEMS WITH INTEGRAL POLYTOPES 83

Then if $\bar{x}$ were a noninteger extreme point of $\bar{A}\bar{x} \geq \bar{a}$, $0 \leq \bar{x} \leq e$, $(\delta, \bar{x})$ would be a noninteger extreme point of $Ax \geq a$. For if $\bar{x}$ is an extreme point, it satisfies $n-1$ independent equations in the system $\bar{A}\bar{x} = \bar{a}$, $\bar{x} = 0$, $\bar{x} = e$. This implies that $(\delta, \bar{x})$ satisfies the corresponding equations of the system $Ax = a$, $x = 0$, $x = e$, plus $x_1 = \delta$ and is therefore an extreme point. □

Since integral satisfiability problems have the unit resolution property, they can be solved not only by linear programming but by the extended unit resolution algorithm described in Section 2.6.4.

Proof of Theorem 18: We first show that if a complete satisfiability problem $Ax \geq a$ defines an integral polytope P, then all maximal set-covering subproblems of it are integral.

Suppose to the contrary, and let $Ax \geq a$ be a counterexample that is minimal with respect to the number of variables. Let x^* be a nonintegral extreme point of a maximal set-covering subproblem $A'x \geq a'$. We will show that x^* is also a nonintegral extreme point of P, which is a contradiction.

Since A' is monotone, one may assume without loss of generality that all its components are nonnegative. Due to maximality it contains all nonnegative rows of A. Since x^* is an extreme point, it is the unique solution of some system $Bx = b$ of n independent equations among $A'x = a'$, $x = 0$, $x = e$.

We first note that due to minimality, all equations in $Bx = b$ are in $A'x = a'$. To see this, suppose to the contrary that $Bx = b$ contains some equation not in $A'x = a'$. Without loss of generality we may assume that this equation has the form $x_1 = \delta$, where δ is 0 or 1. Consider the problem $\bar{A}\bar{x} \geq \bar{a}$ obtained by substituting δ for every occurrence of x_1 in $Ax \geq a$. Let $\bar{A}'\bar{x} \geq \bar{a}'$ and $\bar{B}\bar{x} = \bar{b}$ be the systems similarly obtained from $A'x \geq a$ and $Bx = b$. Then $\bar{A}\bar{x} \geq \bar{a}$ defines an integral polytope, by Lemma 8. Also $\bar{x}^* = (x_2^*, \ldots, x_n^*)$ is a unique solution of $\bar{B}\bar{x} = \bar{b}$ and is therefore a nonintegral extreme point of $\bar{A}'\bar{x} \geq \bar{a}'$. The latter is therefore a counterexample smaller than $Ax \geq a$, which violates minimality.

We are now in a position to show that x^* is a (nonintegral) extreme point of P. For this it suffices to show that x^* satisfies $Ax \geq a$. Suppose to the contrary that x^* fails to satisfy some inequality I, which may be written $\sum_{j \in J} f_j(x_j) \geq 1$. Here each $f_j(x_j)$ is either x_j or $1 - x_j$. We will also write $f(J)$ as shorthand for $\sum_{j \in J} f_j(x_j)$, and $f^*(J)$ as short for $\sum_{j \in J} f_j^*(x_j)$.

We can suppose that $f_t(x_t) = 1 - x_t$ for at least one t, because otherwise I belongs to $A'x \geq a'$ and is satisfied by x^*. There are two cases.

CHAPTER 2   PROPOSITIONAL LOGIC: SPECIAL CASES

Case I. For some t with $f_t(x_t) = 1 - x_t$, no equation in $Bx = b$ represents a clause that can be resolved on x_t with I. But since B is nonsingular, there is an equation of $Bx = b$ containing x_t. Since every equation of $Bx = b$ is, as shown above, an equation of $A'x = a'$, some equation of $Bx = b$ represents a clause containing x_t that cannot be resolved on x_t with I. This equation and I can be written, respectively,

$$\begin{array}{c} x_t + x_{t'} + x(J') = 1 \\ (1 - x_t) + (1 - x_{t'}) + f(J \setminus \{t, t'\}) \geq 1 \end{array} \quad (2.53)$$

for some $t' \neq t$, where $x(J') = \sum_{j \in J'} x_j$. But since x^* satisfies the equation in (2.53), the latter implies that $x_t^* + x_{t'}^* \leq 1$, which in turn implies that

$$(1 - x_t^*) + (1 - x_{t'}^*) \geq 1$$

This means that x^* satisfies I, as desired.

Case II. For every t with $f_t(x_t) = 1 - x_t$, I can be resolved on x_t with a clause represented by some equation of $Bx = b$. We can write I in the form

$$\bar{x}(T) + x(U) \geq 1 \quad (2.54)$$

where $\bar{x}(T) = \sum_{j \in T}(1 - x_j)$. Since B is nonnegative, the equations of $Bx = b$ with which (2.54) can be resolved on x_t for $t \in T$ can be written

$$x_t + x(J_t) = 1 \quad (2.55)$$

where each J_t is disjoint from T. A series of resolutions of the clause corresponding to (2.54) with the clauses corresponding to (2.55) yields the inequality

$$x(\bigcup_{t \in T} J_t \cup U) \geq 1 \quad (2.56)$$

We are given that x^* satisfies each (2.55), which implies $1 - x_t^* = x^*(J_t)$. Substituting this into the left hand side of (2.54) yields the expression

$$\sum_{t \in T} x^*(J_t) + x^*(U) \quad (2.57)$$

But (2.56) must be absorbed by some clause in $Ax \geq a$. Since this clause contains only positive literals, it belongs to $A'x \geq a'$ and is therefore satisfied by x^*. Hence x^* satisfies (2.56). This implies that (2.57), which is the left-hand side of (2.54), is at least 1. Thus x^* satisfies (2.54).

2.7 PROBLEMS WITH INTEGRAL POLYTOPES

To prove the converse of the theorem, suppose that $Ax \geq a$ describes a polytope with a nonintegral extreme point x^* and show that some maximal set-covering subproblem of $Ax \geq a$ also has a nonintegral extreme point. Let us say that a system $A'x = a'$ is monotone if every column of A' is monotone.

Since x^* is an extreme point, an independent subset E of n of the equations $Ax = a$, $x = 0$, $x = e$ has x^* as its unique solution. We will show that an independent *monotone* subset M of n of these same equations has x^* as its unique solution. This will establish that some maximal set-covering subproblem of $Ax \geq a$ has a nonintegral extreme point.

For each j, if $x_j^* = 1$, add the equation $x_j = 1$ to E (if it is not already present), and if $x_j^* = 0$, add $x_j = 0$ to E. Since E has rank n, and the equations in E of the form $x_j = 1$, $x_j = 0$ are independent, the latter are contained in some independent subset of n equations of E. Let this subset be E', of which x^* is the unique solution. We will refer to the equations in E' that are part of the system $Ax = a$ as clausal equations and the remainder as bound equations. We regard each clausal equation i as having the form $f_i(J_i) = 1$. We will also suppose that x_j^* is fractional for $j \in N_1$ and integer for $j \in N_2$.

We will now construct a *monotone* subsystem M of n independent equations from $Ax = a$, $x = 0$, $x = e$ that have the same unique solution x^*. First set M equal to E'. Pick any x_t that changes sign in the clausal equations of M; that is, $f_{it}(x_t) = x_t$ and $f_{i't}(x_t) = 1 - x_t$ for some i, i'. We will show first that x_t^* must be nonintegral. This will allow us to show that clauses i and i' (i.e., the clauses corresponding to equations i and i') can be resolved. Then we show that since they can be resolved, M can be altered so that the sign change in x_t between clauses i and i' is removed, and no sign change is introduced into any column that is already monotone. By repeated application of this procedure, we obtain the desired M.

We begin by showing that x_t^* is neither 0 nor 1. The coefficient matrix B for the equations in M has the structure

$$\begin{pmatrix} C_1 & C_2 \\ D_1 & D_2 \\ 0 & I \end{pmatrix}$$

The left half contains the columns in N_1 and the right half the columns in N_2. (C_1, C_2) consists of rows i and i'. I contains the coefficients of the bound equations.

Suppose first that $x_t^* = 1$. Then since x_j^* is fractional for all $j \in N_1$ and $f_i^*(J_i) = 1$, J_i must be disjoint from N_1. This means that row i in B is a linear combination of rows in $(0, I)$, which contradicts the nonsingularity of B. Similarly, if $x_t^* = 0$, then row i' is a linear combination of rows in $(0, I)$, and the nonsingularity of B is again violated.

Having shown that x_t^* is noninteger, we show that clauses i and i' must have a resolvent. Suppose that they do not. Then equations i and i' can be written, respectively,

$$\begin{aligned} x_t + x_{t'} + f_i(J_i) &= 1 \\ (1 - x_t) + (1 - x_{t'}) + f_{i'}(J_{i'}) &= 1 \end{aligned} \quad (2.58)$$

or

$$\begin{aligned} x_t + (1 - x_{t'}) + f_i(J_i) &= 1 \\ (1 - x_t) + x_{t'} + f_{i'}(J_{i'}) &= 1 \end{aligned} \quad (2.59)$$

Since x^* satisfies equations (2.58), by adding them we obtain that

$$f_i^*(J_i) = f_{i'}^*(J_{i'}) = 0$$

Thus $x_t^* + x_{t'}^* = 1$. Since x_t^* is fractional, this implies that $x_{t'}^*$ is also fractional and $t' \in N_1$. Also, since x_j^* is fractional for all $j \in N_1$, J_i and $J_{i'}$ are disjoint from N_1. Thus the two rows of C_1 are negations of each other.

Similarly, by adding equations (2.59) we obtain that $x_t^* = x_{t'}^*$, which again implies that $t' \in N_1$, and that J_i and $J_{i'}$ are disjoint from N_1. The two rows of C_1 are again negations of each other.

Thus in either case C_1 is not full rank. But

$$\det(B) = \det\begin{pmatrix} C_1 \\ D_1 \end{pmatrix}$$

which is nonzero only if C_1 is full rank. Since B is nonsingular, we have a contradiction.

Since clauses i and i' must have a resolvent, the corresponding equations can be written, respectively,

$$\begin{aligned} x_t + f_i(J_1) + \phantom{f_{i'}(J_2) +} + f_i(J_3) &= 1 \\ (1 - x_t) + f_{i'}(J_2) + f_i(J_3) &= 1 \end{aligned} \quad (2.60)$$

Since $Ax \geq a$ is complete, some clause i'' in $Ax \geq a$ absorbs the resolvent of clauses i and i'. We therefore have

$$f_i(J_1) + f_{i'}(J_2) + f_i(J_3) \geq 1 \quad (2.61)$$

Since x^* satisfies (2.60) and (2.61), by subtracting the sum of equations (2.60) from (2.61) we obtain that

$$f_i^*(J_3) = 0$$

2.8 LIMITED BACKTRACKING

Substituting this into the sum of equations (2.60), we see that x^* satisfies

$$f_i(J_1) + f_{i'}(J_2) + f_i(J_3) = 1 \qquad (2.62)$$

Also, since $f_i^*(J_3) = 0$, the equations $x_j = 0$ or $x_j = 1$ for $j \in J_3$ are part of M.

We now construct a set M' of equations by modifying M as follows: replace either equation in (2.60) by the equality form of clause i'' in $Ax \geq a$, that is, the clause that absorbs (2.61). We claim that since x^* uniquely satisfies the equations in M, it uniquely satisfies the equations in M'. To see this, write clause i'' as follows:

$$f_i(J_1') + f_{i'}(J_2') + f_i(J_3') \geq 1 \qquad (2.63)$$

Here $J_i' \subset J_i$ for $i = 1, 2, 3$ because (2.63) absorbs (2.61). We first note that x^* satisfies (2.63) as an equation; that is, x^* satisfies

$$f_i(J_1') + f_{i'}(J_2') + f_i(J_3') = 1 \qquad (2.64)$$

This is because x^* satisfies both (2.63) and (2.62). Therefore x^* satisfies all the equations in M'. We next show that x^* is the only point that does so. Since $f_i^*(J_3) = 0$, the fact that x^* satisfies (2.62) and (2.64) implies that $f_i^*(J_1 \setminus J_1') = f_{i'}^*(J_2 \setminus J_2') = 0$. Therefore $f_i(J_1 \setminus J_1') = 0$ and $f_{i'}(J_2 \setminus J_2') = 0$ are enforced by equations $x_j = 0$ or $x_j = 1$ belonging to M'. This means that any x that satisfies the equations in M' satisfies (2.62) and therefore both equations in (2.60). So any x that satisfies the equations in M' satisfies the equations in M and must be x^*.

Thus if we let M' be the new M, x^* still uniquely satisfies M. Since (2.61) does not contain x_t, the sign conflict between i and i' involving x_t is removed from M. Furthermore, no sign change is introduced in any monotone column of M.

It is clear that all sign changes involving x_t can be eliminated from M in this manner, without destroying the monotonicity of any column of M. Repeated application of the procedure to each x_t yields the desired M. □

2.8 Limited Backtracking

One final way in which a satisfiability problem can exhibit special structure is for it to be soluble with limited backtracking. That is, when the problem is solved by a branching algorithm, the size of the resulting search tree can be bounded by virtue of special structure in the problem.

The principle of a branching algorithm is very simple. One might begin by branching on x_1; that is, by distinguishing cases in which x_1 is assumed to be true and x_1 is assumed to be false. In each of these cases one simplifies the problem to the extent possible and considers cases in which, say, x_2 is true and false. The process continues, generating a search tree. The leaf nodes of the tree are those at which the problem is simplified to the point that it can easily be checked for satisfiability without branching (i.e., it may be a Horn problem).

The original problem is satisfiable if and only if the problem at some leaf node is satisfiable. One keeps generating nodes until this happens, or the entire tree is generated. Whenever an unsatisfiable leaf node is reached, it is necessary to return to a previous node and take the other branch. This is called *backtracking*. A large amount of backtracking results in the generation of many nodes and retards solution of the problem. The Davis-Putnam-Loveland algorithm, discussed in the next chapter, is a special case of a branching procedure.

There are problem classes in which the amount of backtracking can be limited. This can occur in at least two ways. One is to bound the *depth* of the search tree, or the maximum number of branches required to reach any leaf node. In some problems, for example, the number of branches required to obtain a renamable Horn problem can be bounded. This happens if the problem becomes renamable Horn whenever a certain subset of variables are fixed to any set of values. The task of finding the smallest subset for a given problem instance is NP-hard, but it can be formulated as a set-packing or maximum clique problem, for which several heuristics have been developed.

Another way to limit backtracking is to bound the distance from every node to the *closest* leaf node. This occurs, for example, if taking k or fewer left branches always brings one to a leaf node. Because of the search tree's asymmetrical shape, its size increases polynomially with the number of variables. The tree has this shape if the problem belongs to problem class H_k in a certain type of hierarchy $H_0, H_1, \ldots, H_n$ of problem classes, where H_0 is a class of easy problems, such as renamable Horn or 2-satisfiability problems. This allows one to branch in such a way that left branches always take one to a problem that is lower in the hierarchy, thereby reaching H_0 and therefore a leaf node within k branches.

The idea of such a hierarchy can be traced to Yamasaki and Doshita [302], who defined a class of *generalized Horn* problems. Horn and generalized Horn problems can comprise classes H_0 and H_1 of the hierarchy. After Arvind and Biswas [5] improved the solution algorithm for this class, Gallo and Scutellà [106] recursively defined a particular instance of the hierarchy $H_0, H_1, \ldots, H_n$ in which H_0 and H_1 again consist of Horn and generalized Horn problems. Pretolani [238] later pointed out that, under weak conditions, such a hierarchy can be built on a base other than Horn

2.8 LIMITED BACKTRACKING

problems, such as 2-satisfiability problems. He also defined *splittable* Horn problems as a symmetric version of generalized Horn, and these give rise to an analogous hierarchy. All of Pretolani's hierarchies, except the one based on splittable Horn sets, are special cases of the hierarchy defined here. The satisfiability of any problem belonging to a given level of a hierarchy, as well as whether it belongs to that level, can be checked in polynomial time.

The idea of limited backtracking has been studied in some depth in the constraint satisfaction literature, particularly as it relates to the concept of "consistency" as defined there [282]. Connections between this literature and the ideas presented here are explored in [148].

2.8.1 Maximum Embedded Renamable Horn Systems

It is not uncommon for satisfiability problems in practice to have a large component that is renamable Horn or even Horn. For instance, there may be a small subset of k variables that, when fixed to any set of values, yield a renamable Horn problem. If a branching algorithm branches on these variables, the search tree depth is at most k and therefore contains at most 2^{k+1} nodes. This means that for fixed k, branching solves the problem in polynomial time.

It is best to begin with a precise statement of a generic branching algorithm. Let S be the clause set to be checked for satisfiability. The algorithm generates a search tree whose nodes are associated with simplified subproblems.

Procedure 10. *Branching.*

Step 0. Let A be the set of active nodes, initially containing only the root node, which is associated with the original clause set S.

Step 1. If A is empty, stop; S is unsatisfiable. Otherwise choose a node in A with associated subproblem S' and remove that node from A.

Step 2. If S' is determined without further branching to be satisfiable or to be unsatisfiable, the associated node is a leaf node. If S' is satisfiable, then stop; S is satisfiable. If it is unsatisfiable, go to step 1.

Step 3. Choose a variable x_j that occurs in S'. Create a subproblem S_1 by fixing x_j to true; that is, by deleting all clauses that contain the literal x_j and all occurrences of the literal $\neg x_j$. Create S_0 by fixing x_j to false; that is, by deleting all clauses that contain $\neg x_j$ and all occurrences of the literal x_j. Add child nodes associated with S_1 and S_0 to A and go to step 1.

Let us suppose that leaf nodes are those at which the subproblem is renamable Horn. To bound the tree depth it suffices to identify a subset of variables $x_{j_1}, \ldots, x_{j_k}$ that, when fixed to any values, result in a renamable Horn problem.

Let $x = (x_{j_1}, \ldots, x_{j_k})$, and let $v = (v_1, \ldots, v_k)$ be a true-false vector. Define $S(x, v)$ to be the result of fixing $x = v$ in clause set S; that is, of removing all clauses that contain a true literal, and all false literals from the remaining clauses. S is satisfiable if and only if $S(x, v)$ is satisfiable for some v.

It is desired that $S(x, v)$ be renamable Horn for every possible assignment v. Let $S(x)$ be the result of removing from S every literal that contains a variable in x. Then because $S(x, v) \subset S(x)$ for every assignment v, $S(x, v)$ is renamable Horn for every v if $S(x)$ is renamable Horn.

The problem of finding a shortest possible vector x for which $S(x)$ is renamable Horn can be formulated as a set-packing problem. Recall that a clause set is renamable Horn if zero or more variables can be renamed (i.e., x_j replaced with $\neg x_j$ and vice versa) to obtain a Horn set. For each x_j in S, introduce 0-1 variables y_j and $\bar{y}_j$. Let $y_j = 1$ imply that x_j is not renamed, and $\bar{y}_j = 1$ imply that x_j is renamed. If $y_j = \bar{y}_j = 0$, then x_j is dropped from the problem entirely. Then the set-packing problem is

$$\begin{array}{ll} \max & \sum_j y_j + \bar{y}_j \\ \text{s.t.} & Ay + B\bar{y} \leq e \\ & y_j + \bar{y}_j \leq 1, \text{ all } j \\ & y_j, \bar{y}_j \in \{0, 1\}, \text{ all } j \end{array} \qquad (2.65)$$

Here e is a vector of ones, and A and B are 0-1 matrices given by $a_{ij} = 1$ precisely when the literal x_j occurs in clause i of S, and $b_{ij} = 1$ precisely when $\neg x_j$ occurs in clause i. The formulation (2.65) finds the largest set of variables that, for some renaming, yields a Horn set if all other variables are omitted. It therefore finds a shortest vector of varibles x for which $S(x)$ is renamable Horn, namely, those variables x_j for which $y_j = \bar{y}_j = 0$.

A set-packing problem can always be formulated as a maximum clique problem on a graph. A set-packing problem can in general be written

$$\begin{array}{ll} \max & \sum_j z_j \\ \text{s.t.} & Qz \leq e \\ & z_j \in \{0, 1\}, \text{ all } j \end{array} \qquad (2.66)$$

The associated graph contains a node for each z_j and an arc (z_j, z_k) whenever columns j and k of Q are orthogonal (i.e., $q_{ij} q_{ik} = 0$ for all rows i). A clique is a set of nodes in which every pair is connected by an arc. If C is a clique of maximum size, then z given by $z_j = 1$ if node $z_j \in C$, and $z_j = 0$

2.8 LIMITED BACKTRACKING

otherwise, is an optimal solution of the set-packing problem. In the case of the Horn embedding problem,

$$z = (y, \bar{y}) \quad \text{and} \quad Q = \begin{bmatrix} A & B \\ I & I \end{bmatrix}$$

where I is the identity matrix.

Unfortunately, we have the following theorem.

Theorem 19 ([45]) *The maximum renamable Horn embedding problem (2.65) is NP-hard.*

This can be shown by noting that the general set-packing problem (2.66), which is NP-hard, can always be put into the form (2.65) in polynomial time. Given a set-packing problem (2.66) with $m \times n$ matrix Q, consider the clause set

$$\bigvee_{q_{ij}=1} x_j, \quad i = 1, \ldots, m$$

$$w_1 \vee \neg w_2 \vee \bigvee_{j=1}^{n} \neg x_j \qquad (2.67)$$

$$\neg w_1 \vee w_2 \vee \bigvee_{j=1}^{n} \neg x_j$$

and consider the problem (2.65) of finding a maximum embedded renamable Horn set for (2.67). It suffices to show that (2.66) and (2.67) have the same optimal value. First, one can see as follows that, in (2.67), none of the x_j can be renamed in a maximum renamable Horn embedding. Clearly, at most one can be renamed, and if one is in fact renamed, w_1 and w_2 must be deleted. In this case one could do better by deleting the renamed variable and retaining w_1 and w_2. So the maximum number of variables retained in (2.67) is the maximum number of z_j equal to 1 in a solution of (2.66).

Although (2.65) is NP-hard, it is not necessary in practice to find the largest possible embedding. A reasonably large one will do, if sufficiently few variables $x_{j_1}, \ldots, x_{j_k}$ must be fixed. Remarkably, many exact and heuristic algorithms have been developed for the maximum clique problem; some of them can be found in [12, 13, 39, 93, 100, 110, 189, 230, 260, 301]. The problem can also be solved as a maximum independent set problem on the complementary graph, for which there are several algorithms and heuristics [49, 93, 269].

2.8.2 Hierarchies of Satisfiability Problems

The rationale for defining a hierarchy $H_0, H_1, \ldots, H_n$ of satisfiability problems is to create a lopsided search tree that is limited in size because of

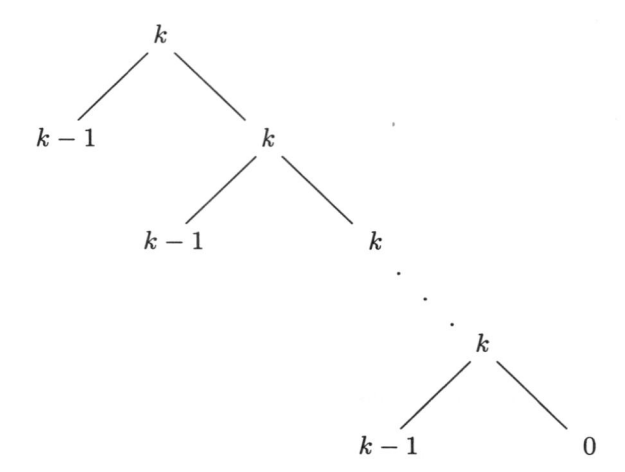

Figure 2.13: Branching pattern for a problem in H_k.

its shape. The subproblems associated with the leaf nodes belong to H_0, which is a class of easily solved problems. Each nonleaf node is associated with a problem in H_k for some k, and at least one of its children should correspond to a problem in H_{k-1}.

This results in a search tree like that of Figure 2.13. The nodes labeled k correspond to problems in H_k, and similarly for nodes labeled $k-1$. The bottom node, labeled 0, corresponds to a problem in which a clause has been falsified or all variables have been fixed. Each node labeled $k-1$ is the root of a similar tree whose nodes are labeled $k-1$ and $k-2$, and so forth. Nodes corresponding to problems in H_0 are leaf nodes, which begin to appear at level k of the tree.

It is easy to bound the size of this search tree. Because each right branch removes at least one variable from the problem, there can be no more than n nodes labeled k, hence no more than n labeled $k-1$. So if the tree for k contains Z_k nodes labeled 0, then $Z_k \leq nZ_{k-1} \leq n^k$. This means that the tree contains at most n^k leaf nodes.

To obtain this type of tree, it suffices for the hierarchy $H_0, \ldots, H_n$ to have the following properties. First, membership in H_0 is closed under variable elimination. That is, if any variable in a clause set belonging to H_0 is fixed to true or false and eliminated from the problem in the usual way, the resulting clause set also belongs to H_0. It is convenient to stipulate that an empty constraint set belongs to H_0.

Second, for each $k > 1$, H_k consists of all constraint sets in H_{k-1}, plus constraint sets S that contain at least one *admissible* variable; that is, S

2.8 LIMITED BACKTRACKING

contains a variable x_j that yields a problem in H_{k-1} when fixed to at least one truth value and yields a problem in H_k when fixed to the other truth value.

H_k, like H_0, is closed under variable elimination.

Lemma 9 *If constraint set $S \in H_k$, then the result of fixing any variable of C to either truth value yields a constraint set in H_k.*

Proof: If $n(S)$ is the number of variables in S, then the lemma can be proved by induction on $n(S)$ and k. The lemma is true for all $S \in H_0$ by definition of H_0. For any k, the lemma is true for any $S \in H_k$ with $n(S) = 0$, because $S \in H_0$ by stipulation and $H_0 \subset H_k$.

Now suppose the lemma is true for (a) all problems $S \in H_{k-1}$ (where $k > 0$) and (b) all problems $S \in H_k$ with $n(S) < n^*$ (where $n^* > 0$). It suffices to show that the lemma is true for any problem $S \in H_k$ with $n(S) = n^*$. For this it must be shown that fixing any x_j to any value v_j yields a problem $S' \in H_k$. But due to the fact that $S \in H_k$, there is a variable x_i in S and a value v_i for which fixing $x_i = v_i$ yields a problem in H_{k-1} and fixing x_i to $\bar{v}_i$ (the opposite of v_i) yields a problem in H_k. There are three cases.

1. $i \neq j$. To show $S' \in H_k$ it suffices to show that fixing $x_i = v_i$ in S' yields a problem S_1 in H_{k-1}, and fixing $x_i = \bar{v}_i$ in S' yields a problem S_0 in H_k. Because $S_1 \in H_{k-1}$, induction hypothesis (a) implies that fixing $x_j = v_j$ in S_1 yields another problem in H_{k-1}. But fixing $x_j = v_j$ in S_1 is equivalent to fixing $x_i = v_i$ in S', which means that the latter yields a problem in H_{k-1}. Also because $S_0 \in H_k$ and $n(S_0) < n^*$, induction hypothesis (b) implies that fixing $x_j = v_j$ in S_0 yields a problem in H_k. But fixing $x_j = v_j$ in S_0 is equivalent to fixing $x_i = \bar{v}_i$ in S', which means that the latter yields a problem in H_k.

2. $i = j$ and $v_j = v_i$. Here fixing x_j to v_j is the same as fixing x_i to v_i, which yields a problem in H_{k-1} and therefore in H_k.

3. $i = j$ and $v_j = \bar{v}_i$. Here fixing x_j to v_j is the same as fixing x_i to $\bar{v}_i$, which yields a problem in H_k. □

At each node of the search tree, an admissible variable x_j on which to branch must be selected. If membership in H_k can be recognized in time T this can be done in at most time nT by simply trying each variable. Because the search tree contains at most n^k leaf nodes and therefore at most n^k nonleaf nodes, a satisfiability problem in H_k can be solved in at most $nT + n^k T'$ time, where T' is the maximum time required to solve a problem in H_0 with n or fewer variables. Given a recognition oracle that operates in polynomial time, the problem can therefore be solved in polynomial time for a given k.

In practice it is best to check for satisfiability and for membership in H_k at the same time. This requires a satisfiability checker for leaf nodes that never gives a wrong answer but may fail to give an answer if the problem does not belong to H_0. For example, if H_0 is the set of renamable Horn problems, then a unit resolution algorithm has this property. (It is always correct if it satisfies all clauses or detects a contradiction. But if it removes all unit clauses before either happens, and the remaining clauses are not all Horn, it fails.) If the satisfiability checker fails when applied to a problem at a leaf node, the problem is not in H_0. If it returns a value, the problem may or may not belong to H_0, but it does not matter, because an answer was obtained in any case.

The branching algorithm can now be written recursively to search at each node for an admissible variable. The process starts with a call to the function procedure **Satk**(S, k) given below to check whether S is satisfiable. This procedure will always succeed if $S \in H_k$ but may fail otherwise. It uses a satisfiability oracle **Sat**(S) that is never wrong but may fail if $S \notin H_0$.

Procedure 11: *Satisfiability Check for H_k*

Function **Satk**(S, k).
    If $k = 0$ then return **Sat**(S).
    If S is empty then return *true*.
    For each x_j in S:
        Set x_j to true to obtain problem S_1.
        If **Satk**$(S_1, k - 1)$ returns *true* then return *true*.
        Else set x_j to false to obtain S_0.
        If **Satk**$(S_0, k - 1)$ returns *true* then return *true*.
        Else if it returns *false* then return *false*.
    Return *fail*.

The algorithm runs in time $n^{k+1}T'$ in the worst case.

2.8.3 Generalized and Split Horn Systems

Generalized Horn systems were originally defined to be those with a nested ones structure [302]. Non-Horn clauses are permitted, but there must be an ordering of them so that all the positive literals in each clause form a subset of those in the next clause. For example, the following clause set is generalized Horn because the first two clauses are Horn and the other three

2.8 LIMITED BACKTRACKING

have the nested ones property:

$$\begin{array}{lllll}
 & \neg x_2 & & \vee & x_5 \\
\neg x_1 & & \vee \neg x_3 & & \\
x_1 \vee & x_2 & & \vee \neg x_5 \\
x_1 \vee & x_2 \vee & x_3 \vee \neg x_4 & \\
x_1 \vee & x_2 \vee & x_3 \vee & x_4 \vee & x_5
\end{array}$$

The pattern is more evident in matrix form:

$$\begin{bmatrix}
0 & -1 & 0 & 0 & 1 \\
-1 & 0 & -1 & 0 & 0 \\
1 & 1 & 0 & 0 & -1 \\
1 & 1 & 1 & -1 & 0 \\
1 & 1 & 1 & 1 & 1
\end{bmatrix}$$

Note that if x_1 is set to true, the problem that remains is Horn, and if it is set to false, the remaining problem still has the nested ones structure. If one branches on $x_1, x_2, \ldots, x_5$, a tree like that of Figure 2.13 (with $k = 1$) results.

The classes of Horn and generalized problems, respectively, qualify for the first two classes H_0, H_1 of a hierarchy. Horn problems clearly remain Horn when any variable is fixed to a value. Furthermore, a generalized Horn problem contains at least one variable x_j (namely, a variable that occurs positively in every non-Horn clause) that when fixed to true yields a Horn set and when fixed to false yields a generalized Horn set. Gallo and Scutellà extended this hierarchy to $H_0, \ldots, H_n$ by defining H_k to be H_{k-1} plus all problems that contain a variable that yields a problem in H_{k-1} when fixed to true and a problem in H_k when fixed to false [106]. This is not the only hierarchy that can be built on a base of Horn problems, however. For instance, H_1 can consist of all generalized Horn problems plus the problem consisting of the clauses

$$\neg x_1 \vee x_2 \vee x_3$$
$$x_1 \vee x_2 \vee \neg x_3$$

This clause set is not generalized Horn, but it satisfies the general conditions for H_1 because a Horn set results when x_1 is fixed to false, and a generalized Horn set results when x_1 is fixed to true.

Pretolani [238] points out than any number of useful problems can be used as the base set H_0, such as renamable Horn problems, 2-satisfiability problems, or extended Horn problems.

Pretolani also generalizes the hierarchy idea in a different way. Given the set $\bar{H}_0$ of Horn problems, he defines $\bar{H}_0, \bar{H}_1, \ldots, \bar{H}_n$ by requiring that $\bar{H}_k$ consist of the problems in $\bar{H}_{k-1}$ plus problems S containing a variable

x_j such that the removal from S of all clauses containing x_j or $\neg x_j$ yields a problem in $\bar{H}_{k-1}$. The problems in $\bar{H}_1$ are *split Horn* problems. Pretolani shows that the recognition and satisfiability problems for $\bar{H}_k$ can be solved in polynomial time, essentially because branching on x_j creates subproblems that together contain fewer non-Horn clauses than the parent problem. Generalization of this idea to other base sets seems to require that one be able to check a clause set for membership in $\bar{H}_0$ by looking at each clause individually, rather than at the entire set. This is true of Horn and 2-satisfiability problems but not, for instance, of renamable Horn or extended Horn sets.

Chapter 3

Propositional Logic: The General Case

We now turn to the general inference or satisfiability problem of propositional logic and discuss solution methods that presuppose no special structure in the problem.

The general satisfiability problem is NP-complete and was in fact the first problem shown to be so [66, 181]. The problem is therefore "hard" in the sense that no known algorithm can always solve it in an amount of time that is a polynomial function of the problem size. (See [108] for an introduction to the idea of NP-completeness.) But in spite of this, there may be solution methods that are effective on most examples that arise.

There is a widespread impression that large NP-complete problems are impossible to solve and that one must settle for a heuristic procedure that may not yield the correct solution. But this impression is wrong. It is not a *problem instance* that is NP-complete, but a *class* of problems. Any class of problems that contains an NP-complete subclass is NP-complete, even if the rest of the problems in the class are easy. A an NP-complete problem class, such as the class of propositional satisfiability problems, may contain many easy problems.

Admittedly, some subclasses of satisfiability problems seem clearly hard. There seems to be no general-purpose algorithm, for instance, that can easily solve the famous pigeonhole problems [128] (see Section 3.2). But computational experience indicates that a wide variety of satisfiability problems can be solved practically in reasonably large instances.

This chapter describes several mathematical-programming-based algorithms that have successfully solved satisfiability problems. They include

(a) branch and bound methods, (b) methods that work with the linear programming "tableau," (c) cutting plane methods related to the resolution method of theorem proving, (d) set covering methods that use facet-defining cuts, and (e) interior-point nonlinear programming methods. These overlap considerably, since the last four can be combined with a tree search. Furthermore, cutting planes allow one to generalize resolution to an inference method for 0-1 inequalities that has application in integer programming.

Inference problems given in the form of a logic circuit can be solved with an integer-programming based method. This algorithm presupposes no special structure in the sense that any satisfiability problem is easily expressed in the form of a logic circuit problem. But it is intended for circuits in which there are many more nodes than inputs, since otherwise it is likely to be very inefficient.

The chapter concludes by exploring a more general inference problem—one that asks not simply, "Can a given proposition be inferred?" but asks, "What propositions relevant to a given question can be inferred?" This is a problem of "logical projection" and is related to polyhedral projection in a way that cutting plane theory illuminates.

One cannot adequately discuss the strengths and weaknesses of the various methods without an understanding of what types of problems tend to be easy or hard. In particular, random problems tend to be quite easy unless they are generated with care, and this must be considered when methods are compared computationally. We therefore discuss these issues before moving to a discussion of the methods themselves.

Finally, a discussion of problem difficulty requires some acquaintance with two classic inference methods, resolution and a simple branching procedure also known as the Davis-Putnam-Loveland algorithm. We begin with a presentation of these.

3.1 Two Classic Inference Methods

Resolution and the Davis-Putnam-Loveland procedure are inference methods well known to computer science and artificial intelligence. We have already encountered in Chapter 1 the resolution method, which is properly called "ground resolution" when applied to propositional logic. It was studied (in another guise) four decades ago by Quine [241, 242], who showed that it is a complete inference method for propositional logic. Chvátal and Szemerédi [61] point out that earlier historical antecedents can be found in the work of Löwenheim [129, 201-203], and Blake [24] (see also [20]). Robinson [244], who used the term 'resolution,' introduced a more general procedure that is a complete inference method for predicate logic.

3.1 TWO CLASSIC INFERENCE METHODS

Before Robinson published his resolution paper, M. Davis and H. Putnam [76] had incorporated resolution into a procedure for solving satisfiability problems in propositional logic. D. W. Loveland later replaced the resolution step with a branching or "splitting" step that often accelerates the procedure [199]. This results in a simple branching method that has also been called the Davis-Putnam-Loveland method.

3.1.1 Resolution for Propositional Logic

Recall that when exactly one variable x_j occurs negated in one clause C and posited in another D, the *resolvent* of C and D is the clause containing all the literals in C and D except x_j and $\neg x_j$. For instance, the resolvent of (3.1) and (3.2) below is (3.3):

$$x_1 \lor x_2 \lor x_3 \tag{3.1}$$

$$\neg x_1 \lor x_2 \lor \neg x_4 \tag{3.2}$$

$$x_2 \lor x_3 \lor \neg x_4 \tag{3.3}$$

The resolvent is implied by the conjunction of its two *parents* but by neither individually. In fact, resolution is reasoning by cases. In this example, if x_1 is false, then x_2 or x_3 must be true, whereas if x_1 is true, then x_2 or $\neg x_4$ is true; it follows that x_2, x_3, or $\neg x_4$ must be true.

The resolution method of theorem proving involves repeated application of resolution to a set S of clauses. If possible, one resolves a pair of clauses in S whose resolvent is not absorbed by a clause already in S, deletes all clauses in S absorbed by the resolvent, and adds the resolvent to S. The process is repeated as many times as possible or until the empty clause is obtained. In the latter case the original set of clauses is unsatisfiable. The resolution procedure is finite, because there is a bound on the number of clauses in S at any one time, and no clause can be returned to S once it is removed. The size of S is bounded because no resolvent contains variables that do not occur in the original set of clauses, and there are finitely many possible clauses using a given set of variables.

W. V. Quine [241, 242] showed four decades ago that resolution is a complete inference method for propositional calculus, in the sense that any clause implied by S is absorbed by one of the clauses obtained in the resolution process. If S is inconsistent, resolution generates the empty clause, which is necessarily false.

In fact, the clauses that remain when the procedure terminates are the *prime implications* of S. A clause is a prime implication if it follows from S but is implied by no other clause that follows from S. Thus the prime implications of S are in a sense the strongest possible implications of S, and every implication of S is absorbed by at least one prime implication.

The conjunction of the clauses in S is equivalent to the conjunction of the prime implications of S. S implies all of its prime implications, since a prime implication is an implication. Conversely, every clause in S is implied by S's prime implications–in fact, by a single prime implication.

Quine actually proved the completeness not of resolution but of *consensus*, a precisely analogous procedure for formulas in disjunctive normal form. For instance, the consensual formula for $(x_1 \wedge x_2) \vee (\neg x_1 \wedge x_3)$ is $(x_1 \wedge x_2) \vee (\neg x_1 \wedge x_3) \vee (x_2 \wedge x_3)$. The completeness proof for resolution is parallel to Quine's proof for consensus.

The motivation for Quine's work was the problem of finding the simplest possible logic circuit that implements a given boolean function. Translating to conjunctive normal form, Quine formulated the problem as that of finding a smallest set of clauses that is equivalent to a given set S of clauses. Unfortunately, the number of prime implications grows exponentially with the number of variables in the worst case [40]. Quine also observed that resolution alone does not completely solve the problem, because some prime implications may be redundant. For instance, if S consists of (3.1), (3.2), and (3.3), then each clause in S is a prime implication of S, but (3.3) can be dropped; the conjunction of (3.1) and (3.2) alone is equivalent to S. Quine formulated the problem of removing redundant clauses as a set-covering problem, but this need not concern us.

It is instructive to review Quine's argument for the completeness of resolution (adapted to formulas in conjunctive normal form). Suppose that we begin with a set S of clauses in which variables $x_1, \ldots, x_n$ appear, and S' is the set that remains when the resolution procedure is finished. Quine's claim is that any implication of S is absorbed by a clause in S'. Suppose to the contrary. Then we can let C be a longest clause implied by S but absorbed by no clause in S'. C is "longest" in the sense that it absorbs no other clause with this property that contains only variables in $\{x_1, \ldots, x_n\}$. Without loss of generality we can suppose that no variables in C are negated, because any negated variable x_j can be replaced in every clause of S by $\neg x_j$ without changing the problem. We need only remember that the interpretation of x_j is the negation of what it was originally. We note first that C cannot contain all the variables $x_1, \ldots, x_n$. If it did, then the only way to falsify C would be to make every atomic proposition false. But because S implies C, this must also falsify some clause C' in S. This means that no variables in C' are negated, which is impossible, since it means that C' absorbs C, contrary to the definition of C.

Thus C must lack some variable x_j. But in this case the clauses $x_j \vee C$ and $\neg x_j \vee C$, which S implies (since S implies C), must be absorbed,

3.1 TWO CLASSIC INFERENCE METHODS

respectively, by clauses in S' that may be named D and $\bar{D}$ (because $x_j \vee C$ and $\neg x_j \vee C$ are longer than C). Furthermore, D must contain x_j, since otherwise it would absorb C, contrary to C's definition, and similarly $\bar{D}$ must contain $\neg x_j$. Thus the resolvent of D and $\bar{D}$ absorbs C, which means that some clause in S' absorbs C, which is again inconsistent with C's definition.

Theorem 20 ([241, 242]) *Resolution is a complete inference method for propositional logic.*

We immediately have the following:

Corollary 5 *The resolution procedure, applied to a set of clauses, generates precisely the prime implications of the set.*

This is because any prime implication C is an implication and is therefore absorbed by some resolvent. That resolvent must be C itself, else C would not be prime. Also, any nonprime implication is eventually deleted because it is absorbed by some prime implication.

3.1.2 A Simple Branching Procedure

A simple branching method, sometimes called the Davis-Putnam-Loveland (DPL) procedure [199], generates a binary search tree in order to find a satisfying solution for a set of clauses.

Branching methods are similar to *tableau* methods discussed in the literature of automated theorem proving. These are not the same, incidentally, as the tableau methods discussed in Section 3.4. In the theorem-proving field, a tableau is essentially a branching tree, whereas in mathematical programming it is a data structure for the simplex method.

Some terminology is helpful here. A *binary tree* consists of a set of nodes, each of which has at most two other nodes as *immediate successors*. Every node except the *root* is the immediate successor of exactly one node, called its *immediate predecessor*. Every node B below node A in the tree is a *successor* of A, and A is a *predecessor* of B.

To generate the search tree, associate the original problem with the root node of the tree and apply the unit resolution procedure (Section 2.2) to simplify the problem. If no inconsistency is detected, set a chosen variable to true and then to false, generating two immediate successor nodes. Then treat the immediate successor nodes in the same way. The process continues until a satisfying solution is found, or until no more successors can be generated, in which case there is no solution.

102   CHAPTER 3   PROPOSITIONAL LOGIC: GENERAL CASE

Some versions of the procedure use *monotone variable fixing*, which simply looks for a variable that is negated in every occurrence or posited in every occurrence. In the former case one can assume the variable is false without affecting the satisfiability of the problem, and in the latter case one can assume it is true. In either case, all clauses containing the variable are removed.

As an example, consider the following set of clauses, which is associated with the root node of the search tree:

$$\begin{array}{ll} 1. & x_1 \vee x_3 \vee \neg x_4 \\ 2. & x_1 \vee x_2 \vee x_3 \vee x_5 \\ 3. & x_1 \vee x_2 \vee x_3 \vee \neg x_5 \\ 4. & \neg x_1 \vee \neg x_2 \vee \neg x_3 \\ 5. & \neg x_2 \vee x_3 \\ 6. & \neg x_1 \vee x_2 \end{array} \quad (3.4)$$

Because x_4 is negated in every occurrence, assume it is false and eliminate clause 1 of (3.4):

$$\begin{array}{c} x_1 \vee x_2 \vee x_3 \vee x_5 \\ x_1 \vee x_2 \vee x_3 \vee \neg x_5 \\ \neg x_1 \vee \neg x_2 \vee \neg x_3 \\ \neg x_2 \vee x_3 \\ \neg x_1 \vee x_2 \end{array} \quad (3.5)$$

Because there are no unit clauses, unit resolution has no effect. Arbitrarily branch on variable x_1 and first set it to true. Add the unit clause x_1 to (3.5) and associate the resulting set with a new immediate successor node. Because there are no monotone variables, immediately perform unit resolution, which yields the empty clause and therefore a contradiction. Since this precludes further branching, backtrack to the root node and take the other branch, adding the unit clause $\neg x_1$ to (3.5) in order to obtain a second successor node. Again there are no monotone variables. Unit resolution fixes x_1 to false, yielding

$$\begin{array}{c} x_2 \vee x_3 \vee x_5 \\ x_2 \vee x_3 \vee \neg x_5 \\ \neg x_2 \vee x_3 \end{array} \quad (3.6)$$

Arbitrarily branch on x_2 next, first adding the unit clause x_2 to (3.6) in order to obtain a successor node. Because x_3 is always posited in the resulting set of clauses, x_3 is fixed to true and all the clauses are eliminated. This reveals a satisfying solution in which x_1 and x_4 are false and x_3 is true, so that no further branching is necessary. The remaining variables x_2 and x_5 can be set to either true or false.

3.1 TWO CLASSIC INFERENCE METHODS

In general the algorithm goes as follows. Begin at the root node, which originally is unprocessed and associated with the original set S of clauses. Start each iteration by picking an unprocessed node that has no successors. If there is no such node, stop, because S is unsatisfiable. Let S' be the set of clauses associated with the node picked. Perform monotone variable fixing and unit resolution to obtain a possibly simplified set S' of clauses. If S' contains the empty clause, declare the node inconsistent, perform another iteration. If S' is empty, stop, because a solution has been found. (To recover the solution, let the variables have the values at which they were fixed by monotone variable fixing or unit resolution at the current node and all its predecessors. Variables that have not been fixed can be set to true or false arbitrarily.) If S' is nonempty and does not contain the empty clause, arbitrarily branch on a variable x_k that occurs in S' by creating two unprocessed successor nodes, one associated with $S' \cup \{x_k\}$ and one associated with $S' \cup \{\neg x_k\}$. Then perform the next iteration.

3.1.3 Branching Rules

The performance of branching methods can be improved by intelligently choosing on which variable to branch, and which value to assign it first. Jeroslow and Wang [171] discovered just how effective a good branching rule can be. Their branching rule says roughly that one should branch on a variable that occurs in a large number of short clauses. If v represents a truth value (0 or 1), define the function

$$w(S', j, v) = \sum_{k=1}^{\infty} N_{jkv} 2^{-k} \qquad (3.7)$$

where N_{jkv} is the number of k-literal clauses in S' in which x_j occurs positively (if $v = 1$) or negatively (if $v = 0$). If (j^*, v^*) maximizes $w(S', j, v)$, then branch on x_{j^*}, taking the $S' \cup \{x_{j^*}\}$ branch first if $v^* = 1$, and otherwise taking the $S' \cup \{\neg x_{j^*}\}$ branch first.

The apparent rationale for (3.7) is twofold.

(a) If S is satisfiable, $w(S', j, v)$ estimates the probability that a random truth assignment will falsify one of the clauses eliminated when one branches on x_j. Thus maximizing $w(S'j, v)$ also maximizes the probability that a random truth assignment will *satisfy* all of the remaining clauses. This can lead to satisfying solution earlier in the search process.

(b) If S is unsatisfiable, then branching on a variable that occurs in short clauses is likely to generate more unit clauses and increase the probability that unit resolution will detect inconsistency.

It is argued in [151] that (a) does not explain the good performance of the Jeroslow-Wang rule on satisfiable problems. Rather, something similar to (b) explains its performance for both satisfiable and unsatisfiable problems. The analysis of [151] leads to a somewhat better "two-sided" branching rule. Namely, branch on a variable x_j for which j maximizes $w(S', j, 0) + w(S', j, 1)$. Explore first the branch in which $x_j = 1$ if $w(S', j, 1) \geq w(S', j, 0)$.

3.1.4 Implementation of a Branching Algorithm

There are various ways to traverse the search tree generated by branching. A *breadth-first search* processes both of a node's immediate successors before processing any other successors. *Depth-first search* does not process the second immediate successor of a node until it has processed all the successors of its first immediate successor.

As the search tree is generated, the set of clauses associated with each node must somehow be stored until the node is processed. The traversal order can therefore have a significant effect on how many problems must be stored. Depth-first traversal tends to require the least storage, because the number of problems to be stored is never greater than the depth d of the tree. Breadth-first traversal, on the other hand, can require that as many as 2^d problems be held in storage.

Even when depth-first search is used, the amount of storage required may be excessive, particularly if the original problem is large. One alternative is not to store the problem at all but to reconstruct it from scratch. That is, to recover the problem at a given node A, begin with the original problem at the root and fix, one at a time, all the variables as they were fixed along the path from the root to A. But this approach can nearly double the computation time, because it repeats the unit resolution procedure at most of the nodes.

A much better alternative is to maintain for each clause i the highest level a_i, in the path from the root to the current node, at which the clause is still in the problem (the root is level 1). In addition, maintain for each variable x_j the highest level b_i at which the x_j still occurs in the problem. A variable that is no longer present at level k may have been removed for either of two reasons: its value was fixed by unit resolution, or all the clauses containing it were removed (or both).

Due to the fundamental importance of branching methods, Procedure 12 shows how this data structure can be implemented in a precise statement of a branching algorithm. Procedure 12 omits monotone variable fixing, because it has a marginal effect on performance. Here L_k is the unit clause added to obtain the left immediate successor of the node last visited at level

3.1 TWO CLASSIC INFERENCE METHODS

k and R_k the unit clause added to obtain the right immediate successor, with $R_k = 0$ if no right successor has been processed. T contains the variables fixed to true at the current node, and F those fixed to false. A procedure **Update** keeps track of the clause set associated with the current node, and a procedure **Restore** restores the problem associated with the current node.

Procedure 12: *A Branching Algorithm (Davis-Putnam-Loveland)*

Begin with a set S of clauses containing variables $x_1, \ldots, x_n$.
Let $S' \leftarrow S$, $k \leftarrow 0$, $T \leftarrow \emptyset$, $F \leftarrow \emptyset$, and perform **Update**.
Set the level $k \leftarrow 1$.
Perform **DPL**.

Procedure **DPL**.
    While $k > 0$:
        Perform **Unit resolution**.
        Perform **Update**.
        If S' is empty, then stop; S is satisfiable.
        Else if S' contains the empty clause, then
            Let $R_k \leftarrow 1$.
            While $k > 0$ and $R_k \neq 0$ let $k \leftarrow k - 1$.
            If $k > 0$ then
                Perform **Restore**.
                Let $R_k \leftarrow \neg L_k$ and $S' \leftarrow S' \cup \{R_k\}$.
                Let $k \leftarrow k + 1$.
        Else
            Pick $j*, v*$ so that $w(S', j*, v*) = \max_{j,v}\{w(S', j, v)\}$.
            If $v* = 1$ then let $L_k \leftarrow x_{j*}$; else let $L_k \leftarrow \neg x_{j*}$.
            Let $R_k \leftarrow 0$, $S' \leftarrow S' \cup \{L_k\}$ and $k \leftarrow k + 1$.
    Stop; S is unsatisfiable.

Procedure **Update**.
    For each clause i in S':
        Let $a_i \leftarrow k$.
        For each variable x_j in clause i: Let $b_j \leftarrow k$.
    End do.

Procedure **Restore**.
    For all j such that $b_j > k$:
        Remove x_j from T or F (whichever contains it).
        Let $b_j \leftarrow k$.
    Let $S' \leftarrow \emptyset$.
    For all clauses i in S:
        If $a_i > k$ then
            Let $a_i \leftarrow k$.
            Delete from clause i all literals involving
                a variable x_j for which $b_j < k$, and add the
                resulting clause to S'.

Procedure **Unit resolution**.
    If S' contains the empty clause, stop.
    While S' contains a unit clause C:
        If S' contains the empty clause, stop.
        Else if the variable x_j in C is positive, then
            Add x_j to T.
            Remove from S' all clauses containing literal x_j
                and all occurrences of literal $\neg x_j$.
        Else
            Add x_j to F.
            Remove from S' all clauses containing literal $\neg x_j$
                and all occurrences of literal x_j.

Despite the simplicity of DPL, Procedure 12 has performed competitively in computational comparisons of several satisfiability algorithms [132, 172, 209, 277].

3.1.5 Incremental Satisfiability

The branching method of the previous section can be modified to solve an important variant of the satisfiability problem, the *incremental* satisfiability problem. It asks: Given that a set S of propositional clauses is satisfiable, is $S \cup \{C\}$ satisfiable for a given clause C? The problem is clearly NP-complete, because one can solve a classical satisfiability problem on m clauses by solving at most m incremental problems.

The incremental problem will play a key role in the treatment of logic circuits in Section 3.9 and first-order predicate logic in Chapter 5. Procedure 13 below states a method for solving the incremental problem that uses the same data structure as Procedure 12 in the previous section and

3.1 TWO CLASSIC INFERENCE METHODS

in fact calls part of it (procedure DPL) as a subroutine. In computational tests, Procedure 13 usually re-solves a satisfiability problem, after adding a clause, in only a small fraction of the time it would take to solve the incremented problem from scratch [143].

When Procedure 13 rechecks for satisfiability after adding a clause, it exploits any information obtained while solving the original problem, much as the dual simplex method does when it re-solves a linear programming problem after adding a new constraint. To see how, consider the search tree that Procedure 12 generates when testing S for satisfiability. When a new clause C is added to S, the clause set S' associated with any fathomed node of this tree remains unsatisfiable. So there is no point in looking again at any part of the tree already generated, except the path from the root to the last node examined. Simply continue to build the tree, beginning at an appropriate point along this path, and keeping in mind that C must also be satisfied. Also, update the data structure to show at what unfathomed nodes C remains in the problem.

More precisely, the algorithm is as follows. When S is initially determined to be satisfiable, save the data structure by saving the numbers a_i, b_j, the sets T, F, the cuts L_k, R_k, and the level k in the search tree at which the search terminates. To test $S \cup \{C\}$ for satisfiability, first check whether the variables fixed at level k satisfy or falsify C (procedure **Check**). If they satisfy C, note that $S \cup \{C\}$ is satisfiable and quit. If they falsify C, then backtrack, along the path from the current node to the root, to the node at which C is first falsified, and resume the DPL algorithm at that point (procedure **Backtrack**). If the fixed variables leave the truth value of C undetermined, branch on one of the variables in C, or if only one variable in C is undetermined, simply fix its value (procedure **Branch**).. Note that $S \cup \{C\}$ is satisfiable and quit. In all of these cases, update the data structure to indicate at which node C is eliminated from the problem.

In Procedure 13, k_0 is the level, along the path from the root to the current node, at which C is first falsified (if at all), and k_1 the level at which it is first satisfied (if at all). The algorithm can be applied repeatedly as new clauses are added.

Procedure 13: *Incremental Satisfiability*

Let $k_0 \leftarrow 0$, $k_1 \leftarrow \infty$, $C' \leftarrow C$. Let C be numbered clause i.
Let $S \leftarrow S \cup \{C\}$.
Perform **Check**.

If $k_1 < \infty$ then
    For each variable x_j in C:
        If $x_j \notin T \cup F$ then let $b_j \leftarrow \max\{b_j, k_1\}$.
    Set $a_i \leftarrow k_1$.
    Stop; $S \cup \{C\}$ is satisfiable.
Else if C' is empty then perform **Backtrack**.
Else perform **Branch**.

Procedure **Check**.
    For each literal l in C:
        Let x_j be the variable in l.
        If $x_j \in T$ then
            If $l = x_j$ then
                Let $k_1 \leftarrow \min\{k_1, b_j\}$.
            Else
                Remove x_j from C' and let $k_0 \leftarrow \max\{k_0, b_j\}$.
        If $x_j \in F$ then
            If $l = \neg x_j$ then
                Let $k_1 \leftarrow \min\{k_1, b_j\}$.
            Else
                Remove $\neg x_j$ from C' and
                    let $k_0 \leftarrow \max\{k_0, b_j\}$.

Procedure **Backtrack**.
    For each variable x_j in C:
        If $x_j \notin T \cup F$ then let $b_j \leftarrow \max\{b_j, k_0\}$.
    Let $k \leftarrow k_0 + 1$, $R_k \leftarrow 1$, $a_i \leftarrow k_0$.
    While $k > 0$ and $R_k \neq 0$ let $k \leftarrow k - 1$.
    If $k > 0$ then
        Perform RESTORE.
        Let $R_k \leftarrow \neg L_k$ and $S' \leftarrow S' \cup \{R_k\}$.
        Let $k \leftarrow k + 1$.
        Perform DPL.
    Else
        Stop; S is unsatisfiable.

Procedure **Branch**.
 Let $a_i \leftarrow k$.
 If C' is a unit clause, then
  Let x_j be the variable in C'.
  If $C' = x_j$ then add x_j to T.
  Else add x_j to F.
 Else
  For each variable x_j in C:
   If $x_j \notin T \cup F$ then let $b_j \leftarrow k$.
  Pick a literal l in C' and
   let x_j be the variable in l.
  If $l = x_j$ then
   Add x_j to T and set $L_k \leftarrow x_j$, $R_k \leftarrow 0$.
  Else
   Add x_j to F and set $L_k \leftarrow \neg x_j$, $R_k \leftarrow 0$.
  Let $a_i \leftarrow k$, $k \leftarrow k + 1$.
 Stop; S is satisfiable.

3.2 Generating Hard Problems

The performance of a satisfiability algorithm can vary enormously from one problem to another. When evaluating an algorithm it is therefore essential to have some idea of the difficulty of the problems on which it is tested.

Unfortunately, there is no general understanding of what makes a problem hard for a given algorithm. But one can obtain a partial understanding by examining some problem classes that are known to be hard for a particular algorithm, and by discussing under what conditions random problems tend to be hard.

At least two problem classes are provably hard for resolution. One is the class of pigeonhole problems, which Haken [128] showed to be hard for resolution and we will show to be hard for branching. The other is a class of problems derived from graphs. Tseitin [278] showed them to be hard for a slightly restricted form of resolution, and Urquhart [280] generalized the result to full resolution. Some families of random problems are also known to be hard, including one shown by Chvátal and Szmerédi [61] to be hard for resolution, and a similar class shown by Franco and Paul [97] to be hard for branching.

3.2.1 Pigeonhole Problems

Haken [128] showed that the "pigeonhole problem" gives rise to a satisfiability problem that is very hard for resolution to solve. The pigeonhole problem is to place n pigeons in $n-1$ holes so that no hole contains more than one pigeon. Since the problem is insoluble, a set of clauses stating otherwise is unsatisfiable. Such a set for $n = 3$ is

$$\begin{aligned}
& x_{11} \vee x_{12} \\
& x_{21} \vee x_{22} \\
& x_{31} \vee x_{32} \\
& \neg x_{11} \vee \neg x_{21} \\
& \neg x_{11} \vee \neg x_{31} \\
& \neg x_{21} \vee \neg x_{31} \\
& \neg x_{12} \vee \neg x_{22} \\
& \neg x_{12} \vee \neg x_{32} \\
& \neg x_{22} \vee \neg x_{32}
\end{aligned} \tag{3.8}$$

Here x_{ij} is true when pigeon i is placed in hole j. Thus the first three clauses in (3.8) assert that each pigeon is placed in a hole. The remaining clauses assert, for each pair of pigeons, that both do not occupy the same hole.

In general, the pigeonhole problem for n pigeons is

$$\bigvee_{j=1}^{n-1} x_{ij}, \text{ all } i \in \{1, \ldots n\} \tag{3.9}$$

$$\neg x_{ik} \vee \neg x_{jk}, \text{ all } k \in \{1, ,\ldots, n-1\}, \; i,j \in \{1, \ldots, n\}, \; i \neq j \tag{3.10}$$

Let us say that the *length* of a proof is the number of resolutions performed.

Theorem 21 ([128]) *No polynomial function of n is an upper bound on the length of a shortest resolution proof of the unsatisfiability of the pigeonhole problem for n pigeons.*

We omit the proof, which is quite difficult, but it is relatively easy to see why a similar result holds for branching.

Theorem 22 *The number of nodes in the smallest search tree (without monotone variable fixing) needed to prove the unsatisfiability of the pigeonhole problem for n pigeons is bounded below by an exponential function of n.*

To see why the theorem is true, note that whenever a variable x_{ij} is fixed to true in a pigeonhole problem with n pigeons ($n \geq 4$), it is reduced

3.2 GENERATING HARD PROBLEMS

to a pigeonhole problem with $n-1$ pigeons. This is because the ith row of (3.9) is eliminated, and unit resolution with rows in (3.10) fixes all the variables x_{kj} to false for $k = 1, \ldots, n$ ($k \neq i$).

So, if p variables have been set to true upon reaching a given node A of depth d, the remaining problem is a pigeonhole problem on $n-p$ pigeons (if $n-p \geq 3$), except that at most $d-p$ variables have been set to false. Unit resolution is possible only if all but one variable in a row of (3.9) have been set to false, and therefore only if at least $n-p-2$ variables have been set to false. Therefore no inconsistency is detected at node A if $d-p \leq n-p-3$, which is to say $d \leq n-3$. This proves that the search tree contains at least 2^{n-3} nodes and therefore grows exponentially with n.

3.2.2 Problems Based on Graphs

Tseitin [278] defined a class of problems for which he showed that a restricted form of resolution requires exponential time in the worst case. Urquhart [280] later showed that full resolution requires exponential time.

Let each arc of an undirected graph G without loops or multiple edges be uniquely associated with an atomic proposition x_j or its negation. Associate with each node i of G a number $t_i \in \{0, 1\}$. Then for each node i generate all clauses having the following form. Each clause is a disjunction of the literals attached to arcs incident to i, where an even number of them are negated if $t_i = 1$, and an odd number are negated if $t_i = 0$. Thus a node with d incident arcs generates 2^{d-1} clauses.

For example, node 3 in Figure 3.1 generates the clauses

$$x_2 \vee x_3 \vee x_5$$
$$\neg x_2 \vee \neg x_3 \vee x_5$$
$$\neg x_2 \vee x_3 \vee \neg x_5$$
$$x_2 \vee \neg x_3 \vee \neg x_5$$

and node 4 generates

$$\neg x_3 \vee x_4 \vee x_7$$
$$x_3 \vee \neg x_4 \vee x_7$$
$$x_3 \vee x_4 \vee \neg x_7$$
$$\neg x_3 \vee \neg x_4 \vee \neg x_7$$

Let $L_1 \oplus \cdots \oplus L_n$ be the sum of the truth values of literals $L_1, \ldots, L_n$ modulo 2. Then if $L_1, \ldots, L_n$ are the literals attached to arcs incident to node i, the clauses generated by i state that

$$L_1 \oplus \cdots \oplus L_n = t_i \tag{3.11}$$

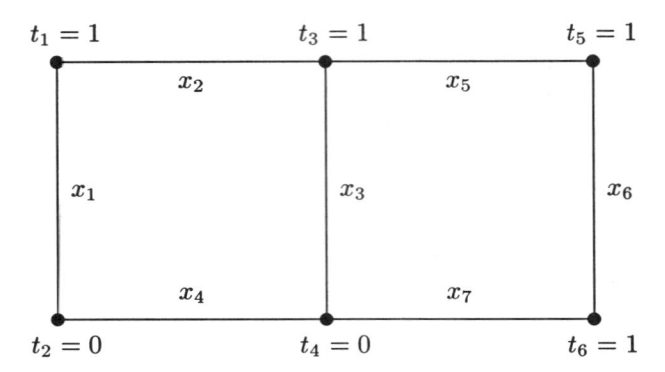

112   CHAPTER 3   PROPOSITIONAL LOGIC: GENERAL CASE

Figure 3.1: A graph that gives rise to a satisfiability problem.

Lemma 10 ([278]) *The set S of clauses generated by the nodes of a connected graph is satisfiable if and only if $\sum_i t_i$ is even.*

It is clear that if $\sum_i t_i$ is odd, S must be unsatisfiable. The sum of the left-hand side of all equations (3.11) is even, because each literal is assigned to an arc incident to two nodes and therefore occurs in two equations. Because the sum of the right-hand sides is odd, the equations and hence the clauses are unsatisfiable.

Conversely, suppose $\sum_i t_i$ is even. If a literal L is assigned to an arc (i, j), one can replace L with $\neg L$ and still satisfy the two equations containing L by complementing t_i and t_j. Thus for any pair of nodes i, j with $t_i = t_j = 1$, one can find a path $t_i, v_1, \ldots, v_m, t_j$ and successively complement the literal on (t_i, v_1), on (v_1, v_2), and so on, with the result that $t_i = t_j = 0$. This reduces $\sum_i t_i$ by 2. Thus if $\sum_i t_i$ is even, one can make each $t_i = 0$ by complementing literals, and the resulting equations (3.11) are satisfied by making all literals false.

Tseitin [278] studied $k \times l$ grid graphs for which $\sum_i t_i$ is odd, such as the 2×3 graph in Figure 3.1. He proved that the time required for *regular* resolution to prove the unsatisfiability of the associated clauses increases exponentially with the minimum of k and l. Regular resolution is that in which the resolution proof can be written as a tree in such a way that (a) the empty clause is the root, (b) each resolvent's immediate predecessors are its parents, and (c) no predecessor of a given clause is obtained by resolving on a variable in the clause.

Urquhart proved a similar result for full resolution, except that he used a different class of graphs, again for which $\sum_i t_i$ is odd.

3.2 GENERATING HARD PROBLEMS

Theorem 23 ([280]) *There is a class of graph-induced satisfiability problems for which the shortest resolution proof of unsatisfiability has length that is exponential in the number of nodes.*

3.2.3 Random Problems Hard for Resolution

Two probability models are generally used for generating random satisfiability problems in n variables. There is some confusion in the literature about what they should be called, but we will use the following denotations.

Fixed density model. Build each clause by randomly choosing k distinct variables from $\{x_1, \ldots, x_n\}$ and negating each with probability $1/2$.

Fixed probability model. Build each clause by letting each variable x_j appear in the clause with probability p, and negate each variable that appears with probability $1/2$. If the resulting clause has length less than $k_{\min}$, generate another.

The minimum clause length $k_{\min}$ in the fixed probability model would be at least one in most applications. It is sometimes useful to set $k_{\min} = 2$, since the presence of unit clauses in the problem may allow unit resolution to simplify the problem substantially.

The fixed density model generates problems that are very hard for resolution, provided the number m of clauses is large relative to n. In particular, resolution requires exponential time if the ratio of the number of clauses to the number of variables is at least about $(0.7)2^k$. The following result is due to Chvátal and Szemerédi.

Theorem 24 ([61]) *If $k \geq 3$ is fixed and $m/n \geq c2^k$, where $c > \ln 2$, there is a positive number ϵ such that, with probability tending to one as n tends to infinity (using the fixed density model), a random set of m clauses of k literals using n variables is unsatisfiable, and a resolution proof of unsatisfiability has complexity at least $(1 + \epsilon)^n$.*

It would not serve our purposes to review the proof of exponential complexity, which is quite involved. But we can easily see why such a random set S of clauses is likely to be unsatisfiable, and this argument is the source of the constant c. A given truth assignment to $x_1, \ldots, x_n$ satisfies a random clause with probability $1 - 2^{-k}$ and therefore satisfies S with probability $(1 - 2^{-k})^m$. Since there are 2^n truth assignments, the probability that at least one of them satisfies S is bounded above by $2^n(1 - 2^{-k})^m < 2^n \exp(-m2^{-k}) = \exp(n \ln 2 - m2^{-k})$. The last expression goes to zero if $m/n \geq (\ln 2)2^k + \delta$ for fixed $\delta > 0$, whence the value of c in Theorem 24.

3.2.4 Random Problems Hard for Branching

The behavior of branching on random problems is particularly important to understand, because a number of satisfiability algorithms use a branching strategy. In this section we will show that a randomly generated problem set is likely to be very easy for branching, or even simpler algorithms, if the parameters are not carefully chosen. This means that a good performance of a branching algorithm on a carelessly generated problem set can be seriously misleading.

A number of asymptotic results have been obtained for algorithms that use some sort of backtracking [50, 51, 96-99, 239, 240], but two results that are particularly relevant are presented here. One of them illustrates that a set of random problems may be trivial to solve even by guesswork. Consider a family of problems generated by the fixed probability model in which the number n of variables and the number m of clauses grow, but the probability p that a variable appears is constant. Franco and Paul showed the following:

Theorem 25 ([97]) *If m is a polynomial function of n, then there is a constant c such that, with probability approaching one as n goes to infinity (on the fixed probability model with constant p), one can find a satisfying solution of a random set S of m clauses in n variables by guessing at most c truth assignments to $x_1, \ldots, x_n$.*

These problems are trivial to solve partly because p is fixed, so that the expected number of literals per clause grows with n. It therefore becomes easier and easier to satisfy the clauses as n increases.

A second asymptotic result shows that when the clause length is fixed, so that p is inversely related to n, the problems become asymptotically hard for branching. This result, also due to Franco and Paul, uses the fixed density model.

Theorem 26 ([97]) *If m/n is a positive constant, then with probability approaching one as n goes to infinity (on the fixed density model with constant k), the number of nodes generated by branching with unit resolution and without monotone variable fixing is greater than $2^{m^{1/4}}$.*

Note that this problem class contains that shown by Chvátal and Szemerédi to be hard for resolution (Theorem 24).

Even though random problems are hard asymptotically for branching, finite instances can be trivial to solve. To see this, recall that in the fixed density model, an arbitrary assignment of truth values to $x_1, \ldots, x_n$ satisfies m random k-literal clauses with probability $(1-2^{-k})^m$. This probability can

3.2 GENERATING HARD PROBLEMS

be surprisingly high. Suppose, for instance, we generate a square system of 100 random clauses in 100 variables, using $k = 5$ literals per clause. Then if one simply *guesses*, say, $2^{10} = 1024$ truth assignments to $x_1, \ldots, x_n$ (which a computer can do in negligible time), one will find $2^{10}(1 - 2^{-k})^m = 43$ solutions on the average!

Similar if less striking results are possible on the fixed probability model. To find the probability that a given truth assignment satisfies a random clause in the fixed probability model, let $b_n(k)$ be the binomial probability $\binom{n}{k} p^k (1-p)^{n-k}$. If $k_{\min}$ is the minimum clause length, then the probability that a clause contains k literals is the truncated binomial probability

$$\bar{b}_n(k) = \frac{b_n(k)}{1 - \sum_{l=0}^{k_{\min}-1} b_n(l)}$$

So the probability that a given truth assignment satisfies a random clause is

$$q = \sum_{k=k_{\min}}^{n} (1 - 2^{-k}) \bar{b}_m(k)$$

It satisfies a set of m clauses with probability q^m.

Suppose again that a square system of 100 random clauses in 100 variables is generated, this time using $p = 0.05$ (about 5 literals per clause) and excluding unit clauses. Then guessing 2^{10} truth assignments finds $2^{10} q^m = 2$ solutions on the average.

Thus a problem may have so many solutions that it is easy to guess one. As the number of clauses increases, however, there are fewer solutions. The expected number of solutions is $2^n(1 - 2^{-k})^m$ under the fixed density model and $2^n q^m$ under the fixed probability model. In the fixed probability example just described, there are an expected number of about 10^{10} solutions for $m = 740$ clauses, but an expected number of only about 0.005 solutions when there are $m = 1200$ clauses. The latter problem is almost certainly unsatisfiable.

Thus as the number of clauses grows, one is less likely to solve a problem by guessing. But at the same time the problem becomes more likely to be unsatisfiable. For a sufficiently large number of clauses, fixing only a few variables is likely to falsify a clause. This means that one may not go very deep in the search tree before backtracking, so that the tree may be small. As the number of clauses continues to increase, then, the problems should begin to get easier again.

To see this in detail, consider a random clause C in the fixed probability model. Let t be a partial truth assignment that fixes the truth values of

$x_1, \ldots, x_i$. Then $Pr(t \text{ falsifies } C)$ is equal to the product of the following probabilities, summed over all possible clause lengths k.

$Pr(t \text{ falsifies } k \text{ literals in } C \mid t \text{ fixes } k \text{ literals in } C, C \text{ has } k \text{ literals}) = 2^{-k}$

$Pr(t \text{ fixes } k \text{ literals in } C \mid C \text{ has } k \text{ literals}) = r_{nik} = \binom{i}{k} / \binom{n}{k}$

$Pr(C \text{ has } k \text{ literals}) = \bar{b}_n(k)$

Therefore $Pr(t \text{ falsifies } C)$ is

$$f = \sum_{k=k_{\min}}^{i} 2^{-k} r_{nik} \bar{b}_n(k)$$

The summation need only be computed up to $k = i$ (rather than $k = n$) because r_{nik} vanishes for $k > i$. The probability that t falsifies at least one of m clauses is $1 - (1-f)^m$.

Consider again the example with $n = 100$ but with, say, 1200 clauses. Fixing 10 variables falsifies at least one clause with probability 0.22. This means that at a depth of 10 in the search tree, there is at least a 0.22 chance of backtracking. The true probability may be much higher than this, because unit resolution is likely to fix additional variables. The problems should therefore get easier as the number of clauses passes 1200.

The foregoing analysis predicts a peak in problem difficulty as the ratio of the number of clauses to the number of variables passes a critical value. In the fixed density model, for example, the expected number of solutions drops from about 10^7 to 10^{-5} as m/n increases from 4 to 6 (when $k = 3$ and $n = 100$). This phenomenon was in fact reported in [138, 149]. It has subsequently come to be known as a *phase transition* and has been discussed extensively in the literature, first for satisfiability problems and more recently for other problems as well (e.g., [54, 74, 112, 190, 218]).

If one wishes to generate nontrivial problems randomly, it is important to set the ratio m/n to something near the critical value for the given k or p. To date, the critical ratios have not been derived analytically, but they may be observed in computational experiments.

3.3 Branching Methods

One simple branching method (Davis-Putnam-Loveland) for solving a satisfiability or inference problem has already been discussed. This section presents four branching methods that are related to or can be viewed as inspired by integer programming: a branch-and-bound method for solving the integer programming formulation, a nonnumeric method of Jeroslow

3.3 BRANCHING METHODS

and Wang that evolved from the branch-and-bound approach, a Horn relaxation method of Gallo and Urbani, and a branching search that uses "bounded" resolution.

The idea of branching occurs in other methods discussed in subsequent sections. It is used in the column subtraction and pivot and complement methods for integer programming, which can be applied to the satisfiability problem (Section 3.4). Branching can also be combined with cutting planes to obtain a branch-and-cut method (Section 3.5).

The branching methods discussed in this section, as well as the branch-and-cut method, are largely characterized by whether and how they use two types of strategies for reducing the size of the search tree.

Sampling heuristic. This is a heuristic device that tries to generate, at each node, a satisfying solution consistent with the truth values that have already been fixed. If one is found, the search can be terminated immediately.

Relaxation. This is a weakening of the satisfiability problem at a node that is generally much easier to solve. The relaxation is satisfiable if the original problem is and may be satisfiable even if the original problem is not. If the relaxation is found to be unsatisfiable, the search backtracks.

The use of sampling heuristics and relaxations in various branching algorithms appears in Table 3.1. We refer to using unit resolution as solving a relaxation of the problem because unit resolution *may* detect unsatisfiability if the problem is unsatisfiable but never detects it if the problem is satisfiable.

3.3.1 Branch and Bound

The most straightforward way to solve an integer programming problem is by *branch and bound*. It seems to have first been applied to the satisfiability problem by Blair, Jeroslow, and Lowe [23].

Consider the following integer programming formulation of the satisfiability problem (3.4):

$$\begin{aligned}
\min \quad & x_0 \\
\text{s.t.} \quad & x_0 + x_1 + x_3 - x_4 \geq 0 \\
& x_0 + x_1 + x_2 + x_3 + x_5 \geq 1 \\
& x_0 + x_1 + x_2 + x_3 - x_5 \geq 0 \\
& x_0 - x_1 - x_2 - x_3 \geq -2 \\
& x_0 - x_2 + x_3 \geq 0 \\
& x_0 - x_1 + x_2 \geq 0 \\
& x_j \in \{0, 1\}, \quad j = 1, \ldots, 5
\end{aligned} \quad (3.12)$$

Table 3.1: Characteristics of Some Branching Methods

Algorithm	Sampling Heuristic	Relaxation
Davis-Putnam-Loveland	None	Use of unit resolution
Integer programming branch-and-bound	Linear relaxation	Linear relaxation
Jeroslow-Wang	Variable-fixing heuristic	Use of unit resolution
Gallo-Urbani	None	Horn relaxation
Billionnet-Sutter	None	Use of bounded resolution
Branch-and-cut	Linear relaxation	Linear relaxation plus cuts

The problem is satisfiable if and only if the minimum value of the objective function in (3.12) is zero.

Branch and bound differs from the Davis-Putnam-Loveland algorithm of Section 3.1.2 in that it solves a linear relaxation of the problem at each node rather than applying unit resolution.

As usual the linear relaxation of (3.12) is obtained by replacing the integrality constraints with $0 \leq x_j \leq 1$ for all j. The relaxation typically has many optimal solutions, but suppose the solution obtained is $(x_0, \ldots, x_5) = (0, 1/3, 1/3, 1/3, 0, 0)$. Because $x_0 = 0$ and the solution is noninteger, it is unclear whether there is an integral solution for which $x_0 = 0$. The next step is to pick a variable with a noninteger value, x_1 for example, and branch on it so as to generate a successor node. That is, first try re-solving the linear relaxation with $x_1 = 1$ as one of the constraints. This yields $(x_0, \ldots, x_5) = (1/4, 1, 3/4, 1/2, 0, 0)$. Since $x_0 > 0$, the corresponding satisfiability problem has no solution, so that there is no point in generating successors of this node. So replace $x_1 = 1$ with $x_1 = 0$ to generate the root's other successor, and re-solve. This might yield the integral solution $(x_0, \ldots, x_5) = (0, 0, 0, 1, 0, 0)$, which solves the satisfiability problem since $x_0 = 0$. If a nonintegral solution with $x_0 = 0$ is obtained, the

3.3 BRANCHING METHODS

next step would have been to branch on one of the nonintegral variables and continue the process.

In general, then, a problem of the following form is associated with every node of the search tree:

$$\begin{aligned}
\min \quad & x_0 \\
\text{s.t.} \quad & ex_0 + Ax \geq b \\
& x_j \in \{0, 1\}, \text{ all } j
\end{aligned} \quad (3.13)$$

where e is a vector of ones. The algorithm begins at the root node, which corresponds to the original satisfiability problem. Upon arriving at any node, there are three possible cases:

(a) The node has no successors yet.

(b) It has one immediate successor.

(c) It has two immediate successors.

In case (a), solve the linear relaxation of (3.13). If $x_0 > 0$ in the optimal solution, move to the immediate predecessor node (i.e., "backtrack"), unless the current node is the root, in which case the original problem is unsatisfiable. If $x_0 = 0$ and the solution is integral, stop, having solved the original problem. Otherwise pick a variable x_j on which to branch, where x_j has a nonintegral value in the solution of the linear relaxation. Generate an immediate successor node by adding the constraint $x_j = 1$ to the current problem and move to the successor node. In case (b), generate and move to the other immediate successor by adding the constraint $x_j = 0$, where x_j is the variable on which we branched to create the first immediate successor. In case (c), backtrack to the node's immediate predecessor.

The linear relaxation therefore serves as both a sampling heuristic and a relaxation. If its solution happens to be integer at a node, then the satisfiability problem is solved and the search stops. If it is infeasible, the node is fathomed.

Ordinarily a branch-and-bound algorithm bounds as well as branches. That is, when it encounters an integral solution with objective function value z, z becomes an upper bound on the value of the optimal solution. So, if at some node A the objective function value of the linear relaxation is greater than or equal to z, there is no point in generating successor nodes. The objective function value at any successor of A can be no better than it is at A. This sort of bounding serves no purpose in a satisfiability algorithm, however. An integer solution with objective function value 1 provides a useless bound, because the linear relaxation cannot have a solution value greater than 1 in any case. An integer solution with value 0 is already a solution of the satisfiability problem.

Solving the linear relaxation at a node is equivalent to applying unit resolution, in the sense that the two detect unsatisfiability in the same instances (Theorem 3). Thus $x_0 > 0$ in the linear relaxation of (3.6) if and only if unit resolution detects no contradiction. In fact, it may be advantageous to apply unit resolution before solving the linear relaxation. If a contradiction is detected, one can backtrack without solving the relaxation. Otherwise one obtains a possibly simplified problem, and its linear relaxation can be solved in hope of obtaining an integer solution with $x_0 = 0$. The advantage of this approach is not only that fewer LP problems are solved, but that the problems become smaller in the lower reaches of the search tree. The disadvantage is that each LP must be solved from scratch. If unit resolution is not applied, the LP is the same as that at the immediate predecessor node except for the addition of one constraint, $x_j = 1$ or $x_j = 0$. It is generally much easier to re-solve an LP after adding a constraint (using, say, the revised simplex method) than to solve a problem of comparable size from scratch.

It may appear that once unit resolution has been performed without finding a contradiction, the linear relaxation provides no information. It can always be solved by setting $x_0 = 0$ and every other $x_j = 1/2$, even if the satisfiability problem it represents is unsatisfiable. (Every constraint contains at least two terms other than x_0 and is therefore satisfied by setting each x_j other than x_0 to $1/2$.) But even though *this* solution provides absolutely no information, it does not follow that there is no point in solving the linear relaxation. Simplex algorithms for linear programming find extreme point solutions, and the solution with all $x_j = 1/2$ is seldom an extreme point solution. An extreme point solution can be quite useful, because it may be an integral solution with $x_0 = 0$.

Unfortunately, solution of the linear relaxation tends to be slow, because it typically has a large number of degenerate solutions. (These are basic solutions in which one or more basic variables have value zero. See the Appendix for background.) In particular, it is usually slower than applying the unit resolution algorithm. For this reason a simple branch-and-bound approach is dominated by other methods on most problems [132, 171]. It becomes competitive when the linear relaxation is augmented with cutting planes or is replaced with a symbolic procedure that achieves a similar effect. We will discuss the former option in Section 3.5.4 and the latter in the next section.

Even if linear-programming based branch and bound is uncompetitive for solving satisfiability problems, it may be useful for the *maximum satisfiability* problem, which requires an optimal as well as feasible solution (Section 2.5.2). This problem asks the maximum number of clauses that can simultaneously be true. If the 0-1 system $Ax \geq b$ represents the clause

3.3 BRANCHING METHODS

set in the usual fashion, the maximum satisfiability problem is

$$\begin{aligned} \max \quad & e^T y \\ \text{s.t.} \quad & y + Ax \geq b \\ & y_i, x_j \in \{0, 1\}, \text{ all } i, j \end{aligned}$$

3.3.2 Jeroslow-Wang Method

In a branch-and-bound approach, the linear relaxation serves as both a sampling heuristic and a relaxation. But as remarked in the previous section, solution of the relaxation tends to be sluggish. This suggests replacing it with something faster that achieves a similar effect. Unit resolution can easily take over the relaxation function, since it is faster than solving the linear relaxation and yet detects unsatisfiability in the same instances. It remains to supply another sampling heuristic.

The approach of Jeroslow and Wang is to design a sampling heuristic that conducts, at each node of the search tree, a secondary search for a satisfying solution consistent with the truth values already fixed. At each node it starts fixing variables (chosen by their branching rule, described in Section 3.1.3) and applies the unit resolution procedure each time a variable is fixed. A satisfying solution may be found by luck; if not, the traversal of the search tree resumes where it left off.

The precise backtracking scheme is rather complex and goes as follows. The "active node" is the current node in the primary search. Node Q is the current node in the secondary search, which begins at the currently active node. At the beginning both Q and the active node are at the root node.

Procedure 14: *Jeroslow-Wang Method*

Step 1. (Try to find a satisfying solution.) Apply the unit resolution algorithm to the problem S' at the current node Q. If a satisfying solution is found, stop; the original problem is satisfiable. If inconsistency is found, go to step 2. Otherwise pick a variable x_j and a truth value v that maximizes $w(S', j, v)$, which is defined in Section 3.1.3. Change Q to a new immediate successor of the current node, and associate $S' \cup \{x_j\}$ with Q if $v = 1$ or $S' \cup \{\neg x_j\}$ with Q if $v = 0$. Repeat step 1.

Step 2. (Backtrack to active node.) If Q is the root node, stop; the original problem is unsatisfiable. Let the active node be Q, unless it is already, in which case both Q and the active node are changed to Q's immediate predecessor. If both of Q's immediate successors have been examined, repeat step 2.

Step 3. (Branch.) If Q has no successors, go to step 1. If Q has two immediate successors, change Q and the active node to the unfathomed immediate successor, and repeat step 3. Otherwise if Q's immediate successor has been examined, generate Q's other immediate successor, let Q be it, and repeat step 3. If it has not been examined, generate Q's other successor, let both Q and the active node be it, and go to step 1.

On close examination of this algorithm one can observe that the nodes generated by the sampling heuristic are afterwards treated as part of the main search tree. Thus Jeroslow-Wang differs from the Davis-Putnam-Loveland algorithm only on the choice of which node to expand next (DPL is generally depth-first). In fact, it turns out that the two methods generate search trees of identical size when the problem is unsatisfiable. The Jeroslow-Wang method may find a solution sooner, however, if the problem is satisfiable.

3.3.3 Horn Relaxation Method

The David-Putnam-Loveland, Jeroslow-Wang, and branch-and-bound algorithms solve essentially the same relaxation of the problem at each node, because they detect unsatisfiability in the same instances. But the Horn relaxation proposed by Gallo and Urbani [107] is weaker in the sense that it detects fewer instances of unsatisfiability. This tends to generate a larger search tree, but the size of the tree may be offset by the rapidity with which the relaxations can be solved. The Horn relaxation can be solved with a variant of unit resolution that resolves only on positive unit clauses (Section 2.3). The Gallo-Urbani method uses no sampling heuristic.

The clause set S' associated with each node of the search tree is used to form a Horn relaxation H and a non-Horn set N. Each Horn clause of S' is placed directly in H. A non-Horn clause in S', such as $x_1 \vee x_2 \vee x_3 \vee \neg x_4 \vee \neg x_5$, is split into a Horn clause $\neg x_4 \vee \neg x_5 \vee y$, which goes into H, and a non-Horn clause $x_1 \vee x_2 \vee x_3 \vee \neg y$, which goes into N, where y is a new variable. If H is unsatisfiable (this can be checked in linear time), then S' is unsatisfiable, and the algorithm backtracks. If H is found to be satisfiable, then the satisfying solution obtained is examined. If it makes every new variable y false, it satisfies S', and the algorithm terminates. Otherwise the algorithm branches on a shortest clause in N as follows. Suppose this clause is $x_1 \vee x_2 \vee x_3 \vee \neg y$. Form three branches, the first of which is associated with clause set $S' \cup \{x_1\}$, the second with $S' \cup \{\neg x_1, x_2\}$, and the third with $S' \cup \{\neg x_1, \neg x_2, x_3\}$.

3.4 TABLEAU METHODS

Weak Horn relaxation seems to result in one of the fastest methods for relatively easy satisfiability problems [132]. It can run quite slowly for harder problems, however, since the search tree can grow explosively.

3.3.4 Bounded Resolution Method

Whereas the Horn relaxation is weaker than using unit resolution, one can obtain a stronger relaxation by using *bounded resolution*. By this we mean resolution in which a resolvent is generated only if its length does not exceed some bound.

Billionnet and Sutter [21] proposed an algorithm with a weaker bound and therefore a stronger relaxation. Their algorithm is essentially the same as the Davis-Putnam-Loveland method except that the resolvent is required to be no longer than the maximum parent length. Thus two clauses are resolved if and only if one would absorb the other if the variable on which resolution takes place were removed. (Their particular implementation uses consensus rather than resolution and is applied only to problems with at most three literals per term.) It is unclear what effect this relaxation has on the size of the search tree.

3.4 Tableau Methods

The simplex method for solving linear programming problems provides a framework for two approaches to solving satisfiability problems. The simplex method essentially works by finding a sequence of nonnegative solutions for a system of linear equations. The array of numbers that are manipulated to obtain these solutions is called a *tableau*. When the linear relaxation of a satisfiability problem is solved in this way, the resulting tableau can sometimes be further manipulated to find an *integral* solution of the problem, thereby solving the satisfiability problem.

The best known tableau method is the *pivot and complement* method of Balas and Martin [10], which is designed for any 0-1 linear programming problem. Because it is a heuristic, when applied to a satisfiability problem it can only try to find a solution; if it fails, the satisfiability question is unresolved. But the method can be embedded in a branch-and-bound method by applying it at each node of the search tree. This may lead to a more rapid discovery of an integer solution if one exists.

The motivation for the pivot and complement method may be reconstructed as follows. The linear relaxation of a satisfiability problem defines a polytope that in general has many extreme points (vertices), one of which, say x^*, is found by the simplex method. Point x^* is likely to have nonintegral components, but if the problem is satisfiable, the polytope has one or more integral extreme points. If the integral extreme points are fairly well distributed among the nonintegral ones, an integral point may lie close to x^*. Because each iteration (i.e., each *pivot*) of the simplex method moves from an extreme point to an adjacent one, a few more iterations allow one to explore the vicinity of x^* and possibly to find an integral solution. If the search fails, it may be useful to *complement* one or more variables (change an integral x_j^* to $1 - x_j^*$) and search again.

Another tableau-based approach is the *column subtraction* method, originally developed by Harche and Thompson [133] for set-covering and packing problems and adapted in [132] to the satisfiability problem. It is based on the principle that when the system of equations represented by a tableau is solved for some variables (basic variables) in terms of others (nonbasic variables), every integral solution may be obtained by setting the nonbasic variables to appropriate integral values. In the solution of the linear programming relaxation, every nonbasic variable is set to zero (which generally results in a nonintegral solution). A nonbasic variable can be flipped to one by performing a simple column subtraction operation in the tableau. The possible flips can be enumerated by a branching scheme until an integral solution is found, or until it is proved than none exists.

Although the column subtraction method involves branching, it differs from branch-and-bound. The former fixes a complete and unchanging set of nonbasic variables at every node of the search tree. Branch-and-bound fixes only the variables on which the search has branched so far, and the set of nonbasic variables is determined anew at each node.

Based on tests reported in [132], the column subtraction method is unusually robust. It may be slower than some other methods on problems of easy to moderate difficulty but solves several problems that appear to be intractable for other methods.

It is best to begin with a brief review of the tableau form of the simplex method, as it applies to the linear relaxation of satisfiability problems.

3.4.1 The Simplex Method in Tableau Form

The simplex method rests on the fact that a linear programming problem can be written as a system of equations subject to nonnegativity constraints and, optionally, nonnegative upper bounds on the variables.

$$\begin{aligned} \min \quad & cx \\ \text{s.t.} \quad & Ax = b \\ & 0 \leq x \leq h \end{aligned} \qquad (3.14)$$

3.4 TABLEAU METHODS

If there are m (independent) constraints, one can solve the equations for a vector x_B of m basic variables in terms of the remaining nonbasic variables x_N. So if (3.14) is written

$$\begin{aligned} \min \quad & c_B x_B + c_N x_N \\ \text{s.t.} \quad & B x_B + N x_N = b \\ & 0 \leq x_B, x_N \leq h \end{aligned} \quad (3.15)$$

and if B is nonsingular, one obtains the following after solving the constraint for x_B:

$$x_B = B^{-1}b - B^{-1}N x_N \quad (3.16)$$

Every feasible solution (x_B, x_N) of (3.15) can be obtained by setting x_N to some value between 0 and h. The particular solution obtained by setting $x_N = 0$, namely $(x_B, x_N) = (B^{-1}a, 0)$, is the *basic solution* corresponding to B. The basic solution is feasible if $0 \leq x_B \leq h$.

If the basic solution $(x_B, 0)$ is not optimal, a better solution can be obtained by making some nonbasic variable positive. It is straightforward to determine which nonbasic variable(s) will have this effect by substituting (3.16) into the objective function of (3.15), to yield

$$c_B B^{-1} b + (c_N - c_B B^{-1} N) x_N$$

So the objective function value for the basic solution is $c_B B^{-1} b$. Increasing a nonbasic variable x_j in x_N decreases the objective function when the corresponding component of $c_N - c_B B^{-1} N$, called the *reduced cost* of x_j, is negative.

The simplex method picks a nonbasic variable x_j with negative reduced cost and increases it until one of the basic variables x_i hits a bound (0 or h_i). If x_i hits its upper bound, it is replaced with $h_i - x_i$ so that it is now at zero. x_i therefore becomes nonbasic, x_j becomes basic, and the equations are re-solved for the new set of basic variables. This comprises one iteration. The iterations continue until all the reduced costs are nonpositive, whereupon the basic solution is optimal. (See [60] for a fuller description of the simplex method.)

The re-solution for a new basis is readily carried out when the problem is written as a tableau. (The problem can also be stored in factored form for purposes of efficient solution.) So (3.15) is displayed as

c_B	c_N	0
B	N	b

Gauss-Jordan elimination (a series of row operations) is applied to solve the equations for x_B. If the objective function row is carried along in the

126   CHAPTER 3   PROPOSITIONAL LOGIC: GENERAL CASE

calculations, the tableau becomes

0	$c_N - c_B B^{-1} N$	$-c_B B^{-1} b$
I	$B^{-1} N$	$B^{-1} b$

0	r	d_0
I	D	d

(3.17)

The alternative notation on the right will presently become useful. Note that the values $B^{-1}b$ of the basic variables in the basic solution can be read from the rightmost column and that the reduced costs appear above the nonbasic columns.

To see this concretely, consider again the satisfiability problem (3.4), whose integer programming form is given in (3.12). The linear programming relaxation replaces the integrality constraints with $0 \leq x_j \leq 1$. The problem in equality form is

$$
\begin{aligned}
\min \quad & x_0 \\
\text{s.t.} \quad & x_0 + x_1 \phantom{{}+x_2} + x_3 - x_4 \phantom{{}+x_5} - s_1 \phantom{{}-s_2-s_3-s_4-s_5-s_6} = 0 \\
& x_0 + x_1 + x_2 + x_3 \phantom{{}-x_4} + x_5 \phantom{{}-s_1} - s_2 \phantom{{}-s_3-s_4-s_5-s_6} = 1 \\
& x_0 + x_1 + x_2 + x_3 \phantom{{}-x_4} - x_5 \phantom{{}-s_1-s_2} - s_3 \phantom{{}-s_4-s_5-s_6} = 0 \\
& x_0 - x_1 - x_2 - x_3 \phantom{{}-x_4-x_5-s_1-s_2-s_3} - s_4 \phantom{{}-s_5-s_6} = -2 \\
& x_0 \phantom{{}+x_1} - x_2 + x_3 \phantom{{}-x_4-x_5-s_1-s_2-s_3-s_4} - s_5 \phantom{{}-s_6} = 0 \\
& x_0 - x_1 + x_2 \phantom{{}-x_3-x_4-x_5-s_1-s_2-s_3-s_4-s_5} - s_6 = 0 \\
& 0 \leq x_j \leq 1, \ j = 1, \ldots, 5 \\
& x_0, s_i \geq 0, \ i = 1, \ldots, 6
\end{aligned}
$$
(3.18)

(The surplus variable names $s_1, \ldots, s_6$ may be viewed as alternate names for $x_6, \ldots, x_{11}$.) The tableau form of the problem is

x_0	x_1	x_2	x_3	x_4	x_5	s_1	s_2	s_3	s_4	s_5	s_6	RHS
1	0	0	0	0	0	0	0	0	0	0	0	0
1	1	0	1	-1	0	-1	0	0	0	0	0	0
1	1	1	1	0	1	0	-1	0	0	0	0	1
1	1	1	1	0	-1	0	0	-1	0	0	0	0
1	-1	-1	-1	0	0	0	0	0	-1	0	0	-2
1	0	-1	1	0	0	0	0	0	0	-1	0	0
1	-1	1	0	0	0	0	0	0	0	0	-1	0

(3.19)

To obtain a basic solution, one must pick a set of six basic variables and solve for them. If they are picked at random, however, the solution may be infeasible. A "phase I" technique is ordinarily used to find a set of basic variables that give rise to a feasible basic solution, but in a satisfiability problem, such a set can be found by inspection. Since all the constraints are satisfied by making the artificial variable $x_0 = 1$, the surplus variables $s_1, \ldots, s_6$ are nonnegative if the remaining variables are regarded as nonbasic and set to zero. Since one additional nonbasic variable is needed, the surplus variable that occurs in a constraint with a right-hand side of 1 is

3.4 TABLEAU METHODS

regarded as nonbasic, as there will be no surplus in such a constraint. One may safely assume that there is at least one such constraint, because otherwise the satisfiability problem is trivially solved by setting each $x_j = 0$.

So the matrix B in the tableau (3.19) consists of the columns marked x_0, s_1, s_3, s_4, s_5, s_6. (In practice one does not bother to permute the columns so that the B columns are adjacent.) A few iterations of Gauss-Jordan elimination solve for the basic variables by producing a tableau in the form of (3.17).

	x_0	x_1	x_2	x_3	x_4	x_5	s_1	s_2	s_3	s_4	s_5	s_6	RHS
	1	−1	−1	−1	0	−1	0	1	0	0	0	0	−1
s_1	0	0	1	0	1	1	1	−1	0	0	0	0	1
x_0	1	1	1	1	0	1	0	−1	0	0	0	0	1
s_3	0	0	0	0	0	2	0	−1	1	0	0	0	1
s_4	0	2	2	2	0	1	0	−1	0	1	0	0	3
s_5	0	1	2	0	0	1	0	−1	0	0	1	0	1
s_6	0	2	0	1	0	1	0	−1	0	0	0	1	1

Note that each constraint row is labeled by the basic variable that occurs in the row. The objective function has value 1, which indicates that the clause set has not been satisfied.

One next allows the simplex method to solve this linear relaxation of the satisfiability problem. A variable with negative reduced cost, such as x_1, becomes basic. The basic variable that first hits a bound is determined by noting that as a nonbasic variable x_j increases, each basic variable x_i becomes $x_i = d_i - d_{ij}x_j$. For instance, $s_4 = 3 - 2x_1$. Since we want $0 \le d_i - d_{ij}x_j \le h_i$ for all basic variables x_i, we pick the basic variable that achieves the following minimum:

$$\min\left\{ \min_{\substack{i \\ d_{ij} > 0}} \left\{\frac{d_i}{d_{ij}}\right\}, \min_{\substack{i \\ d_{ij} < 0}} \left\{\frac{h_i - d_i}{-d_{ij}}\right\} \right\} \qquad (3.20)$$

where $h_i = 0$ for logical variables and ∞ for surplus variables. So if x_1 becomes basic, the ratios in (3.20) are

$$\{\{\tfrac{1}{1}, \tfrac{3}{2}, \tfrac{1}{1}, \tfrac{1}{2}\}, \{\}\}$$

corresponding, respectively, to basic variables x_0, s_4, s_5, s_6. So s_6 hits a bound (zero) first and becomes nonbasic. Row operations are carried out to solve for x_1, yielding

	x_0 x_1 x_2 x_3 x_4 x_5 s_1 s_2 s_3 s_4 s_5 s_6	RHS
	0 0 -1 $-\frac{1}{2}$ 0 $-\frac{1}{2}$ 0 $\frac{1}{2}$ 0 0 0 $\frac{1}{2}$	$-\frac{1}{2}$
s_1	0 0 1 0 1 1 1 -1 0 0 0 0	1
x_0	1 0 1 $\frac{1}{2}$ 0 $\frac{1}{2}$ 0 $-\frac{1}{2}$ 0 0 0 $-\frac{1}{2}$	$\frac{1}{2}$
s_3	0 0 0 0 0 2 0 -1 1 0 0 0	1
s_4	0 0 2 1 0 0 0 0 0 1 0 -1	2
s_5	0 0 2 $-\frac{1}{2}$ 0 $\frac{1}{2}$ 0 $-\frac{1}{2}$ 0 0 1 $-\frac{1}{2}$	$\frac{1}{2}$
x_1	0 1 0 $\frac{1}{2}$ 0 $\frac{1}{2}$ 0 $-\frac{1}{2}$ 0 0 0 $\frac{1}{2}$	$\frac{1}{2}$

Two more iterations ("pivots") of this sort deliver the optimal solution of the linear programming problem:

	x_0 x_1 x_2 x_3 x_4 x_5 s_1 s_2 s_3 s_4 s_5 s_6	RHS	
	1 0 0 0 0 0 0 0 0 0 0 0	0	
s_1	$-\frac{1}{3}$ 0 0 0 1 $\frac{2}{3}$ 1 $-\frac{2}{3}$ 0 0 $-\frac{1}{3}$ $\frac{1}{3}$	$\frac{2}{3}$	
x_3	$\frac{4}{3}$ 0 0 1 0 $\frac{1}{3}$ 0 $-\frac{1}{3}$ 0 0 $-\frac{2}{3}$ $-\frac{1}{3}$	$\frac{1}{3}$	(3.21)
s_3	0 0 0 0 0 2 0 -1 1 0 0 0	1	
s_4	-2 0 0 0 0 -1 0 1 0 1 0 0	1	
x_2	$\frac{1}{3}$ 0 1 0 0 $\frac{1}{3}$ 0 $-\frac{1}{3}$ 0 0 $\frac{1}{3}$ $-\frac{1}{3}$	$\frac{1}{3}$	
x_1	$-\frac{2}{3}$ 1 0 0 0 $\frac{1}{3}$ 0 $-\frac{1}{3}$ 0 0 $\frac{1}{3}$ $\frac{2}{3}$	$\frac{1}{3}$	

The satisfiability question remains undecided, because the solution

$$(x_1, \ldots, x_5) = (\frac{1}{3}, \frac{1}{3}, \frac{1}{3}, 0, 0)$$

is not integral.

3.4.2 Pivot and Complement

The 0-1 satisfiability problem (3.13) may be written in equality form,

$$\begin{aligned} \min \quad & x_0 \\ \text{s.t.} \quad & ex_0 + Ax - s = b \\ & x_j \in \{0, 1\}, \text{ all } j \\ & s_i \geq 0 \text{ all } i \end{aligned}$$

The pivot and complement method is based on the observation that this problem may be solved by finding an optimal basic solution of the linear

3.4 TABLEAU METHODS

relaxation,

$$\begin{aligned}
\min \quad & x_0 \\
\text{s.t.} \quad & ex_0 + Ax - s = b \\
& 0 \leq x_j \leq 1, \quad \text{all } j, \\
& s_i \geq 0, \quad \text{all } i
\end{aligned} \qquad (3.22)$$

in which all of the logical variables x_j are nonbasic. This is because a variable that lies at one of the bounds 0,1 can be nonbasic in a basic solution. The method therefore begins with an optimal solution of the linear relaxation and tries to carry out further pivots in such a way that the logical variables become nonbasic.

The original pivot and complement method has an initial phase that tries to find a feasible 0-1 solution, followed by an improvement phase that tries to decrease the objective function value. Since only a feasible solution is desired here (where $x_0 > 0$ is taken as a sign of infeasibility), only the first phase is relevant.

The method may be described as follows. Note that a pivot maintains $x_0 = 0$ if the variable that it selects to become basic has a zero reduced cost. The pivot maintains feasibility if the variable that becomes nonbasic is selected according to the ratio test (3.20).

Procedure 15: *Pivot and Complement*

Step 1. Solve the linear relaxation (3.22). If the solution is integral, stop with success.

Step 2. Search for a pivot that maintains feasibility and $x_0 = 0$ and that reduces the number of logical variables that are basic (i.e., replaces a basic logical variable with a basic surplus variable). If such a pivot exists, perform it; if the resulting solution is integral, stop with success, and otherwise repeat this step. (If no such pivot exists, continue.)

Step 3. Search for a pivot that a) maintains feasibility and $x_0 = 0$, b) does not change the number of logical variables that are basic, and c) reduces the sum of integer infeasibilities, defined to be

$$\sum_{i \geq 1} \min\{d_i, h_i - d_i\},$$

where $h_i = 0$ for logical variables and ∞ for surplus variables. If such a pivot exists, perform it; if the resulting solution is integral, stop with success, and otherwise go to step 2. (If no such pivot exists, continue.)

130   CHAPTER 3   PROPOSITIONAL LOGIC: GENERAL CASE

Step 4. Check whether rounding the solution produces a feasible solution with zero objective function value. If so, stop with success.

Step 5. Among pivots that make a surplus variable basic and positive and make a logical variable nonbasic, do the one that minimizes the sum of infeasibilities, $z = \sum_{i>0} \max\{0, -d_i\}$. (A basic variable that exceeds the upper bound $\bar{1}$, as well as one that achieves the bound, should be complemented before pivoting.)

Step 6. If there is a nonbasic logical variable that can be complemented to reduce z, complement the one that yields the largest reduction in z. Then if $z = 0$, go to step 4. (If there is no such variable, continue.)

Step 7. If there is a pair of nonbasic logical variables than can simultaneously be complemented to reduce z, complement them. Then if $z = 0$, go to step 4, and otherwise go to step 6. If there is no such pair, stop with failure.

Consider again problem (3.4), whose linear relaxation in equality form is (3.18). The optimal tableau for the linear relaxation is (3.21). Surplus variables s_2, s_5, and s_6 can become basic without increasing x_0. If s_2 becomes basic, then the variable that becomes nonbasic is the one corresponding to the minimum of the ratios (3.20):

$$\left\{\left\{\begin{matrix}1\\1\end{matrix}\right\}, \left\{\frac{\infty-\frac{2}{3}}{\frac{2}{3}}, \frac{1-\frac{1}{3}}{\frac{2}{3}}, \frac{\infty-1}{1}, \frac{1-\frac{1}{3}}{\frac{1}{3}}, \frac{1-\frac{1}{3}}{\frac{1}{3}}\right\}\right\}$$

where the ratios correspond, respectively, to $s_4, s_1, x_3, s_3, x_2, x_1$. So if s_3 becomes basic, s_4 becomes nonbasic. Similarly, if s_5 becomes basic, then x_3, x_2, or x_1 could become nonbasic, and if s_6 becomes basic, x_1 becomes nonbasic. Because step 1 requires a pivot that makes a logical variable nonbasic, use the first such pivot found: exchange s_5 for x_3. Since x_3 will hit its upper bound, first replace x_3 with $\bar{x}_3 = 1 - x_3$.

	x_0	x_1	x_2	$\bar{x}_3$	x_4	x_5	s_1	s_2	s_3	s_4	s_5	s_6	RHS
	1	0	0	0	0	0	0	0	0	0	0	0	0
s_1	$-\frac{1}{3}$	0	0	0	1	$\frac{2}{3}$	1	$-\frac{2}{3}$	0	0	$-\frac{1}{3}$	$\frac{1}{3}$	$\frac{2}{3}$
x_3	$\frac{4}{3}$	0	0	-1	0	$\frac{1}{3}$	0	$-\frac{1}{3}$	0	0	$-\frac{2}{3}$	$-\frac{1}{3}$	$-\frac{2}{3}$
s_3	0	0	0	0	0	2	0	-1	1	0	0	0	1
s_4	-2	0	0	0	0	-1	0	1	0	1	0	0	1
x_2	$\frac{1}{3}$	0	1	0	0	$\frac{1}{3}$	0	$-\frac{1}{3}$	0	0	$\frac{1}{3}$	$-\frac{1}{3}$	$\frac{1}{3}$
x_1	$-\frac{2}{3}$	1	0	0	0	$\frac{1}{3}$	0	$-\frac{1}{3}$	0	0	$\frac{1}{3}$	$\frac{2}{3}$	$\frac{1}{3}$

3.4 TABLEAU METHODS

Now the pivot operation yields

	x_0	x_1	x_2	$\bar{x}_3$	x_4	x_5	s_1	s_2	s_3	s_4	s_5	s_6	RHS
	1	0	0	0	0	0	0	0	0	0	0	0	0
s_1	-1	0	0	$\frac{1}{2}$	1	$\frac{1}{2}$	1	$-\frac{1}{2}$	0	0	0	$\frac{1}{2}$	1
s_5	-2	0	0	$\frac{3}{2}$	0	$-\frac{1}{2}$	0	$\frac{1}{2}$	0	0	1	$\frac{1}{2}$	1
s_3	0	0	0	0	0	2	0	-1	1	0	0	0	1
s_4	-2	0	0	0	0	-1	0	1	0	1	0	0	1
x_2	1	0	1	$-\frac{1}{2}$	0	$\frac{1}{2}$	0	$-\frac{1}{2}$	0	0	0	$-\frac{1}{2}$	0
x_1	0	1	0	$-\frac{1}{2}$	0	$\frac{1}{2}$	0	$-\frac{1}{2}$	0	0	0	$\frac{1}{2}$	0

The solution
$$(x_1, \ldots, x_5) = (0, 0, 1, 1, 0)$$
is already integral, and there is no need to go beyond step 1.

3.4.3 Column Subtraction

Like the simplex method, the column subtraction method rests on the fact that any solution of a linear programming problem may be obtained by setting nonbasic variables to appropriate values. In particular, an integral solution may be obtained by setting nonbasic variables to integral values.

When a basic solution of the linear relaxation is derived, the nonbasic variables are all set to zero. The column subtraction method simply tries setting subsets of them to one, in order to obtain integer values of the basic variables. Obviously, setting a variable to one is equivalent to crossing out its column in the tableau and subtracting that column from the right-hand side, whence the name of the method.

Nonbasic surplus variables require a slightly different treatment, as they can take integral values greater than one. In fact, the value of a surplus variable can be as large as one less than the number of literals in its clause. Each integral value must be enumerated, and the corresponding multiple of that variable's column subtracted from the right-hand side.

Consider, for instance, the optimal tableau (3.21). The surplus variables $s_1, \ldots, s_6$ can be as large as 2, 3, 3, 2, 1, 1, respectively. Any of the nonbasic variables x_5, s_2, s_5, s_6 may be set to a positive value. (x_0 is not changed, because it must be zero.) A search tree is created in which the root node has six immediate successors, at which variables are fixed as follows:

$$x_5 = 1 \quad s_2 = 1 \quad s_2 = 2 \quad s_2 = 3 \quad s_5 = 1 \quad s_6 = 1$$

Harche and Thompson recommend a breath-first search down to a certain depth, followed by a complete depth-first search if necessary.

A breadth-first search begins by visiting the six successors of the root. Setting $x_5 = 1$ subtracts the x_5 column from the right-hand side and makes $s_3 = -1 < 0$, an infeasible solution. The $x_5 = 1$ node is therefore removed. Setting $s_2 = 1$ yields another nonintegral solution, and the node remains in the tree so that its successors may be created if necessary. Setting $s_2 = 2, 3$ is infeasible, and these two nodes are removed. But setting $s_5 = 1$ results in an integral right-hand side, with solution $(x_1, \ldots, x_5) = (0, 1, 1, 0, 0)$.

	x_0	x_1	x_2	x_3	x_4	x_5	s_1	s_2	s_3	s_4	s_5	s_6	RHS
	1	0	0	0	0	0	0	0	0	0	0	0	0
s_1	$-\frac{1}{3}$	0	0	0	1	$\frac{2}{3}$	1	$-\frac{2}{3}$	0	0	0	$\frac{1}{3}$	1
x_3	$\frac{4}{3}$	0	0	1	0	$\frac{1}{3}$	0	$-\frac{1}{3}$	0	0	0	$-\frac{1}{3}$	1
s_3	0	0	0	0	0	2	0	-1	1	0	0	0	1
x_2	$\frac{1}{3}$	0	1	0	0	$\frac{1}{3}$	0	$-\frac{1}{3}$	0	0	0	$-\frac{1}{3}$	1
x_1	$-\frac{2}{3}$	1	0	0	0	$\frac{1}{3}$	0	$-\frac{1}{3}$	0	0	0	$\frac{2}{3}$	0

The search can therefore stop.

Branching is simplified if each surplus variable is replaced by a sum of binary surplus variables. If s_i has a maximum value less than 2^{k+1}, it can be replaced by $\sum_{j=0}^{k} 2^j s_{ij}$.

Procedure 16: *Column Subtraction*

Step 1. Replace the surplus variables with sums of binary variables as indicated above, and solve the linear relaxation. If $x_0 > 0$, stop; the problem is unsatisfiable. If the solution is integral with $x_0 = 0$, stop; a solution has been found.

Step 2. Associate the linear relaxation with the root node of a search tree. The immediate successors of each node of the tree correspond to the nonbasic columns (the columns of D) that remain in the tableau (3.17), except the x_0 column. Each successor is obtained by subtracting one of these columns from the right-hand side d and removing that column from the tableau.

Step 3. Conduct a breadth-first search of the tree down to a predetermined depth. Remove a node if $0 \leq d \leq e$ is not satisfied; do not generate its successors. Stop with a solution if d consists of 0's and 1's.

Step 4. Conduct an exhaustive depth-first search of the tree, backtracking when $0 \leq d \leq e$ is not satisfied, and stopping with a solution if d consists of 0's and 1's. If the search backtracks to the root node, the problem is unsatisfiable.

3.5 Cutting Plane Methods

Some of the more remarkable connections between logic and optimization are revealed by cutting plane theory. They begin with the observation that the resolution procedure can be interpreted as a cutting plane algorithm. Even this elementary fact has an important practical implication. Treating resolvents as cutting planes accelerates resolution enormously, because the integer programming context provides a guide as to which resolvents should be generated [139].

An algorithm competitive with those discussed in the previous section, however, requires a deeper analysis. One such analysis reveals that a restricted form of resolution powerful enough to solve the inference problem for Horn clauses is equivalent to a restricted cutting plane procedure that generates "rank 1" cutting planes [142]. This leads to a cutting plane method for non-Horn problems that can be combined with branch-and-bound enumeration to obtain a "branch-and-cut" method for inference [149]. (Effective cutting planes for the inference problem can also be obtained with the help of polyhedral theory that does not rely on connections with logic, as we will see in Section 3.7.)

The cutting planes described here are useful not only for satisfiability problems but also for any optimization problem in which logical clauses are among the constraints.

A different type of analysis reveals that if resolution is extended by a provision for adding new variables, the resulting algorithm is as powerful an inference method as *any* cutting plane procedure—in the sense that the former solves a problem in polynomial time whenever the latter does [68]. This theoretically interesting fact has not, however, led to an improvement in inference algorithms.

A third type of analysis develops a logic of inequalities and uses cutting plane theory to generalize resolution to a complete inference method for them [140, 144]. It provides a good illustration of how logic and optimization can enjoy a symbiotic relationship. The cutting plane results contribute to logic by providing, at least in principle, a complete inference method for any type of logical formulas that can be expressed as linear 0-1 inequalities. Logic contributes to optimization by providing a new class of cutting planes ("logic cuts") that are useful in integer and mixed integer/linear programming [154]. Due to the complexity of this third type of analysis, it is given a section of its own (Section 3.6).

3.5.1 Resolvents as Cutting Planes

Recall from Section 2.2.1 that a satisfiability problem can be formulated:

$$\begin{aligned} \min \quad & x_0 \\ \text{s.t.} \quad & ex_0 + Ax \geq b \\ & x_j \in \{0,1\}, \quad \text{all } j \end{aligned} \qquad (3.23)$$

where e is a vector of ones. One approach to solving (3.23) is by cutting planes. A *cut* or *cutting plane* for (3.23) is an inequality satisfied by all integral solutions of the constraints. To solve (3.23), begin by finding a solution x^* for its linear relaxation, which is

$$\begin{aligned} \min \quad & x_0 \\ \text{s.t.} \quad & ex_0 + Ax \geq b \\ & 0 \leq x_j \leq 1 \quad \text{all } j \end{aligned} \qquad (3.24)$$

If $x_0^* = 0$ and x^* is nonintegral, add one or more *separating cuts* to (3.24) and solve it again, where a separating cut is a cut that x^* violates. The process continues until $x_0^* > 0$ or x^* is integral.

One way to obtain a cut is to take a positive linear combination of one or more constraints and round up any nonintegers that result. Such a cut is a *Chvátal cut*. Chvátal cuts that can be generated using the original constraints are *rank 1 cuts*. Rank 2 cuts can be generated using the rank 1 cuts, and so on. Chvátal [59] proved that if the feasible set for (3.23) is nonempty and bounded (it is clearly bounded), all cuts can be generated in this fashion. So Chvátal cuts suffice in theory to solve any integer programming problem.

It is easy to see that resolvents are rank 1 cuts generated from their parents and bounds of the form $0 \leq x_j \leq 1$. As an example, consider the resolution of the first two clauses below to obtain the third:

$$\begin{array}{l} x_1 \vee x_2 \vee x_3 \\ \neg x_1 \vee x_2 \qquad \vee \neg x_4 \\ x_2 \vee x_3 \vee \neg x_4 \end{array} \qquad (3.25)$$

We can obtain the resolvent by computing the weighted sum:

$$\begin{array}{ll} (1/2) & x_1 + x_2 + x_3 \qquad \geq \quad 1 \\ (1/2) & -x_1 + x_2 \quad - x_4 \geq -1 \\ (1/2) & \qquad x_3 \qquad \geq \quad 0 \\ (1/2) & \qquad \underline{\quad - x_4 \geq -1} \\ & x_2 + x_3 - x_4 \geq -1/2 \end{array}$$

where the weights appear on the left. Rounding up the fraction $-1/2$ on the right yields the resolvent.

3.5 CUTTING PLANE METHODS

The connection between resolvents and cutting planes was observed by Cook, Coullard, and Turán [68], Hooker [140], Williams [293], and probably others. Also, the cutting plane procedure of C. Blair [22], when applied to clauses, is resolution.

Resolution is therefore a means of generating cuts. Only separating cuts are useful, however, and some way is needed to find them. It is impractical to generate all resolvents and simply use the ones that are separating, because the number of resolvents can explode exponentially. Also, if we were going to generate all resolvents, we could just as well solve the problem by resolution.

There is, however, a very useful condition the parents must satisfy before the resolvent can be separating. Given a solution x^* of the relaxation (3.24), let the *truth value* of a positive literal x_j be x_j^*, and the truth value of a negative literal $\neg x_j$ be $1 - x_j^*$. Then a resolvent is separating only if the sum of the truth values of its literals is strictly less than one. Now consider either parent of the resolvent. All of its literals, other than the literal on which we resolve, occur in the resolvent. So the sum of the truth values of all its literals must be strictly less than two, or else the resolvent will not be separating.

Furthermore, because the sum of these truth values must be at least one (since x^* satisfies the constraint), the truth value of the literal (say, x_j) on which we resolve must be greater than zero. Applying the same argument to the other parent, the truth value of $\neg x_j$ must be greater than zero, which means the truth value of x_j is strictly between zero and one.

Lemma 11 *A clause can be the parent of a separating resolvent only if the sum of the truth values of its literals is strictly less than 2, and only if at least one literal (including one on which resolution takes place) is nonintegral.*

This suggests a cutting plane strategy. Solve the relaxed problem (3.25) and generate all possible resolvents using only those parents that satisfy the conditions of Lemma 11. Add to the problem the resolvents that are separating and repeat. Continue until an integer solution is found or $x_0^* > 0$.

This method can be applied to problem (3.12). The initial LP relaxation yields the solution $(x_0^*, \ldots, x_5^*) = (0, 1/3, 1/3, 1/3, 0, 0)$, as before. Constraints 1, 2, 4, and 5 of (3.12) pass the truth value test of Lemma 11. They yield three resolvents on variables with fractional values: $x_1 + x_3 + x_5 \geq 1$, $x_2 + x_3 + x_5 \geq 1$, and $-x_1 + x_3 \geq 0$. Only the first two are separating. When we add them to the constraint set of (3.12) and re-solve the relaxation, we get an integer solution, $(x_0^*, \ldots, x_5^*) = (0, 0, 1, 1, 0, 0)$.

Computational tests [139] have shown that this simple cutting plane method, with occasional recourse to branch and bound, can solve some

CHAPTER 3 PROPOSITIONAL LOGIC: GENERAL CASE

Theorem 29 *Given a set S of clauses and a clause C, let $Ax \geq b$ be the linear relaxation of the inequalities representing S. Then C is a rank 1 cut for $Ax \geq b$ if it is the result of an input proof from S.*

Consider first an illustration of the theorem. The input proof (3.27) is reproduced below in inequality form, where $x = (x_1, \ldots, x_6)$. For the moment ignore the inequalities in italics.

$$(1) \quad [\,1\;0\;1\;1\;1\;0\,]x \geq 1$$

$$
\begin{array}{ll}
(1) \;[\;\;1\;\;\;0\,{-}1\;\;\;0\;\;\;1\;\;\;0\,]x \geq 0 & [\,1\;0\;0\;1\;1\;0\,]x \geq 1 \\
 & [\,\mathit{2\;0\;0\;1\;2\;0}\,]x \geq \mathit{1} \\[4pt]
(2) \;[{-}1\;\;\;1\;\;\;0\;\;\;0\;\;\;1\;\;\;1\,]x \geq 0 & [\,0\;1\;0\;1\;1\;1\,]x \geq 1 \\
 & [\,\mathit{0\;2\;0\;1\;4\;2}\,]x \geq \mathit{1} \qquad (3.28) \\[4pt]
(2) \;[\;\;\;0\,{-}1\;\;\;0\;\;\;1\;\;\;0\;\;\;1\,]x \geq 0 & [\,0\;0\;0\;1\;1\;1\,]x \geq 1 \\
 & [\,\mathit{0\;0\;0\;3\;4\;4}\,]x \geq \mathit{1} \\[4pt]
(3) \;[\;\;0\;\;\;0\;\;\;0\,{-}1\;\;\;0\;\;\;1\,]x \geq 0 & [\,0\;0\;0\;0\;1\;1\,]x \geq 1 \\
 & [\,\mathit{0\;0\;0\;0\;4\;7}\,]x \geq \mathit{1}
\end{array}
$$

To show that the equation $[000011]x \geq 1$ (i.e., $x_5 + x_6 \geq 1$) is a rank 1 cut, recover it by taking a positive linear combination of the premises and the bounds $0 \leq x_j \leq 1$, and rounding up the nonintegers. Note that (3.28) shows how. Each inequality in italics below a resolvent is a weighted sum of the premises used to obtain the resolvent (as parents, parents of parents, etc.) where the weights are indicated on the left in parentheses. Note that the weights are chosen to cancel the variable on which the resolution takes place. Begin by giving premise 1 weight 1 and then give premise 2 weight 1 as well so that its $-x_3$ will cancel x_3 in premise 1. Then give premise 3 weight 2 so that its $-x_1$ will cancel $2x_1$, and so on. Now the weighted sum of the premises is $4x_5 + 7x_6 \geq 1$:

$$
\begin{array}{rl}
(1) & [\;\;\;1\;\;\;0\;\;\;1\;\;\;1\;\;\;1\;\;\;0\,]x \geq 1 \\
(1) & [\;\;\;1\;\;\;0\,{-}1\;\;\;0\;\;\;1\;\;\;0\,]x \geq 0 \\
(2) & [{-}1\;\;\;1\;\;\;0\;\;\;0\;\;\;1\;\;\;1\,]x \geq 0 \\
(2) & [\;\;\;0\,{-}1\;\;\;0\;\;\;1\;\;\;0\;\;\;1\,]x \geq 0 \\
(3) & [\;\;\;0\;\;\;0\;\;\;0\,{-}1\;\;\;0\;\;\;1\,]x \geq 0 \\ \hline
 & [\;\;\;0\;\;\;0\;\;\;0\;\;\;0\;\;\;4\;\;\;7\,]x \geq 1
\end{array}
\qquad (3.29)
$$

$$
\begin{array}{rl}
(5) & [\;0\;\;0\;\;0\;\;0\;\;1\;\;0\,]x \geq 0 \\
(2) & [\;0\;\;0\;\;0\;\;0\;\;0\;\;1\,]x \geq 0 \\ \hline
 & [\;0\;\;0\;\;0\;\;0\;\;9\;\;9\,]x \geq 1
\end{array}
$$

3.5 CUTTING PLANE METHODS

By adding to $4x_5 + 7x_6 \geq 1$ multiples of the bounds $x_5 \geq 0$ and $x_6 \geq 0$ as shown in (3.29), dividing the sum by 9 (the sum of the weights), and rounding up the $1/9$ on the right, one obtains the desired $x_5 + x_6 \geq 1$, which is therefore a rank 1 cut.

To generalize this argument, let the ith premise be $a^i x \geq a_0^i$, where $a_0^i = 1 + n(a^i)$. Then the ith premise serves as a parent in the following resolution:

$$b^{i-1} x \geq b_0^{i-1}$$
$$u^{i-1} x \geq u_0^{i-1}$$
$$(w_i) \quad a^i x \geq a_0^i \qquad b^i x \geq b_0^i$$
$$u^i x \geq u_0^i$$

Here $b^{i-1} x \geq b_0^{i-1}$ and $a^i x \geq a_0^i$ resolve to yield $b^i x \geq b_0^i$ (where $b^0 x \geq b_0^0$ is the first premise $a^1 x \geq a_0^1$). The inequalities with u^{i-1} and u^i correspond to the italicized inequalities in (3.29). Let $\text{sgn}(\alpha) = 1$ if $\alpha > 0$, -1 if $\alpha < 0$, and 0 otherwise. First show inductively that

$$b_j^i = \text{sgn}(u_j^i) \text{ for } j = 1, \ldots, n \qquad (3.30)$$
$$u_0^i = 1 + n(u^i) \qquad (3.31)$$

which is trivially true for $i = 1$. Assume, then, that (3.30) and (3.31) hold for $i - 1$.

To show (3.30), let $x_{j(i)}$ be the variable on which we resolve. The weight w_i needed to make $a_{j(i)}$ and $u_{j(i)}^{i-1}$ cancel is clearly a positive number (namely, $|u_{j(i)}^{i-1}|$), because the induction hypothesis and the fact that we are resolving on $x_{j(i)}$ imply that $a_{j(i)}$ and $u_{j(i)}^{i-1}$ have opposite signs. For the same reasons u_j^{i-1} and a_j^i do not have opposite signs for each $j \neq j(i)$, and (3.30) follows.

To show (3.31), note that by definition,

$$u_0^i = w_i a_0^i + u_0^{i-1}$$

which by the definition of a_0^i and the induction hypothesis implies

$$u_0^i = w_i(1 + n(a^i)) + (1 + n(u^{i-1}))$$
$$= w_i + n(w_i a^i + u^{i-1}) - w_i + 1$$
$$= 1 + n(u^i)$$

Here the second equation follows from the fact that one negative component cancels in the sum $w_i a^i + u^{i-1}$.

Suppose that the final resolvent is $b^m x \geq b_0^m$. To obtain it from the weighted sum $u^m x \geq u_0^m$, let $W = w_1 + \cdots + w_m$ and add to $u^m x \geq$

140   CHAPTER 3   PROPOSITIONAL LOGIC: GENERAL CASE

u_0^m the inequalities $(W - u_j^m)x_j \geq 0$ for each j for which $u_j^m > 0$ and $(-W - u_j^m)x_j \geq -W - u_j^m$ for each j for which $u_j^m < 0$. Due to (3.30) and (3.31) we obtain the inequality

$$Wb^m x \geq 1 + n(u^m) - \sum_{u_j^m < 0} (W + u_j^m) = 1 + Wn(b^m)$$

Dividing by W and rounding up on the right yields the desired inequality $b^m x \geq 1 + n(b^m)$.

The following corollary of Theorem 29 will be useful in Section 3.9.

Corollary 7 *Let S, C, and $Ax \geq b$ be as in Theorem 29. Then if C is the result of an input proof from S, it is equivalent to an inequality that can be obtained as a nonnegative linear combination of the inequalities of $Ax \geq b$, excluding bounds $0 \leq x_j \leq 1$.*

The term "equivalent" means that C and the inequality are satisfied by exactly the same 0-1 points. The desired inequality can be obtained just by omitting the bounds from the linear combination described above. In the example, it is $(4/9)x_5 + (7/9)x_6 \geq 1/9$.

The other direction of the implication in Theorem 28 may now be proved in a strengthened form. Let a literal L containing variable x_j be *monotone* in a set of clauses if x_j always occurs in the set with the same sign as in L.

Theorem 30 *Given a set S of clauses and a clause C, let $Ax \geq b$ be the linear relaxation of the inequalities representing S. Then the inequality representing C is a rank 1 cut for $Ax \geq b$ only if some clause that absorbs C is the result of an input proof from a subset T of S in which the literals of C are monotone.*

The proof of Theorem 30 relies on the following lemma.

Lemma 12 *Any rank 1 clausal cut for the inequalities representing a set S of clauses is a rank 1 cut of the inequalities representing some subset of S in which the literals of C are monotone.*

(There is an interesting connection here with the idea of regular resolution, discussed in Section 3.2.2.) To illustrate the lemma, note that a linear combination of the following four clausal inequalities and two bounds, using the weights indicated, yields (after rounding) the rank 1 clausal cut

3.5 CUTTING PLANE METHODS

$x_3 + x_4 + x_5 \geq 1$.

$$\begin{array}{rl}
(1/4) & [0 1 0 1 1]x \geq 1 \\
(1/4) & [1 0 1 1 0]x \geq 1 \\
(2/4) & [-1 0 1 0 1]x \geq 0 \\
(1/4) & [1 -1 -1 0 1]x \geq -1 \\
(2/4) & [0 0 1 0 0]x \geq 0 \\
(2/4) & [0 0 0 1 0]x \geq 0 \\ \hline
& [0 0 1 1 1]x \geq \tfrac{1}{4}
\end{array}$$
(3.32)

One can delete the fourth clause, which spoils the monotonicity of x_3, by adding the last four bounds below to compensate:

$$\begin{array}{rl}
(1/4) & [0 1 0 1 1]x \geq 1 \\
(1/4) & [1 0 1 1 0]x \geq 1 \\
(2/4) & [-1 0 1 0 1]x \geq 0 \\
(2/4) & [0 0 1 0 0]x \geq 0 \\
(2/4) & [0 0 0 1 0]x \geq 0 \\
\\
(1/4) & [1 0 0 0 0]x \geq 0 \quad \text{(i)} \\
(1/4) & [0 -1 0 0 0]x \geq -1 \quad \text{(ii)} \\
(1/4) & [0 0 0 1 0]x \geq 0 \quad \text{(iv)} \\
(2/4) & [0 0 0 0 1]x \geq 0 \quad \text{(iii)} \\ \hline
& [0 0 \tfrac{5}{4} \tfrac{5}{4} \tfrac{5}{4}]x \geq \tfrac{1}{4}
\end{array}$$
(3.33)

After multiplying by 4/5 and rounding up, the desired clause is obtained.

To see how to do this in general, suppose without loss of generality that no negative literals occur in the conclusion C. (If they do, complement the variables.) Let $ux \geq u_0$ be the inequality that results from the given linear combination. Then if a variable x_k in C is negated in some inequality that has weight, say, w in the linear combination, remove this inequality and compensate by adding inequalities as follows. Here N_2 is the set of indices of variables in C, and $N_1 = \{1, \ldots, n\} \setminus N_2$.

(i) For all $j \in N_1$ such that $a_j = 1$, add the inequality $wx_j \geq 0$.

(ii) For all $j \in N_1$ such that $a_j = -1$, add $-wx_j \geq -w$.

(iii) For all $j \in N_2$ such that $a_j = 1$, add $2wx_j \geq 0$.

(iv) For all $j \in N_2$ such that $a_j = 0$, add $wx_j \geq 0$.

(The application of (i)–(iv) in the above example is indicated in (3.33).) Let $u'x \geq u'_0$ be the resulting linear combination. From (i) and (ii), $u'_j = u_j = 0$ for $j \in N_1$ and $u'_j = u_j + w$ for $j \in N_2$. Also, if s is the number of negative literals in the N_2 portion of the deleted inequality, one can check that

$u'_0 = u_0 + (s-1)w \geq u_0$, where the inequality follows from the fact that $s \geq 1$. Dividing $u'x \geq u'_0$ by $1+w$, we obtain $ux \geq u'_0/(1+w) \geq u_0/(1+w)$, which after rounding yields C.

Theorem 30 can now be derived from Lemma 12. The argument can clearly be indicated with the above example. Supposing that we have the rank 1 cut $x_3 + x_4 + x_5 \geq 1$ for the inequalities in (3.32), the aim is to show that it is the result of an input proof from a subset of these inequalities in which x_3, x_4, and x_5 are monotone. First use Lemma 12 to obtain $x_3 + x_4 + x_5 \geq 1$ as a rank 1 cut from such a subset, as in (3.33). If x_3, x_4 and x_5 are removed from (3.33), the same linear combination yields an unsatisfiable inequality:

$$\begin{array}{r} (1/4) \quad [\ 0 \quad 1\,]\bar{x} \geq 1 \\ (1/4) \quad [\ 1 \quad 0\,]\bar{x} \geq 1 \\ (2/4) \quad [-1 \quad 0\,]\bar{x} \geq 0 \\ (1/4) \quad [\ 1 \quad 0\,]\bar{x} \geq 0 \\ \underline{(1/4) \quad [\ 0 -1\,]\bar{x} \geq -1} \\ 0 \geq 1/4 \end{array} \quad (3.34)$$

Here $\bar{x} = (x_1, x_2)$. This means by Theorem 3 in Chapter 2 that the three clauses x_2, x_1, and $\neg x_1$ represented by the first three inequalities of (3.34) are unsatisfiable and have a unit refutation. So, by Theorem 27, they have an input refutation, namely:

$$\begin{array}{cc} & x_1 \\ \neg x_1 & \emptyset \end{array}$$

To this add the variables x_3, x_4, and x_5 as they occur in (3.33) and obtain a unit proof of a clause $x_3 \vee x_4 \vee x_5$ that (trivially) absorbs $x_3 \vee x_4 \vee x_5$:

$$\begin{array}{cc} & x_1 \vee x_3 \vee x_4 \\ \neg x_1 \vee x_3 \vee x_5 & x_3 \vee x_4 \vee x_5 \end{array}$$

The development is now reached a point at which Corollary 6 can be demonstrated. Given an input proof of C, Theorem 29 implies that C is a rank 1 cut. So by Theorem 30 C is absorbed by some clause that is the result of an input proof from clauses in which the literals of C are monotone. Removing the literals in C from this proof obtains an input refutation, because these literals are monotone. Thus by Theorem 27 there is a unit refutation of the same clauses. Restoring the literals in C yields a unit proof of a clause absorbing C, whence Corollary 6.

3.5.3 A Separation Algorithm for Rank 1 Cuts

Lemma 11 of Section 3.5.1 provides a useful condition that a clause must satisfy before it can be the parent of a separating resolvent. This forms the basis of a simple separation algorithm: generate all resolvents whose parents satisfy the truth value condition stated in Lemma 11, and pick out the separating resolvents.

This algorithm can be generalized to rank 1 clausal cuts. Recall that one condition in Lemma 11 for the parent of a separating resolvent is that the truth values of its literals must sum to a number strictly less than two. This is generalized in the following easy lemma.

Lemma 13 *If the inequality representing clause C is a rank 1 separating cut for the inequalities representing a set S of clauses in which the literals of C are monotone, then within any clause of S the truth values of literals also in C sum to a number strictly less than one.*

The lemma is easy because the truth values of literals in C must sum to less than one if C is to be separating. Since in any clause of S the literals also in C occur with the same sign as in C, their truth values must sum to less than one if C is to be separating.

Another condition of Lemma 11 is that the resolution must take place on a variable with a nonintegral truth value. This is generalized in step 2 of the following algorithm, which finds a separating rank 1 clausal cut if one exists. In the algorithm K is a set of variables on which resolutions are to take place, and $k = |K|$.

Procedure 17: *Separation Algorithm for Rank 1 Clausal Cuts*

Step 0. Begin with a set S of clauses and delete all clauses absorbed by others. Solve the linear relaxation (3.24) and set $k = 0$.

Step 1. Set $k = k + 1$. If $k > n$, stop.

Step 2. Set $T = S$, and pick a set K of k variables (a set not already picked) with nonintegral truth values in the solution of (3.24). If no such sets remain, go to step 1.

Step 3. (Apply Lemma 13.) Pick a clause D in T with exactly one variable x_j in K, such that the sum of the truth values of the other literals in D is strictly less than one. If there is no such D, go to step 2.

Step 4. (Apply unit resolution.) Derive all possible resolvents on x_j of D with other clauses in T, and add the resolvents to T. Go to step 3, unless one of these resolvents is separating (and no more separating cuts are desired), in which case the algorithm stops.

Theorem 31 ([142]) *The above algorithm finds a separating rank 1 clausal cut for S if one exists.*

Proof: Let a K-subset of S be a subset of S in which all the variables that do not occur in C belong to K. Let A be the algorithm stated above, and let A' be the result of removing the truth value conditions in steps 2 and 3. For a given K in either algorithm, say that the algorithm *resolves on K*.

Note that while resolving on K, A' generates all clauses that are the result of unit proofs from K-subsets of S, until a separating cut is found. Thus by Theorems 27 and 29, A' generates all rank 1 cuts of K-subsets of S until a separating cut is found. If follows that if some K-subset of S has a rank 1 separating cut but no K'-subset of S has such a cut, where K' is any proper subset of K, then A' generates a separating rank 1 cut while resolving on K. Thus it suffices to show the following:

Claim: If A' generates a separating cut while resolving on K but on no proper subset of K, then A generates a separating cut while resolving on K.

To prove the claim, suppose that there is a separating rank 1 cut C that A' generates while resolving on K but on no proper subset of K. To show that A also generates C, it suffices by Lemma 13 to show that all variables in K are nonintegral. Suppose to the contrary that x_i^* is integral for some $x_i \in K$. Since C is a rank 1 cut for some K-subset T of S, it is easy to see that both $x_i \vee C$ and $\neg x_i \vee C$ are rank 1 cuts for T. Thus by Theorems 27 and 30, both $x_i \vee C$ (or a clause that absorbs it) and $\neg x_i \vee C$ (or a clause that absorbs it) are the result of unit proofs from a $K \setminus \{x_i\}$-subset of T. This means that A' generates $x_i \vee C$ (or a clause absorbing it) and $\neg x_i \vee C$ (or a clause absorbing it) while resolving on a proper subset $K \setminus \{x_i\}$ of K, and thus by hypothesis neither can be separating. But since x_i^* is integral, it is 0 or 1. If $x_i^* = 0$, then $x_i \vee C$ (and any clause that absorbs it) is separating, and if $x_i^* = 1$, then $\neg x_i \vee C$ (and any clause that absorbs it) is separating. It follows that each x_i^* is nonintegral. □

3.5.4 A Branch-and-Cut Algorithm

Section 3.3.1 observed that the branch-and-bound approach to solving satisfiability problems is handicapped by the sluggishness of known algorithms for solving the linear relaxation at each node. One way to offset this handicap, at least partially, is to use a *branch-and-cut* approach—that is, to

3.5 CUTTING PLANE METHODS

augment the linear relaxation at each node with separating cuts. Since separating cuts produce a stronger relaxation, they tend to reduce the size of the search tree.

The cutting plane algorithm developed in the last two sections can form the basis for a branch-and-cut approach. But it is necessary first to modify the separation algorithm of the previous section to make it more practical. Afterwards the implementation of a branch and cut algorithm may be discussed.

The difficulty with the separation algorithm of Section 3.5.3 is that one must enumerate all possible subsets K of the n variables before being assured of finding any separating rank 1 cuts that exist. As a result, one repeatedly checks the same pairs of clauses for possible resolution. A more practical strategy is to drop the set K and accumulate resolvents rather than resetting T to S in step 2. This allows one to benefit from previous checking. The resulting algorithm, below, generates any separating cut that is generated by Procedure 17.

Procedure 18: *Accelerated Separation Algorithm for Rank 1 Clausal Cuts*

Step 0. Begin with a set S of clauses. Solve the linear relaxation (3.24) and set $k = 0$.

Step 1. Set $k = k + 1$. If $k > n$, stop.

Step 2. *(Apply Lemma 13.)* Pick from S a clause D containing a literal with a nonintegral truth value in the solution of (3.24), such that the sum of the truth values of the other literals in D is strictly less than one. If there is no such D, go to step 1.

Step 3. *(Apply resolution.)* Derive all possible resolvents of D with other clauses in S that take place on a variable with a nonintegral value. Add the resolvents to S. Go to step 2, unless one of these resolvents is separating (and no more separating cuts are desired), in which case the algorithm stops.

The algorithm may be illustrated by deriving a separating rank 1 cut from the clausal inequalities below. A hypothetical solution

$$x^* = (0, 1/5, 2/5, 1/5, 4/5, 2/5, 2/5)$$

146   CHAPTER 3   PROPOSITIONAL LOGIC: GENERAL CASE

is shown in the first row for easy reference:

$$\begin{array}{llllllll}
0 & 1/5 & 2/5 & 1/5 & 4/5 & 2/5 & 2/5 & \\
x_0 + x_1 & & + x_3 + x_4 + x_5 & & \geq 1 & \text{(a)} \\
x_0 + x_1 & & - x_3 & + x_5 & & \geq 0 & \text{(b)} \\
x_0 - x_1 + x_2 & & & + x_5 + x_6 \geq 0 & \text{(c)} \\
x_0 & - x_2 & + x_4 & & + x_6 \geq 0 & \text{(d)} \\
x_0 & & - x_4 & & + x_6 \geq 0 & \text{(e)}
\end{array} \quad (3.35)$$

Since the truths values of literals other than $-x_3$ in (3.35b) sum to 3/5, which is less than 1, (a) and (b) resolve to obtain (3.36a) below. Similarly, (a) and (e) resolve to obtain (3.36b), and (d) and (e) resolve to obtain (3.36c).

$$\begin{array}{lllll}
x_0 + x_1 & & + x_4 + x_5 & \geq 1 & \text{(a)} \\
x_0 + x_1 & + x_3 & + x_5 + x_6 & \geq 1 & \text{(b)} \\
x_0 & - x_2 & + x_6 & \geq 0 & \text{(c)}
\end{array} \quad (3.36)$$

We can now resolve (3.36c) and (3.35c) to produce (3.37a) below, and resolve (3.36a) and (3.35e) to obtain (3.37b).

$$\begin{array}{ll}
x_0 - x_1 + x_5 + x_6 \geq 1 & \text{(a)} \\
x_0 + x_1 + x_5 + x_6 \geq 1 & \text{(b)}
\end{array} \quad (3.37)$$

These two resolve to produce

$$x_0 + x_5 + x_6 \geq 1 \quad (3.38)$$

which is a separating cut. It is a rank 1 cut, as can be verified by taking a linear combination in which (3.35a–e) and the bound $x_5 \geq 0$, respectively, have weights 1/7, 1/7, 2/7, 2/7, 3/7, 3/7. By comparison, the algorithm of Section 3.5.3 must enumerate 15 sets K and generate 33 resolvents before discovering a separating cut, namely, (3.38), when $K = \{x_1, x_2, x_3, x_4\}$.

Theorem 32 ([149]) *Procedure 18 finds all separating cuts found by Procedure 17 and therefore generates a rank 1 clausal cut if one exists.*

Proof: We will show that Procedure 18 generates all the clauses that Procedure 17 does. Let a resolvent obtained in Procedure 17 have *depth* $d + 1$ if the maximum depth of its parents is d, where all of the original clauses have depth zero. The proof is by induction on depth.

Obviously Procedure 18 generates all clauses of depth zero that Procedure 17 does. Now, supposing that Procedure 18 generates all clauses of depth d that Procedure 17 does, we will prove the same for depth $d + 1$.

Let R be any resolvent of depth $d + 1$ that Procedure 17 obtains. Let D^* be the clause D that Procedure 17 uses in step 4 to obtain R, and E^*

3.5 CUTTING PLANE METHODS

a clause with which it resolves D^* to obtain R. Then D^* and E^* have depth d or less. Also D^* contains exactly one variable x_j in K, and the truth values of variables in D^* other than x_j sum to less than one. By the induction hypothesis Procedure 18 generates D^* and E^*, which at some point have been added to S in step 3. Thus in some execution of step 2, Procedure 18 picks D^* as the clause D and resolves it on x_j with E^*, thus obtaining R. □

Procedure 18 is actually more powerful than Procedure 17, because it can generate cuts that have rank greater than 1. Consider the clauses,

$$x_0 + x_1 + x_2 + x_3 \geq 1$$
$$x_0 + x_1 - x_2 + x_3 \geq 0$$
$$x_0 - x_1 + x_2 + x_3 \geq 0$$
$$x_0 - x_1 - x_2 + x_3 \geq -1$$

and suppose that $(x_0^*, \ldots, x_3^*) = (0, 1/2, 1/2, 0)$. The above algorithm produces the separating cut $x_0 + x_3 \geq 1$, which is not a rank 1 cut and is not produced by the algorithm of Section 3.5.3.

The branch-and-cut algorithm goes as follows. At each node of the search tree, apply unit resolution to the associated set S' of clauses before solving the linear relaxation, since unit resolution is generally faster. If this finds S' to be satisfiable, the search terminates, and if it finds S' to be unsatisfiable, the search backtracks. Otherwise obtain the solution x^* of the linear relaxation. x^* exists, because otherwise unit resolution would have detected unsatisfiability. If x^* is integer, the search terminates with a determination of satisfiability; this provides a sampling heuristic. Otherwise try to generate one or more separating cuts using the above algorithm. As each cut is generated, every clause it absorbs is deleted from S', and the cut is added to S'. Next apply unit resolution to S'. If it detects satisfiability or unsatisfiability, terminate or backtrack as before. Otherwise use the Jeroslow-Wang branching rule (Section 3.1.3) to choose a variable x_j for branching for which x_j^* is nonintegral. Create a new immediate successor node associated with $S' \cup \{x_j\}$ or $S' \cup \{\neg x_j\}$ and continue the process at that node.

In practice it may be advantageous to generate cuts only down to a certain level of the search tree. Cuts added near the top of the tree have a greater impact that is more likely to offset the cost of generating them. Below this level the search can proceed as in the Davis-Putnam-Loveland method.

There are also a number of time-saving devices that are important in the implementation of the separation algorithm. One should put an upper bound $k_{\max}(< n)$ on the number of iterations, or else it will run too long.

It is easy to check that a resolvent generated in the last iteration k_{max} can be separating only if parents we consider in iteration k have literals whose truth values sum to less than $k_{max} - k + 2$. Furthermore, experience shows that an excessive number of resolvents tend to be generated before the algorithm begins to obtain the strongest (i.e., shortest) separating cuts. To prevent this one can put an upper bound L on the length of clauses generated in iteration k_{max}, where perhaps $L = 1$. Since a parent cannot be more than one literal longer than its resolvent, it is enough to consider parents in iteration k with length at most $k_{max} - k + L + 1$. This results in the following, more practical separation algorithm.

Procedure 19: *Restricted Separation Algorithm*

Step 0. Let $T = S'$, $k = 0$.

Step 1. Set $= k + 1$. If $k > k_{max}$, stop.

Step 2. Pick from T a clause D of length at most $k_{max} - k + 2$ that contains a variable with a nonintegral value in the linear relaxation, such that the truth values of the other literals in D sum to less than one. If there is no such D, go to step 1.

Step 3. Derive all possible resolvents of D with other clauses E in T that take place on a variable with a nonintegral value, where E has length at most $k_{max} - k + L + 1$, and the truth values of its literals sum to less than $k_{max} - k + 2$. Add to T the resolvents containing at least one variable with a nonintegral value. Add to S' the resolvents that are separating. If more separating cuts are desired, go to step 2; otherwise stop.

3.5.5 Extended Resolution and Cutting Planes

A curious phenomenon occurring in both logic and polyhedral theory is that one can sometimes make a problem easier by adding more variables. Tseitin [278] pointed out, for instance, that conversion to conjunctive normal form requires only linear time if additional variables are used but can require exponential time otherwise. Doubling the number of variables so as to formulate the satisfiability problem as a set-covering problem makes possible the polyhedral analysis of Section 3.7.

Tseitin also realized that resolution proofs can be shortened by adding new variables. He defined *extended resolution* to be a proof system that not only permits the addition to a set S of clauses the resolvent of two clauses in S, but also permits the addition of clauses $x_i \vee x_j$, $x_i \vee x_k$ and

3.5 CUTTING PLANE METHODS

$\neg x_i \vee \neg x_j \vee \neg x_k$, where the variable x_i does not occur in S. This has the effect of introducing a new variable x_i that is equivalent to $\neg x_j \vee \neg x_k$.

In fact, the addition of new variables gives resolution the same theorem-proving power as cutting planes, in the sense that extended resolution can prove unsatisfiability in polynomial time whenever cutting planes can. A cutting plane proof consists of repeatedly adding a Chvátal cut (Section 3.5.1) to a set of inequalities until $0 > 1$ is obtained. Cook, Coullard, Turán proved the following result.

Theorem 33 ([68]) *There exists a polynomial $p(N, M)$ such that for any unsatisfiable set S of clauses containing M literals, if the unsatisfiability of S has a cutting plane proof of length N, then it has an extended resolution proof of length $p(N, M)$.*

Here the length of a resolution proof is the number of resolvents, and the length of a cutting plane proof is the number of cuts. The proof is based on a result of S. A. Cook [67] (distinct from the W. Cook who helped prove Theorem 33) that would take the discussion much too far afield if reviewed here.

Cutting plane proofs are more powerful than resolution because they can solve the pigeonhole problems (Section 3.2.1) in polynomial time, whereas resolution cannot (Theorem 21). In fact, a problem with n pigeons can be solved with a proof of length n^3. For instance, the clause set (3.8) for three pigeons can be refuted by computing the cut $-x_{1j} - x_{2j} - x_{3j} \geq -1$ (i.e., $x_{1j} + x_{2j} + x_{3j} \leq 1$) for $j = 1, 2$ as follows:

$$\begin{array}{rl}
-x_{1j} - x_{2j} \phantom{- x_{3j}} \geq -1 & (1/2) \\
-x_{1j} \phantom{- x_{2j}} - x_{3j} \geq -1 & (1/2) \\
\phantom{-x_{1j}} - x_{2j} - x_{3j} \geq -1 & (1/2) \\
\hline
-x_{1j} - x_{2j} - x_{3j} \geq -3/2 &
\end{array}$$

where the weights are on the right and the right-hand side $-3/2$ is rounded up to -1. These two cuts can now be added to the first three clauses of (3.8) to obtain a contradiction.

For an n-pigeon problem one obtains a series of cuts in stages. The rth stage ($r = 1, \ldots, n-1$) obtains for each $j \in \{1, \ldots, n-1\}$ the cut

$$\sum_{k=i}^{i+r} x_{kj} \leq 1 \qquad (3.39)$$

This cut is derived by taking a linear combination of the following three inequalities, each with weight 1/2,

$$\sum_{k=i}^{i+r} x_{kj} \leq 1, \quad \sum_{k=i+1}^{i+r+1} x_{kj} \leq 1, \quad x_{ij} + x_{i+r+1,j} \leq 1$$

and rounding down the right-hand side. The cuts (3.39) obtained in the final stage ($r = n-1$) are then added to the premises of the form $\sum_{j=1}^{n-1} x_{ij} \geq 1$ for $i = 1, w \ldots, n - 1$.

Thus cutting plane proofs and extended resolution have the same power on pigeonhole problems, and extended resolution is at least as powerful as cutting planes on arbitrary problems.

The operation resulting in each cut (3.39) is an instance of a "diagonal summation" operation obtained in Section 3.6 by generalizing resolution to inequalities. This suggests that generalized resolution may be as powerful as cutting plane proofs when applied to sets of clauses, even though it uses only two particular types of cuts. But again it is unknown whether this is true.

More generally, the advantages of "lifting" a problem into a space of higher dimension by adding variables, followed by a "projection" onto the original space, are not well understood. This type of strategy has recently been investigated for integer programming [8, 258], but its full implications for logical inference are apparently unknown.

3.6 Resolution for 0-1 Inequalities

Section 3.5 showed how certain inequalities are equivalent to logical clauses and can be studied from a logical point of view. This section extends logical analysis to all linear inequalities in 0-1 variables.

An inequality in 0-1 variables can be viewed as a logical proposition that is true if and only if the inequality is satisfied. One inequality implies another if all the 0-1 points satisfying the first satisfy the second. Two inequalities are equivalent if they are satisfied by the same 0-1 points.

A complete inference method for inequalities can be obtained by extending the connection between resolution and cutting planes. Recall from Section 3.1.1 that resolution generates all "prime" or undominated implications of a set of clauses. It is therefore a complete inference method for clauses, in the sense that any clause implied by a set of clauses is absorbed by some prime implication generated by resolution.

In addition, Section 3.5.1 established that a resolvent is a particular type of cutting plane. Resolution is therefore a cutting plane operation that, when applied repeatedly, finds all clausal cuts for a given set of clausal

3.6 RESOLUTION FOR 0-1 INEQUALITIES

inequalities, in the sense that any clausal cut is implied by a cut generated by this operation.

This cutting plane operation may be extended to find all cuts for any set of linear 0-1 inequalities. Or more precisely, the problem of finding the implications of a *set* of inequalities, that is, the problem of finding all cuts, may be reduced to that of finding the implications of a single inequality. We will describe two cutting plane operations, one of which is essentially resolution, that generate all undominated cuts, up to equivalence. A cut is undominated if it is a "strongest possible" cut in a logical sense, meaning that it is implied only by equivalent cuts. Then a given inequality is a cut if and only if it is implied by one of the undominated cuts. These two cutting plane operations therefore comprise a *generalized resolution algorithm* that computes implications of 0-1 inequalities.

The algorithm applies to subsets of inequalities as well as the set of all inequalities. Let T be any class of inequalities that satisfies an innocuous technical property that might be called monotonicity. The algorithm finds for a given set of inequalities all undominated cuts in T, up to equivalence. If T is the set of clausal inequalities, the algorithm reduces to resolution.

This means that the algorithm makes a contribution to logic: it provides, at least in principle, a complete inference method for any class of formulas that can be represented as linear inequalities in 0-1 variables. Or more precisely, it reduces the problem of finding all implications of a set of formulas to that of finding all implications of a single formula.

Finding the implications of a single formula is trivial in the case of clauses, because implication for clauses is absorption. But it may be nontrivial for other types of formulas. The generalized resolution algorithm is therefore practical only when applied to classes of formulas for which it is easy to recognize when one formula implies another. At the end of the section it will be shown how to specialize the algorithm to certain types of logical formulas with this property.

The algorithm also has relevance to integer programming. Section 3.3 noted that when an integer programming problem is solved by branch and bound, the search tree can be reduced by adding cuts to the linear relaxation of the problem. Cuts are generally regarded as devices that strengthen the relaxation by cutting off fractional solutions. But they can also reduce the search tree by explicitly representing logical relations that are only implicit in the original constraints. For instance, if no solution in which $x_j = x_k = 1$ is feasible, one could add a cut representing the logical fact that $\neg x_j \vee \neg x_k$, such as $-x_j - x_k \geq -1$.

This sort of cut might be called a *logic cut*. (Logic cuts increase the degree of *consistency* of a constraint set, a concept that has been extensively discussed in the constraint programming literature [148, 282, 287, 288].)

Two cuts that rule out the same integer points are the same logic cut, because the function of a logic cut is to impose logical relations among the integer variables. An interesting feature of logic cuts is that they need not be inserted into the linear relaxation as inequalities. They can perform their function externally if no branch is taken in the branch-and-bound tree without first checking whether the resulting assignment of values to integer variables would violate a logic cut.

The relevance of the generalized resolution algorithm is that it generates, at least in principle, all valid logic cuts for a given set of inequality constraints. The cuts are generated in the form of inequalities, but as just noted, they need not be used as inequality constraints within the linear relaxation. Generalized resolution is therefore a logical counterpart of Chvátal's cutting plane algorithm, since the latter in principle generates all valid cuts—in the traditional sense whereby two cuts are equivalent if they define the same half-space. Since any valid logic cut, expressed as an inequality, is a cut in the traditional sense, it is not surprising that only two types of Chvátal cutting plane operations are needed to generate all valid logic cuts.

There is no simple relation, however, between undominated logic cuts (the output of our algorithm) and the output of Chvátal's algorithm, which are facet cuts for the convex hull of the set of feasible integer solutions. An inequality that is an undominated logic cut need not be facet-defining, and we will see that, curiously, a facet-defining inequality may be strictly dominated by other logic cuts.

The notion of a logic cut can be naturally extended to mixed integer/linear programming, in which some variables are integral and others are continuous. In this context a logic cut may actually cut off feasible integer solutions. It is defined to be a constraint on the values of the 0-1 variables that does not change the projection of the problem's epigraph onto the space of continuous variables, whenever the integer variables have nonnegative coefficients. (See [154] for an explanation of these ideas). These generalized logic cuts have proved quite useful in practical problems, as noted in [154].

This raises the question as to how generalized logic cuts may in principle be generated. Since it turns out that generalized logic cuts are necessarily valid logic cuts in a pure 0-1 integer programming problem (a problem in which all variables are 0-1), generalized resolution partially answers the question by answering it for pure 0-1 programming problems. The question remains unanswered for general mixed integer/linear programming.

3.6.1 Inequalities as Logical Formulas

An inequality $ax \geq a_0$ can be regarded as a boolean function of $x_1, \ldots, x_n$ that is true if and only if $x_1, \ldots, x_n$ satisfy the inequality. It can therefore be regarded as a logical formula. Boolean functions that are representable with inequalities are known as *threshold functions* and have been studied intensely [157, 196, 221]. The problem of characterizing threshold functions is quite difficult and has never been completely solved. But it is clear that a wide variety of boolean functions, and therefore a wide variety of logical formulas, can be written as inequalities.

The focus here is on inequalities $ax \geq a_0$ in which a_0 and the components of a are integers. These in effect encompass all inequalities with rational coefficients, as rational numbers can be converted to integers by multiplying the inequality by an appropriate constant.

It will often be convenient to write an inequality $ax \geq a_0$ in the form $ax \geq \beta + n(a)$, where $n(a)$ is the sum of the negative components of a, and β is the *degree* of the inequality. For instance, $2x_1 - 3x_2 \geq -1$ becomes $2x_1 - 3x_2 \geq 2 - 3$, which has degree 2. Thus a clausal inequality has degree 1. An inequality is satisfiable when $\sum_j |a_j| \geq \beta$ and tautologous when $\beta \leq 0$. The coefficients can be made nonnegative by complementing variables with negative coefficients, that is, by replacing x_j by $1 - \bar{x}_j$. This makes the degree identical to the right-hand side. In the example this yields $2x_2 + 3\bar{x}_2 \geq 2$. It is useful to think of the degree as the "true" right-hand side, which is "disguised" by the presence of negative coefficients.

It is natural to deal with $\geq$ inequalities because of their connection with logical clauses. The results can be restated for $\leq$ inequalities by complementing every variable and multiplying every inequality by -1. Thus an inequality $ax \geq \beta + n(a)$ becomes $a\bar{x} \leq p(a) - \beta$, where $p(a)$ is the sum of the positive components of a, and β is again the degree.

Absorption can be generalized to inequalities. One inequality $ax \geq a_0$ *absorbs* another when the latter is the sum of $ax \geq a_0$ and zero or more inequalities of the form $x_j \geq 0$ (if $a_j \geq 0$), $-x_j \geq -1$ (if $a_j \leq 0$), and $0 \geq -1$. For instance, $2x_1 - 3x_2 \geq 2 - 3$ absorbs $4x_1 - 3x_2 \geq 2 - 3$, as well as $4x_1 - 3x_2 \geq 1 - 3$ and $4x_1 - 4x_2 \geq 1 - 4$. Because the bounds $0 \leq x_j \leq 1$ are valid for 0-1 inequalities, an inequality implies any inequality it absorbs.

One inequality $ax \geq a_0$ *reduces* to another if the latter (the *reduction*) is the sum of $ax \geq a_0$ and zero or more inequalities of the form $x_j \geq 0$ (if $a_j < 0$) and $-x_j \geq -1$ (if $a_j > 0$). For instance, $2x_1 - 3x_2 \geq 4 - 3$ reduces to $x_1 - 3x_2 \geq 3 - 3$ and to $x_1 - x_2 \geq 1 - 1$.

Unsurprisingly, an undominated cut for a set S of inequalities need not be a facet-defining inequality for the convex hull of the 0-1 solutions. Suppose, for instance, that $S = \{x_1 + x_2 \geq 1, \ x_1 + x_3 \geq 1, \ 0 \leq x_j \leq$

154 CHAPTER 3 PROPOSITIONAL LOGIC: GENERAL CASE

1, all j}. The 0-1 solutions are $(1,0,0), (1,0,1), (0,1,1), (1,1,0), (1,1,1)$. Then $2x_1 + x_2 + x_3 \geq 2$ is an undominated cut, but it does not define a facet of the convex hull of these points. It is somewhat surprising, however, that a facet-defining inequality need not be undominated. This is true of $x_1 + x_2 \geq 1$, which is strictly dominated by $2x_1 + x_2 + x_3 \geq 2$ because the latter inequality is satisfied by all points satisfying the former except $(0,1,0)$.

3.6.2 A Generalized Resolution Algorithm

The generalized resolution algorithm repeatedly applies two rank 1 cutting plane operations to a given set S of inequalities. One of the operations is essentially resolution. That is, when clausal inequality C is implied by an inequality in S, and similarly for clausal inequality D, then the resolvent of C and D (if there is one) is generated as a cut. Note that it is necessary to recognize when an inequality is implied by another. Section 3.5.1 observes that the resolvent is a rank 1 cut.

The other operation is a "diagonal sum" that may be illustrated as follows. Suppose that each of the following inequalities is implied by an inequality in S:

$$
\begin{aligned}
x_1 + 5x_2 + 3x_3 + x_4 &\geq 4 \quad \text{(a)} \\
2x_1 + 4x_2 + 3x_3 + x_4 &\geq 4 \quad \text{(b)} \\
2x_1 + 5x_2 + 2x_3 + x_4 &\geq 4 \quad \text{(c)} \\
2x_1 + 5x_2 + 3x_3 &\geq 4 \quad \text{(d)}
\end{aligned}
\tag{3.40}
$$

Note that each is a reduction of the following inequality,

$$2x_1 + 5x_2 + 3x_3 + x_4 \geq 5 \tag{3.41}$$

obtained by reducing one coefficient at a time in a diagonal pattern. Assigning weight 2/10 to (3.40a), 5/10 to (b), 3/10 to (c), and 1/10 to (d) yields the nonnegative linear combination

$$2x_1 + 5x_2 + 3x_3 + x_4 \geq 44/10 \tag{3.42}$$

Rounding up the right-hand side produces (3.41), the *diagonal sum* of (3.40a–d). The diagonal sum is therefore a rank 1 cut and so is implied by (3.40a–d) and S.

In general, a satisfiable inequality $ax \geq \beta + n(a)$ is the diagonal sum of inequalities $a_i x \geq \beta - 1 + n(a^i)$ for $i \in J \subset \{1,\ldots,n\}$ when $a_j \neq 0$ for all $j \in J$, $a_j = 0$ for all $j \notin J$, and

$$a^i_j = \begin{cases} a_j - 1 & \text{if } j = i \text{ and } a_j > 0 \\ a_j + 1 & \text{if } j = i \text{ and } a_j < 0 \\ a_j & \text{otherwise} \end{cases} \tag{3.43}$$

3.6 RESOLUTION FOR 0-1 INEQUALITIES

To verify that $ax \geq \beta + n(a)$ is a rank 1 cut (when $n \geq 2$), assign each $a_i x \geq \beta - 1 + n(a^i)$ weight $|a_i|/(W-1)$, where $W = \sum_j |a_j|$. The weighted sum of the inequalities is $ax \geq (\beta - 1)W/(W-1) + n(a)$. Because satisfiability implies $W \geq \beta$, $\beta - 1 < (\beta-1)W/(W-1) \leq \beta$. So the desired $ax \geq \beta + n(a)$ is obtained after rounding up the right-hand side.

Before stating the algorithm precisely, a *monotone* class T of inequalities must be defined. T is monotone if it contains all clausal inequalities, and given any inequality $ax \geq \beta + n(a)$ in T, T contains all inequalities $a'x \geq \beta' + n(a')$ for which $|a'_j| \leq |a_j|$ for all j, and $0 \leq \beta' \leq \beta$. The set of clausal inequalities is obviously monotone, as is the set of all inequalities with integral coefficients.

The following algorithm generates, for a given set S of inequalities in T, all undominated cuts in T.

Procedure 20: *Resolution for 0-1 Inequalities*

Step 0. Let $S' = S$. Remove inequalities from S' to ensure that no inequality in S' implies another.

Step 1. If possible, find clausal inequalities C and D that have a resolvent R that no inequality in S' implies, such that C and D are each implied by some inequality in S'. Remove from S' all inequalities that R implies, and add R to S'.

Step 2. If possible, find inequalities $I_1, \ldots, I_m \in T$ that have a diagonal sum $I \in T$ that no inequality in S' implies, such that $I_1, \ldots, I_m$ are each implied by some inequality in S'. Remove from S' all inequalities that I implies, and add I to S'.

Step 3. If inequalities were added to S' in either step 1 or step 2, return to step 1. Otherwise stop.

The algorithm is finite because there are finitely many inequalities in variables $x_1, \ldots, x_n$ with integral coefficients, S' contains no pairs of equivalent inequalities, and a cut is never added to S' once it is removed.

Theorem 34 ([144]) *The above algorithm generates all undominated cuts in T, up to equivalence, for a given set S of inequalities in T, if T is monotone.*

Two lemmas help to prove the theorem. Let the *length* of an inequality be the sum of the absolute values of its coefficients. An inequality has maximal length with respect to a given property if it has the property and would lose it if one or more coefficients were increased in absolute value.

156   CHAPTER 3   PROPOSITIONAL LOGIC: GENERAL CASE

Lemma 14 *Let $ax \geq 1 + n(a)$ be a cut of maximal length for S that is implied by no inequality in S, and suppose that $a_k = 0$ for some k. Then $ax \geq 1 + n(a)$ is the resolvent of two inequalities, each of which is implied by an inequality in S.*

Proof: The clausal inequalities $x_k + ax \geq 1 + n(a)$ and $-x_k + ax \geq 1 + n(a) - 1$ are cuts for S, because they are absorbed and therefore implied by $ax \geq 1 + n(a)$. Since they are longer than $ax \geq 1 + n(a)$, each is implied by some inequality in S. But their resolvent is $ax \geq 1 + n(a)$, and the lemma follows. □

The next lemma will form the initial step of an inductive argument.

Lemma 15 *If T is monotone, any clausal cut for S is implied by some cut generated by the above algorithm.*

Proof: Suppose otherwise, and let $ax \geq 1 + n(a)$ be a clausal inequality in T of maximal length that is a cut for S but implied by no cut the algorithm generates. Since T is monotone and therefore contains all clausal inequalities, $ax \geq 1 + n(a)$ is a clausal inequality, *simpliciter*, of maximal length that is implied by no cut the algorithm generates. Since variables can be complemented, suppose without loss of generality that $a \geq 0$. Then $a_k = 0$ for some k. To see this, note that otherwise the only point x violating $ax \geq 1 + n(a)$ is the origin, which means that any clause violated by the origin implies $ax \geq 1 + n(a)$. If the origin violated no clause in S, $ax \geq 1 + n(a)$ would not be a cut for S. Therefore the origin violates some clause in S, and this clause implies $ax \geq 1 + n(a)$, contrary to hypothesis. Thus some $a_k = 0$. Given this, Lemma 14 implies that the algorithm generates a resolvent that implies $ax \geq 1 + n(a)$, contrary to hypothesis. □

Proof of Theorem 34: It will be proved that any cut $ax \geq \beta + n(a)$ for S in T is implied by some cut the algorithm generates. The proof is by induction on the degree β.

First suppose $\beta = 1$. It is easy to see that any inequality $ax \geq 1 + n(a)$ of degree 1 is equivalent to the clausal cut $a'x \geq 1 + n(a')$, where $a'_j = 1$ if $a_j > 0$, -1 if $a_j < 0$, and 0 if $a_j = 0$. But since T is monotone, it follows from Lemma 15 that $a'x \geq 1 + n(a')$, and therefore $ax \geq 1 + n(a)$ is implied by a cut the algorithm generates.

We now assume that the theorem is true for all inequalities in T of degree $\beta - 1$ and show that it is true for inequalities of degree β. Suppose otherwise. Let $ax \geq \beta + n(a)$ be a cut for S of maximal length that is implied by no cut the algorithm generates. For all $i \in \{j | a_j \neq 0\} = J$, let

3.6 RESOLUTION FOR 0-1 INEQUALITIES

a^i be defined by (3.43). Then $ax \geq \beta + n(a)$ is the diagonal sum of the inequalities $a^i x \geq (\beta - 1) + n(a^i)$ for $i \in J$. The following statements can be made about $a^i x \geq (\beta - 1) + n(a^i)$ for each $i \in J$: (a) it is a reduction of $ax \geq \beta + n(a)$ and is therefore a cut for S; (b) it belongs to T, since T is monotone; (c) since it has degree $\beta - 1$, (a), (b), and the induction hypothesis imply that it is implied by some cut the algorithm generates. But (c), together with step 2 of the algorithm, implies that $ax \geq \beta + n(a)$ is likewise implied by a cut the algorithm generates, contrary to hypothesis. The theorem follows. □

3.6.3 Some Examples

The algorithm of the previous section may be adapted to some classes of logical formulas in which it is easy to check whether one formula implies another.

If the algorithm is specialized to clauses, the result is ordinary resolution. But one can also apply it to *set packing* clauses; that is, clauses that assert that *at most* one of a set of atomic propositions is true. They might be called "set packing clauses" because they can be written as set packing inequalities. For instance, a clause stating that at most one of x_1, x_2, and x_3 is true can be written $x_1 + x_2 + x_3 \leq 1$.

In general, set packing inequalities have the form $ax \leq 1$, where each $a_j \in \{0, 1\}$. After complementing variables the inequality becomes $a\bar{x} \geq p(a) - 1$. Also, $ax \leq 1$ implies $bx \leq 1$ if and only if $a \geq b$. Resolution (cancelation) does not apply here, because there are no negative coefficients. Fortuitously, diagonal summation is quite simple, because the diagonal sum of set packing inequalities is a set packing inequality. The algorithm of the previous section simplifies to the following.

Procedure 21: *Resolution for Set Packing Inequalities*

Step 0. Let S be a set of set packing inequalities, and set $S' = S$. Remove from S' every inequality that is implied by a longer one in S'.

Step 1. If possible, find a set $J \subset \{1, \ldots, n\}$ of at least two indices and, for each $i \in J$, an inequality $a^i x \leq 1$ in S' such that $a^i_j = 1$ for all $j \in J \setminus \{i\}$, and such that no inequality in S' dominates the diagonal sum $ax \leq 1$, where $a_i = 1$ for $j \in J$ and $a_j = 0$ for $j \notin J$. (Assume without loss of generality that $a^i_i = 0$.) Remove from S' all inequalities that $ax \leq 1$ dominates, and add $ax \leq 1$ to S'.

Step 2. If a cut was added to S' in step 1, return to step 1. Otherwise stop.

For instance, the following set packing inequalities

$$\begin{aligned} x_2 + x_3 + x_4 &\leq 1 \\ x_1 \phantom{{}+{}} + x_3 \phantom{{}+x_4} &\leq 1 \\ x_1 + x_2 \phantom{{}+x_3} + x_4 &\leq 1 \end{aligned} \qquad (3.44)$$

respectively absorb the following inequalities

$$\begin{aligned} x_2 + x_3 &\leq 1 \\ x_1 \phantom{{}+{}} + x_3 &\leq 1 \\ x_1 + x_2 \phantom{{}+x_3} &\leq 1 \end{aligned}$$

which have the diagonal sum

$$x_1 + x_2 + x_3 \leq 1 \qquad (3.45)$$

So if S' contains inequalities (3.44), the algorithm generates (3.45). This is a well-known cut for set packing problems.

The algorithm also solves the maximum cardinality set packing problem. The cardinality of a maximum set packing (maximum of $\sum_j x_j$ subject to S) is the minimum of $n - p(a) + 1$ over all undominated cuts $ax \leq 1$.

Another class of formulas to which one can apply the algorithm of the previous section consists of "extended clauses," which assert that at least β of a set of literals are true. For ordinary clauses, $\beta = 1$. Extended clauses are easily written as inequalities, since $x_1 - x_2 + x_3 \geq 2 - 1$, for instance, says that at least two of $x_1, \neg x_2, x_3$ are true. In general, an extended clause has the form $ax \geq \beta + n(a)$, where each $a_j \in \{0, 1, -1\}$. These are studied in [9], which calls them "canonical cuts."

It is shown in [140] that one extended clause implies another if and only if the former reduces to an extended clause that absorbs the latter. Equivalent extended clauses are identical. Finally, it is clear that extended clauses form a monotone set. The algorithm takes the following form:

Procedure 22: *Resolution for Extended Clauses*

Step 0. For a set S of extended clauses, let $S' = S$. Remove from S' every clause implied by a distinct one in S'.

Step 1. If possible, find clausal inequalities C and D that have a resolvent R that no inequality in S' implies, such that C and D are each implied by some inequality in S'. Remove from S' all inequalities that R implies, and add R to S'.

Step 2. If possible, find extended clauses $a^i x \geq \beta + n(a^i)$ for $i \in J \subset \{1, \ldots, n\}$, each of which is implied by an extended clause in S',

3.7 A SET-COVERING FORMULATION WITH FACET CUTS

such that the diagonal sum $ax \geq \beta + 1 + n(a)$ is implied by no clause in S'. Here $a_i^i = 0$ for each $i \in J$, $a_j^i = a_j = 0$ for each $j \in \{1, \ldots, n\}$ and each $i \in J$, and $a_j = 1$ for each $j \in J$. Remove from S' all extended clauses that $ax \geq \beta + 1 + n(a)$ implies, and add $ax \geq \beta + 1 + n(a)$ to S'.

Step 3. If inequalities were added to S' in either step 1 or step 2, return to step 1. Otherwise stop.

For instance, suppose that S' contains the extended clauses

$$\begin{aligned} x_1 - x_2 + x_3 + x_4 - x_5 &\geq 4 - 2 \\ x_2 + x_3 - x_4 - x_5 &\geq 3 - 2 \\ -x_1 \qquad\qquad x_4 - x_5 &\geq 2 - 2 \end{aligned}$$

If we choose $J = \{3, 4, 5\}$, these imply, respectively,

$$\begin{aligned} x_4 - x_5 &\geq 1 - 1 \\ x_3 \qquad - x_5 &\geq 1 - 1 \\ x_3 + x_4 \qquad &\geq 1 \end{aligned}$$

So we add the diagonal sum $x_3 + x_4 - x_5 \geq 2 - 1$ to S'.

We noted in Section 3.5.5 that diagonal sums of this sort can solve pigeonhole problems in polynomial time.

Barth [16] has written a careful study of how generalized resolution may be efficiently applied to extended clauses. His aim is to solve 0-1 programming problems by inferring extended clauses from the constraints and processing the clauses with various forms of generalized resolution. Ohyanagi et al. [225] propose an alternate method for solving the satisfiability problem for extended clauses.

Finally, a generalized resolution algorithm is given in [144] for first and second degree inequalities whose coefficients belong to $\{0, \pm 1, \pm 2\}$.

3.7 A Set-Covering Formulation with Facet Cuts

The collective wisdom in integer programming is that the use of cutting planes that are "deep" or facet-inducing can have a remarkable impact on controlling the size of the enumeration trees (cf. [222, 229]) in branch-and-cut methods. This requires detailed understanding of these facet cuts and techniques for their generation. A *facet cut* is a cutting plane that exposes a facet of the convex hull of integer solutions. For readers who are unfamiliar with this terminology of polyhedral theory, the Appendix contains a brief review.

There are many examples in integer programming where introducing redundancy in formulations has been extremely useful in gaining insight about the polyhedral structure of the convex hull of integer solutions. In this section we introduce such an "extended formulation" of satisfiability in propositional logic. In this formulation we have as many decision variables as there are literals in the proposition (twice the number of atomic propositions). This additional freedom allows us to formulate satisfiability as a set-covering problem, a class of integer programs for which some theory of facet cuts has already been developed.

After introducing the set-covering formulation we describe recent results on facet cuts for the set-covering formulation of satisfiability in propositional logic [4]. The main result is that *resolvent facet cuts* are precisely the *prime implications* of the proposition. This is an intuitively pleasing result since prime implications represent the strongest possible logical implications of a proposition and facet cuts represent the strongest possible cutting planes of the integer hull. Thus we have another example of a remarkable connection between logic and optimization.

It is unfortunate that demonstrating this connection requires the rather involved and sometimes technical derivations that are developed in this section (this is a warning to the unsuspecting reader). However, results on facet cuts are invariably complicated to derive since arguments based on the affine dimension of polyhedra are central to this subject. To a large extent, though, we finesse these dimensionality issues by resorting to the set-covering formulation.

A natural question to ask is whether the facet cuts defined by prime implications are all that we need. Unfortunately, this is not the case. Recall (Section 3.5.5) that to obtain a polynomial length cutting plane proof of the pigeonhole principle, we had to resort to cutting planes that were not resolvents. In fact, we will exhibit an example in this section for which all the prime implications are not sufficient to separate a fractional and optimal extreme point of the linear programming relaxation from the integer hull. As a partial remedy, we will also describe a method for generating facet cuts that do not arise as resolvents.

3.7.1 The Set-Covering Formulation

Given a set of clauses $S = \{C_1, C_2, \ldots, C_m\}$ on the atomic propositions $\{x_1, x_2, \ldots, x_n\}$, the satisfiability problem is to determine if there exists a set of truth assignments (a model) that satisfies all clauses in S. We define

3.7 A SET-COVERING FORMULATION WITH FACET CUTS

a set of $2n$ variables $\{y_1, z_1, y_2, z_2, \ldots, y_n, z_n\}$ that have the interpretation

$$y_j = \begin{cases} 1 & \text{if } x_j \text{ is true} \\ 0 & \text{otherwise} \end{cases}$$

$$z_j = \begin{cases} 1 & \text{if } x_j \text{ is false} \\ 0 & \text{otherwise} \end{cases}$$

Hence there is a 1-1 correspondence between a set of truth assignments for $\{x_1, x_2, \ldots, x_n\}$ and nonnegative integers y_j, z_j meeting the condition

$$y_j + z_j = 1 \text{ for } j = 1, 2, \ldots, n \tag{3.46}$$

Let $T(S)$ denote the subset of 0-1 vectors (y, z) that meet (3.46) and whose corresponding truth assignments satisfy all clauses in S. In $\Re^{2n}$ coordinatized by (y, z) we can interpret $T(S)$ as a collection of extreme points of the unit hypercube. The y components of these extreme points are precisely incidence vectors of satisfying truth assignments of S. Having embedded truth assignments in Euclidean space, we now construct a polyhedron around them by letting P be the convex hull $conv(T(S))$ of the feasible set $T(S)$. The polyhedron P is empty if and only if $T(S)$ is empty (i.e., S is unsatisfiable).

Given a clause C_i of S that is not a tautology, that is, it is nonempty and does not contain a literal and its negation, we write down a clausal inequality

$$a^i y + b^i z \geq 1$$

where

$$a^i_j = \begin{cases} 1 & \text{if } x_j \text{ is in } C_i \\ 0 & \text{otherwise} \end{cases}$$

$$b^i_j = \begin{cases} 1 & \text{if } \neg x_j \text{ is in } C_i \\ 0 & \text{otherwise} \end{cases}$$

Satisfiability of S can be resolved by the following integer programming problem.

$$\begin{aligned} \min \quad & \sum_{j=1}^{n}(y_j + z_j) \\ \text{s.t.} \quad & a^i y + b^i z \geq 1, \text{ for } C_i \in S \\ & y_j + z_j \geq 1, \text{ for } j = 1, 2, \cdots, n \\ & (y, z) \in \{0, 1\}^{2n} \end{aligned} \tag{3.47}$$

Notice that n is a lower bound on the minimum value that can be obtained in (3.47) since each pair (y_j, z_j) must add to at least 1 in any solution.

Clearly S is satisfiable if and only if the optimal objective value of (3.47) is n. Indeed, the solutions attaining an objective value of n correspond precisely to the set of satisfying truth assignments $T(S)$. If we denote as P_S3, the convex hull of the 0-1 solutions to (3.47), the polytope P is the face of P_S defined by the valid inequality $\sum_{j=1}^{n}(y_j + z_j) \geq n$.

Let us define the *clause set* of clause C_i as the subset of the variables $\{y_1, \cdots, y_n, z_1, \cdots, z_n\}$ corresponding to the literals in C_i (y_j if x_j is in C_i, z_j if $\neg x_j$ is in C_i and neither otherwise). Let the *ground set* E contain the pairs $\{y_j, z_j\}$ for $j = 1, 2, \cdots, n$ and the clause sets of all clauses $\{C_i\}$. Corresponding to each y_j is a subset Y_j of E (the "pair set" $\{y_j, z_j\}$ and the clause sets containing y_j). Similarly for each z_j there is a subset Z_j of E. The optimization problem (3.47) therefore seeks a minimum cardinality subcollection of the sets $Y_1, \ldots, Y_n, Z_1, \ldots, Z_n$ whose union covers E. This is the "set-covering" interpretation of (3.47).

It is convenient in the discussion of facets to work with a modification of the set-covering formulation (3.47). The modification is simply to replace the 0-1 restriction with nonnegative and integer restriction for all the variables.

$$\begin{aligned} \min \quad & \sum_{j=1}^{n}(y_j + z_j) \\ \text{s.t.} \quad & a^i y + b^i z \geq 1, \quad \text{for } C_i \in S \\ & y_j + z_j \geq 1, \quad \text{for } j = 1, 2, \ldots, n \\ & (y, z) \in \Re_{+}^{2n} \text{ and integer} \end{aligned} \quad (3.48)$$

Clearly, it is still true that S is satisfiable if and only if the optimal objective value of (3.48) is n. Let Q_S denote the convex hull of the set of feasible solutions to (3.48). P is the face of Q_S as well as P_S that is defined by the valid inequality $\sum_{j=1}^{n}(y_j + z_j) \geq n$. In fact Q_S is simply the dominant of P_S, i.e. Q_S contains all points in $\Re^{2n}$ whose components are at least as large as the components of some point in P_S. This may be denoted as

$$Q_S = P_S + \Re_{+}^{2n}$$

where the addition "+" is of two sets. Because we will study the structure of Q_S, it is appropriate also to define the dominant of P, which we denote by Q. Both P and Q are empty if and only if $T(S)$ is empty (i.e., S is unsatisfiable). We do not know the complexity of computing the dimension of P even under the assumption that S is satisfiable. It is clearly NP-hard to compute the dimension of P (or for that matter Q) if the satisfiability of S is not predetermined. However, if S is satisfiable then $T(S)$ is nonempty and Q has full dimension. Also, P is the face of Q defined by the valid inequality $\sum_{j=1}^{n}(y_j + z_j) \geq n$.

3.7 A SET-COVERING FORMULATION WITH FACET CUTS

The polyhedra Q and Q_S are by construction *upper comprehensive*. This means that for any point $(\tilde{y}, \tilde{z})$ in Q, all points (y, z) satisfying $(y, z) \geq (\tilde{y}, \tilde{z})$ are also in Q, and similarly for Q_S. In contrast with P and Q, Q_S is nonempty even if S is unsatisfiable since the vector of $2n$ ones is always in Q_S. The following lemma is a summary of the relationship between Q_S, P, and Q.

Lemma 16 *For the polyhedra P, Q and Q_S as defined above:*

(i) $P \subseteq Q \subseteq Q_S$.

(ii) P *is the face of Q (and of Q_S) defined by the valid inequality*

$$\sum_{j=1}^{n}(y_j + z_j) \geq n$$

(iii) Q_S *is of full dimension (even if S is unsatisfiable).*

We do not have a thorough understanding of the structure of the linear programming relaxation of (3.48) (i.e., of the polyhedron obtained if we require (y, z) to be nonnegative reals rather than nonnegative integers). However, we do know that if every clause in S has at least two literals, then a vector of $1/2$'s is feasible in the linear programming relaxation of (3.48). Thus solving the linear program does not necessarily give any information about the satisfiability of S. In fact, we can construct simple examples of satisfiable S such that the linear programming relaxation of (3.48) has fractional extreme points.

For instance, consider the family of Horn clauses

$$\begin{array}{ll} x_1 \vee \neg x_2 \vee \neg x_3 & \neg x_1 \vee x_2 \\ x_1 \vee \neg x_2 \vee \neg x_4 & \neg x_1 \vee x_3 \\ x_1 \vee \neg x_3 \vee \neg x_4 & \neg x_1 \vee x_4 \end{array}$$

for which we can write the linear programming relaxation of (3.48) as

$$\begin{array}{rl} \min & \sum_{i=1}^{4}(y_i + z_i) \\ \text{s.t.} & y_1 + z_2 + z_3 \geq 1 \\ & y_1 + z_2 + z_4 \geq 1 \\ & y_1 + z_3 + z_4 \geq 1 \\ & z_1 + y_2 \geq 1 \\ & z_1 + y_3 \geq 1 \\ & z_1 + y_4 \geq 1 \\ & z_i + y_i \geq 1, \quad i = 1, \ldots, 4 \\ & z_i, y_i \geq 0, \quad i = 1, \ldots, 4 \end{array}$$

The feasible region has a fractional extreme point given by $y_1 = 0$, $z_1 = 1$, and $y_i = z_i = 1/2$ for $i = 2, 3, 4$, which is an optimal solution for (3.7.1).

3.7.2 Elementary Facets of Satisfiability

We shall use two different approaches to prove that a valid inequality $ax \geq a_0$ defines a facet of a full-dimensional polyhedron K in $\Re^{2n}$. Either we exhibit $2n$ affinely independent points in K satisfying the inequality as an equality; or we simply assume that the face of K defined by $ax \geq a_0$ is contained in the face defined by some other inequality $\alpha x \geq \alpha_0$, and proceed to use some of the points in K satisfying $ax = a_0$ to show that $\alpha = \lambda a, \alpha_0 = \lambda a_0$ for some positive scalar λ.

The process described above can be simplified in the case of an upper comprehensive polyhedron, as a consequence of the following lemma.

Lemma 17 *Let $K \subseteq \Re^{2n}$ be a nonempty upper comprehensive polyhedron. Then:*

(i) *K is full-dimensional.*

(ii) *Every valid inequality has nonnegative coefficients.*

(iii) *If $ax \geq a_0$ is a valid inequality defining a nonempty face F of K, and $\alpha x \geq \alpha_0$ induces a facet of K containing F, then $a_i = 0$ implies $\alpha_i = 0$, for all i.*

Proof: Part (i) follows from considering the points $x^0, x^0 + e_1, \ldots, x^0 + e_n$, where $x^0 \in K$ and e_i is the unit vector with 1 on the ith component and 0 elsewhere. Part (ii) follows from considering that $x^0 + \lambda e_i \in P$ for all $\lambda \geq 0$. If, in addition, $x^0 \in F$, for all i with $a_i = 0$ we have that $x^0 + \lambda e_i \in F$. Hence $\alpha(x^0 + \lambda e_i) = \alpha_0$, implying that $\alpha_i = 0$. □

We know that Q_S is full-dimensional whether the set of clauses in S is satisfiable or not. The following theorem summarizes some basic results on the facets of Q_S.

Theorem 35 ([4]) *The elementary facets of Q_S are characterized below.*

(i) *The inequality $y_i \geq 0$ ($z_i \geq 0$) defines a facet of Q_S if and only if the clause x_i ($\neg x_i$) is not in S.*

(ii) *The inequality $y_i + z_i \geq 1$ defines a facet of Q_S if and only if neither x_i nor $\neg x_i$ is a clause in S.*

3.7 A SET-COVERING FORMULATION WITH FACET CUTS

(iii) The clausal inequality associated to $C_i \in S$ defines a facet of Q_S if and only if C_i is not absorbed by any other clause in S.

Proof: (i) We only consider the clause x_1; the other cases are similar. If x_1 is a clause in S then $y_1 \geq 1$ is a valid inequality for Q_S. Hence $y_1 \geq 0$ is not facet inducing. On the other hand, if x_1 is not a clause in S then the point $(\bar{y}, \bar{z})$, with $\bar{y}_1 = 0$ and 1 elsewhere, is in Q_S. If e_i is a unit vector in $\Re^{2n}$, then the $2n$ affinely independent vectors $(\bar{y}, \bar{z}), (\bar{y}, \bar{z}) + e_2, \ldots, (\bar{y}, \bar{z}) + e_{2n}$ are in Q_S and satisfy $y_1 = 0$.

(ii) If x_1 or $\neg x_1$ is a clause in S then either $y_1 \geq 1$ or $z_1 \geq 1$. Any one of those inequalities implies $y_1 + z_1 \geq 1$. Now, if x_1 and $\neg x_1$ are not clauses in S then (y^1, z^1), with $\bar{y}_1 = 0$ and 1 elsewhere, and (y^2, z^2), with $\bar{z}_1 = 0$ and 1 elsewhere, are both in Q_S. The result follows from considering the vectors $(y^1, z^1), (y^2, z^2), (y^1, z^1) + e_i, (y^2, z^2) + e_i$ for $i = 2, \ldots, n$, which are affinely independent and satisfy $y_1 + z_1 = 0$.

(iii) If C_1 is absorbed by C_2 then the clausal inequality for C_1 is dominated by the clausal inequality for C_2, namely,

$$a^1 y + b^1 z \geq a^2 y + b^2 z \geq 1$$

and cannot induce a facet of Q_S. On the other hand, for every literal in C_1, we can define a vector that is 1 on the variable corresponding to that literal, 0 on the variables corresponding to other literals in C_1, and 1 elsewhere; that is, if x_a is a literal in C_1 we define a vector (y^a, z^a) with components

$$y_i^a = \begin{cases} 1 & \text{if } i = a \\ 0 & \text{if } x_i \text{ is in } C_1 \text{ but } i \neq a \\ 1 & \text{if } x_i \text{ is not in } C_1 \end{cases}$$

$$z_i^a = \begin{cases} 0 & \text{if } \neg x_i \text{ is in } C_1 \\ 1 & \text{if } \neg x_i \text{ is not in } C_1 \end{cases}$$

Likewise, if $\neg x_a$ is a literal in C_1 we define a similar vector having $y_a^a = 0$ and $z_a^a = 1$ instead. If C_1 is not absorbed by any clause in S then $(y^a, z^a) \in Q_S$ and satisfies the clausal inequality for C_1 with equality. By Lemma 17, if $\alpha y + \beta z \geq \alpha_0$ defines a facet of Q_S containing the face defined by the clausal inequality $a^1 y + b^1 z \geq 1$, then $\alpha_i = \beta_j = 0$ for any literals $x_i, \neg x_j$ not in C_1. Now, if x_a is a literal in C_1 we then have $\alpha y^a + \beta z^a = \alpha_a = \alpha_0$, and if $\neg x_a$ is a literal in C_1 we obtain $\alpha y^a + \beta z^a = \beta_a = \alpha_0$, which allows us to conclude that the inequality $\alpha y + \beta z \geq \alpha_0$ is equal to the clausal inequality times the scalar α_0. □

3.7.3 Resolvent Facets Are Prime Implications

From the definitions it follows that for a nonempty family of nonempty clauses S we have that

$$P \subseteq Q \subseteq Q_S$$

and that if S is unsatisfiable then $P = Q = \emptyset$. Hence we focus on the relationship between Q and Q_S when S is satisfiable. The last part of Theorem 35 indicates that we can discard from S those clauses which are absorbed. However, this is not enough to obtain $Q = Q_S$, as follows from the next example.

Let $S = \{x_1 \vee x_2, \ x_1 \vee \neg x_2, \ \neg x_1 \vee \neg x_2\}$ with $T(S) = \{(1,0,0,1)\}$. Then $(0,1,1,1)$ is in Q_S since it satisfies the system

$$\begin{aligned} y_1 + y_2 &\geq 1 \\ y_1 + z_2 &\geq 1 \\ z_1 + z_2 &\geq 1 \\ y_i + z_i &\geq 1, \quad i = 1,2 \\ y_i, z_i &\geq 0, \quad i = 1,2 \\ y_i, z_i \ &\text{integer} \end{aligned} \qquad (3.49)$$

but it is not in $Q = \{(1,0,0,1)\} + \Re_+^4$, hence $Q \neq Q_S$.

The resolution method is used in logic to obtain all the implications from a set of clauses. Let C_0 denote a nonempty clause obtained after applying resolution on two parent clauses in S, and assume that C_0 is not absorbed by any clause in S. If we take $S' = S \cup \{C_0\}$, we can show that the 0-1 vector (y,z) defined as zero on all the components corresponding to variables associated with the literals in C_0 and one elsewhere is a vector in Q_S but not in $Q_{S'}$. Hence $Q_{S'} \subset Q_S$.

We thus interpret the effect of a resolution step from a polyhedral point of view. When a new set of clauses S' is obtained by resolution on S, we change the polyhedron Q_S to a polyhedron $Q_{S'}$ which is closer to Q since

$$Q \subseteq Q_{S'} \subset Q_S$$

The last inclusion is proper when the resolution algorithm generates at least one clause that is not absorbed by any of the initial ones. Now if we purge absorbed clauses, we are just eliminating clauses whose clausal inequalities define faces but not facets of the polyhedron. Those inequalities are dominated and redundant. We could use resolution combined with the elimination of any absorbed clauses to try to obtain a family of nonempty clauses S', that reduces the gap between Q and $Q_{S'}$. It is then natural to think that we should replace S by the best family of clauses S' that results

3.7 A SET-COVERING FORMULATION WITH FACET CUTS

in Q and $Q_{S'}$ being as close as possible. In example (3.49), if we replace S by its prime implications, $S' = \{x_1, \neg x_2\}$, then $Q = Q_{S'}$. This observation is true in general.

Theorem 36 ([4]) *If the family of clauses S is unsatisfiable, then $Q = \emptyset$. If S is satisfiable and S' is the set of prime implications of S then Q is nonempty and $Q = Q_{S'}$.*

We shall prove Theorem 36 by noting that $T(S)$ is the set of extreme points of Q. Although they are also extreme points of Q_S, this latter polyhedron might have some other extreme points given by 0-1 vectors which do not correspond to truth assignments, that is, satisfy $y_i = z_i = 1$ for some i. These "undesirable" extreme points are cut off by clausal inequalities of prime implications of S. Thus if S' is the set of prime implications of S, then $T(S)$ becomes the set of extreme points of $Q_{S'}$. From that we immediately have Theorem 36. We give proofs of all the necessary facts in the following lemma.

Lemma 18 *If S is satisfiable then the points in $T(S)$ are extreme points of P, Q, and $Q_{S'}$. In fact, they are all the extreme points of P and Q.*

Proof: By definition the extreme points of P and Q are a subset of $T(S)$; then it suffices to show that any point $(y, z) \in T(S)$ is an extreme point of $Q_{S'}$. We do it by assuming that (y, z) is the midpoint along the line connecting two points in Q_S, and prove that those two points must be (y, z). Hence assume that

$$(y, z) = \tfrac{1}{2}(y^1, z^1) + \tfrac{1}{2}(y^2, z^2), \qquad (y^i, z^i) \in Q_{S'}$$

Note that $y_j = 0$ implies $y_j^1 = y_j^2 = 0$, $z_j = 0$ implies $z_j^1 = z_j^2 = 0$, and $y_j + z_j = 1$ implies $y_j^1 + z_j^1 = y_j^2 + z_j^2 = 1$. From those relations it follows that $(y, z) = (y^1, z^1) = (y^2, z^2)$, so (y, z) is an extreme point of $Q_{S'}$. □

In example (3.49), $(0, 1, 1, 1)$ is an extreme point of Q_S that is not in Q. That point does not satisfy the inequality for the prime implication x_1. We now essentially have the proof of Theorem 36.

Proof Theorem 36: We only need to prove any extreme point of $Q_{S'}$ is in $T(S)$. So assume $(\tilde{y}, \tilde{z})$ is an extreme point of $Q_{S'}$ not in $T(S)$. Since all the inequalities in the set-covering formulation have 0-1 coefficients, it follows that $(\tilde{y}, \tilde{z})$ is a 0-1 vector. Now $(\tilde{y}, \tilde{z})$ must have at least a zero component, because the vector of ones $(\mathbf{1}, \mathbf{1})$ is strictly larger than any vector in $T(S)$

and cannot be an extreme point. Hence we can construct a (nonempty) clause C_0 using the literals corresponding to the zero components; that is, the literal x_a is used in C_0 if $\tilde{y}_a = 0$, and the literal $\neg x_a$ is used if $\tilde{z}_a = 0$. Note that $\tilde{y}_a + \tilde{z}_a \geq 1$ implies that C_0 does not contain a literal and its negation.

Now consider the clausal inequality for C_0, $\alpha_0 y + \beta_0 z \geq 1$. If a point $(\bar{y}, \bar{z}) \in T(S)$ satisfies
$$\alpha_0 \bar{y} + \beta_0 \bar{z} = 0$$
then $(\bar{y}, \bar{z}) \leq (\tilde{y}, \tilde{z})$, and $(\tilde{y}, \tilde{z})$ cannot be an extreme point of $Q_{S'}$. Thus for all $(\bar{y}, \bar{z}) \in T(S)$
$$\alpha_0 \bar{y} + \beta_0 \bar{z} \geq 1$$
and C_0 is an implication of S. Then there exists a prime implication absorbing C_0 and whose clausal inequality cuts $(\tilde{y}, \tilde{z})$ off, contradicting that $(\tilde{y}, \tilde{z}) \in Q_{S'}$. □

Now we pay some attention to the relationship between P and Q. It is clear that we are interested in solving the satisfiability problem by transforming it into an optimization problem, minimizing $\sum_{i=1}^{n}(y_i + z_i)$. Clearly, the problem can be solved optimizing over either P or Q. Since Q is a more tractable polyhedron than P, it seems advantageous to optimize over it. In fact, for a satisfiable set S, Q is full-dimensional while the dimension of P is hard to determine. In addition, we shall see that prime implications induce facets of Q, while the same is not true for P. The following example illustrates that the result in Theorem 36 does not apply to P. The example is visualized better by projecting onto $\Re^3$ defined by the y-variables.

As an example, let $C = \{x_1 \vee x_2, x_2 \vee x_3, x_1 \vee x_3\}$, which is satisfied by truth assignments where at least two of x_1, x_2, x_3 are true. Then
$$T(S) = \{(1,1,0,0,0,1), (1,0,1,0,1,0), (0,1,1,1,0,0), (1,1,1,0,0,0)\}$$
and dim $P = 3$. The inequality for the prime implication $x_1 \vee x_2$ is satisfied by just two points in $T(S)$; thus it just defines a face of P of dimension 1.

Now if we add the clause $\neg x_1 \vee \neg x_2 \vee \neg x_3$ we obtain a family that is satisfied by truth assignments having exactly two variables true. In this case dim $P = 2$, the inequalities associated to the clauses $x_i \vee x_j$ define facets, but the clause $\neg x_1 \vee \neg x_2 \vee \neg x_3$ defines an implicit equation for P.

We are clearly interested in the case $Q \neq \emptyset$, that is, when the set of clauses S is satisfiable. Under this condition, $Q = Q_{S'}$, where S' is the family of prime implications of S. We can generalize to the polyhedron Q all the results about Q_S presented in Theorem 35.

3.7 A SET-COVERING FORMULATION WITH FACET CUTS

Theorem 37 ([4]) *If S is a satisfiable family of clauses then:*

(i) *The inequality $y_i \geq 0$ ($z_i \geq 0$) defines a facet of Q if and only if the clause x_i ($\neg x_i$) is not a prime implication of S.*

(ii) *The inequality $y_i + z_i \geq 1$ defines a facet of Q if and only if neither x_i nor $\neg x_i$ is a prime implication of S.*

(iii) *The clausal inequality associated with a clause C_i defines a facet of Q if and only if C_i is a prime implication of S.*

Unfortunately, the theorem does not provide a complete description of the polyhedron Q. But this is to be expected since the set-covering polytope has a complex facial structure (cf. [11], [71], [224], and [246]). Even in the case of Horn clauses, a description of Q might require some nonclausal inequalities.

Consider the family of Horn clauses given in a previous example (3.7.1). The prime implications of S have the form

$$\neg x_1 \vee x_i \vee \neg x_j, \quad i \neq j, \ i,j \in \{2,3,4\}$$
$$x_i \vee \neg x_j \vee \neg x_k, \quad i,j,k \text{ a permutation of } 2,3,4$$
$$\neg x_1 \vee x_i, \quad i = 2,3,4$$

However, there are facet-inducing inequalities for Q which do not correspond to clausal inequalities. Consider

$$\begin{aligned} y_1 + z_2 + z_3 &\geq 1 \\ y_1 + z_2 + z_4 &\geq 1 \\ y_1 + z_3 + z_4 &\geq 1 \\ \hline 3y_1 + 2z_2 + 2z_3 + 2z_4 &\geq 3 \end{aligned}$$

Dividing by 2 and rounding up we obtain the inequality

$$2y_1 + z_2 + z_3 + z_4 \geq 2$$

which can be proved to be facet-inducing for Q. That inequality is clearly nonclausal since it does not have 0-1 coefficients. In addition, none of the prime implications yields a clause violated by the fractional solution described in example (3.7.1), while our nonclausal inequality is violated and thus solves the separation problem.

The following theorem imposes some structure on the general facet-inducing inequalities of Q. It is interesting to point out that part (iii) implies the y- and z-variables for the same literal cannot appear in the same inequality, except for the inequalities $y_i + z_i \geq 1$. Hence the facet-inducing inequalities can be thought as generalized clauses, with coefficients other than 1.

Theorem 38 ([4]) *Let Q be defined by a satisfiable family of clauses. If an inequality*
$$\alpha y + \beta z \geq \alpha_0 \qquad (3.50)$$
is facet-inducing for Q then:

(i) *All coefficients are nonnegative.*

(ii) $\alpha_0 \geq \max_i\{\alpha_i, \beta_i\}$, *except when (3.50) corresponds to a nonnegativity constraint.*

(iii) *For each i, only one of the two coefficients α_i, β_i may be strictly positive, unless the inequality (3.50) is a positive multiple of $y_i + z_i \geq 1$.*

Proof: Part (i) follows from Lemma 17. Now if $\alpha_i > \alpha_0$ ($\beta_i > \alpha_0$), then all the extreme points of Q satisfying (3.50) with equality also satisfy $y_i = 0$ ($z_i = 0$), so if (3.50) defines a facet, it is the same facet defined by $y_i \geq 0$ ($z_i \geq 0$). This proves part (ii) of the theorem.

To show part (iii) we assume, with no loss of generality, that $\alpha_1 \geq \beta_1$. We shall prove that either $\beta_1 = 0$, or both $\alpha_1 = \beta_1$ and $\sum_{i=2}^{n}(\alpha_i + \beta_i) = 0$. We argue by contradiction. If $\alpha_1 > \beta_1$ or $\sum_{i=2}^{n}(\alpha_i + \beta_i) > 0$, then the inequality
$$(\alpha_1 - \beta_1)y_1 + \sum_{i=2}^{n}(\alpha_i y_i + \beta_i z_i) \geq \alpha_0 - \beta_1 \qquad (3.51)$$
has a nonzero left-hand side and is well defined. Moreover, inequality (3.51) is valid for all points in P and, since it has nonnegative coefficients, for all points in Q. If $\beta_1 > 0$, we proceed to sum inequality (3.51) and inequality $\beta_1(y_1 + z_1) \geq \beta_1$ to obtain (3.50). Hence (3.50) is implied by two valid inequalities and cannot define a facet of Q. □

The last part of Theorem 38 does not apply to Q_S, when S is not the set of prime implications. Consider again the family of clauses given in example (3.49), where we find a facet-inducing inequality containing both y_2 and z_2. Taking the weighted sum

$$\begin{array}{r} y_1 + y_2 \phantom{{}+z_2} \geq 1 \\ y_1 \phantom{{}+y_2} + z_2 \geq 1 \\ {} + y_2 + z_2 \geq 1 \\ \hline y_1 + y_2 + z_2 \geq \frac{3}{2} \end{array}$$

and rounding up we obtain the inequality

$$y_1 + y_2 + z_2 \geq 2$$

3.7 A SET-COVERING FORMULATION WITH FACET CUTS

We can prove it defines a facet of Q_S by considering the vectors

$$(0,1,1,1),\ (1,0,1,1),\ (1,1,1,0),\ (1,0,0,1)$$

which are linearly independent points in Q_S and satisfy the inequality with equality.

We can strengthen Theorem 38 to characterize all 0-1 facet-inducing inequalities of Q. From Theorem 38(iii), a 0-1 inequality that is neither a nonnegativity constraint nor a constraint of the type $y_i + z_i \geq 1$ must correspond to a clausal inequality. By Theorem 37, only prime implications define facets. Hence we have proved the following theorem.

Theorem 39 ([4]) *If S is satisfiable, then $Q \neq \emptyset$ and every 0-1 inequality (with 0's and 1's in the left- and right-hand side coefficients) that is facet-inducing for Q is either a nonnegativity constraint, or $y_i + z_i \geq 1$, or corresponds to a prime implication of S.*

3.7.4 A Lifting Technique for General Facets

We have noted above that resolvent facet cuts were not quite enough to even solve the separation problem for the simple example (3.7.1) of a Horn proposition with four atomic propositions. So we perceive the need for more general (nonclausal) facet cuts. Although Theorem 38 gives some insight on the structure of general facets, we still do not have a methodology for generating them (in contrast with resolvent facets, which are generated by resolution). In the brief discussion below we describe one constructive approach for general facets. This approach exploits special combinatorial substructures in the underlying set-covering problem (3.47). For convenience, let us abbreviate the set-covering problem as follows:

$$\begin{aligned} \min \quad & c^T x \\ \text{s.t.} \quad & Ax \geq e \\ & x \in \{0,1\}^{2n} \end{aligned} \quad (3.52)$$

where A is a matrix with 0-1 coefficients, and **1** is a vector with s ones. The set-covering polytope P_A is the convex hull of the feasible solutions to (3.52). Similarly, we can define Q_A as the convex hull of the set

$$\{x \mid Ax \geq e, x \geq 0 \text{ and integer}\}$$

There is considerable literature on the facial structure of the set-covering polytope. Many of the important results are cited in Sassano [246] and Nobili and Sassano [224]. We observe below that many of these results apply directly to the dominant polyhedron Q_A.

First we define a (deletion) minor of A. Let $D \subseteq \{1, \ldots, 2n\}$. We denote by A_D the submatrix of A obtained by deleting all the rows with a nonzero coefficient in those columns indexed by D, followed by the columns in D, and finally any remaining rows that are dominated. We call A_D the minor of A obtained by the deletion of the columns in D.

Nobili and Sassano [224] prove that P_{A_D} is the projection of P_A onto the subspace $\Re^{\{1,\ldots,2n\}\setminus D}$, and that an inequality $\sum_{i \notin D} a_i x_i \geq a_0$ is valid for P_{A_D} if and only if it is valid for P_A. In addition, if that inequality defines a facet of P_A then it defines a facet of P_{A_D}, but the converse is not true. However, a nontrivial inequality $\sum_{i \notin D} a_i x_i \geq a_0$ defining a facet for a full-dimensional polytope P_{A_D} can be lifted to a nontrivial facet-inducing inequality of P_A of the form

$$\sum_{i \notin D} a_i x_i + \sum_{i \in D} b_i x_i \geq a_0 + b_0$$

with nonnegative coefficients. In general, the lifting coefficients b_i are not unique.

We extend this lifting result to the dominant Q_A (and hence to the satisfiability problem). First we show that facet-inducing inequalities other than $x_i \leq 1$ can be trivially lifted from P_A to Q_A, that is, all the lifting coefficients are zero, and then we extend the result to a lifting from Q_{A_D} to Q_A.

Lemma 19 *Assume the set-covering polytope P_A is full-dimensional and the inequality*

$$\sum_i a_i x_i \geq a_0 \tag{3.53}$$

has nonnegative coefficients. If (3.53) defines a facet of P_A, then it defines a facet of Q_A.

Proof: Since inequality (3.53) has nonnegative coefficients then it is valid for Q_A. Since $P_A \subset Q_A$ and $\dim P_A = \dim Q_A$, it follows that (3.53) also defines a facet of Q_A. □

Theorem 40 ([4]) *Let A_D be the minor of A obtained by deletion of the columns in D, and let*

$$\sum_{i \notin D} a_i x_i \geq a_0 \tag{3.54}$$

be an inequality with nonnegative coefficients, and $a_0 > 0$. If (3.54) defines a facet of Q_{A_D}, then it defines a facet of Q_A.

Proof: From the definition of minor, a valid inequality for Q_{A_D} is also valid for Q_A. Now any $x^D \in Q_{A_D}$ can be extended to a vector in Q_A by setting the components in D to 1, that is

$$x_i = \begin{cases} 1 & \text{if } i \in D \\ x_i^D & \text{if } i \notin D \end{cases}$$

Clearly, if x^D satisfies (3.54) with equality, so does x. Now let $k = \dim Q_{A_D}$. Since (3.54) defines a facet of Q_{A_D}, there exist k affinely independent vectors $x^{i,D} \in Q_{A_D}$ satisfying (3.54) with equality. Those vectors can be extended as indicated above. Then we can check that the set

$$\{x^i \mid i = 1, \ldots, k\} \cup \{x^1 + e_j \mid j \in D\}$$

contains $\dim Q_A$ affinely independent vectors in Q_A satisfying (3.54) with equality. Hence (3.54) defines a facet of Q_A. □

The two lifting results imply the following strategy of constructing facets for the polyhedron Q_S arising from the satisfiability problem. We pick subsets L of the set of literals $L \subset \{x_1, \neg x_1, \ldots, x_n, \neg x_n\}$, and then consider only the y, z inequalities encoding the clauses in S with literals in L. In case both $x_i, \neg x_i \in L$, we also throw in the inequality $y_i + z_i \geq 1$. These subsets must be chosen so that (a) the set-covering problem defined by those inequalities has at least two nonzero coefficients on the left-hand side (so the set-covering polytope is full-dimensional [246]) and more importantly, (b) there are known facet-inducing inequalities for this set-covering problem. From the theorem we know that these can be used directly as facets of Q_S.

Of particular interest is the case where S is a family of prime implications. In this case, we know (Theorem 36) that facets of Q_S will also be facets of Q. Generating all prime implications of a family S is harder, however, than solving the satisfiability problem. What could be done in practice is to use any set of clauses representing a formula, overlooking the fact that they may not be prime implications. Any set-covering inequalities obtained from minors of the embedded set-covering problem are certainly valid inequalities for Q and are likely to improve the representation of Q. This approach will make available a large set of nonclausal valid inequalities that may help in "branch-and-cut" procedures for the satisfiability problem. We believe this may be important in adding strength to linear programming proofs of the unsatisfiability of propositions.

3.8 A Nonlinear Programming Approach

The satisfiability problem can be given a nonlinear programming formulation and attacked in that form with an interior point algorithm. This approach was proposed by A. P. Kamath, N. K. Karmarkar, K. G. Ramakrishnan, and M. G. C. Resende [174]. As originally presented in [174] the method was used only as a heuristic for finding a solution of the satisfiability problem, as it tried to find a solution by rounding intermediate solutions found in the process of solving the nonlinear problem.

If no solution was found this way, the procedure simply stopped without determining whether a solution really exists. The procedure can readily be incorporated in a branching or cutting plane method, however, resulting in a decision algorithm. The articles [175, 176] describe some experiments along this line. Here the focus is on the nonlinear programming aspect, because it is the distinguishing feature of the method.

The interior point method described here evolved from Karmarkar's polynomial-time "projective scaling" algorithm for linear programming [178] and from subsequent "affine scaling" algorithms [14, 288] (anticipated in [83, 84]). Yet it is based on the classical Levenberg-Marquardt approach to solving nonlinear programming problems.

3.8.1 Formulation as a Nonlinear Programming Problem

The nonlinear programming formulation of the satisfiability problem goes as follows. Start with the usual integer programming formulation of a satisfiability problem, which may be written

$$\begin{aligned} a^i x \geq 1 + n(a^i), \quad & i = 1, \ldots, m \\ x_j \in \{0, 1\}, \quad & j = 1, \ldots, n \end{aligned} \quad (3.55)$$

where $n(a^i)$ is, as usual, the sum of the negative components of row a^i. Next, change variables so that -1 and 1, rather than 0 and 1, respectively, signify false and true:

$$y_j = 2x_j - 1 \quad \text{or} \quad x_j = \tfrac{1}{2}(1 + y_j)$$

The problem (3.55) becomes

$$\begin{aligned} a^i y \geq b_i, \quad & i = 1, \ldots, m \\ y_j \in \{-1, 1\}, \quad & j = 1, \ldots, n, \end{aligned} \quad (3.56)$$

where $b_i = 2 + n(a^i) - p(a^i)$, and $p(a^i)$ is the sum of the positive components of a^i.

Now consider the nonlinear programming problem

$$\begin{aligned} \max \quad & y^T y = \sum_{j=1}^n y_j^2 \\ \text{s.t.} \quad & a^i y \geq b_i, \quad i = 1, \ldots, m \\ & -1 \leq y_j \leq 1, \quad j = 1, \ldots, n \end{aligned} \quad (3.57)$$

3.8 A NONLINEAR PROGRAMMING APPROACH

It is clear that y solves the integer programming problem (3.56) if and only if $y^T y = n$, which is to say, if and only if it is an optimal solution of (3.57).

3.8.2 An Interior Point Algorithm

The following interior point algorithm solves the nonlinear problem (3.57). It was designed for any right-hand side b [179, 180], but here it is applied to satisfiability problems, for which $b_i = 2 + n(a^i) - p(a^i)$.

Rather than maximize $y^T y$ as in (3.57), minimize a "potential function" that is actually a modified barrier function inspired by Karmarkar's work in linear programming:

$$\phi(y) = \log \sqrt{n - y^T y} - (1/n) \sum_{i=1}^{m} \log d_i(y) \quad (3.58)$$

where

$$d_i(y) = b_i - a^i y, \quad i = 1, \ldots, m$$

Note that the second term of (3.58) is a barrier term that grows without bound as one approaches the boundary of the feasible region.

The idea is to minimize (3.58) subject to the constraints in (3.57) by minimizing successive quadratic approximations of (3.58) over a "trust region" inside the feasible region. For convenience, write the constraints in (3.57) as $Bx \leq c$. The quadratic approximation of ϕ at the current iterate y^k is

$$Q(y) = \tfrac{1}{2}(y - y^k)^T H^2(y - y^k) + h^T(y - y^k) + \text{constant}$$

where h is the gradient and H the Hessian of ϕ at y^k. If $D = \text{diag}(d_1(y), \ldots, d_n(y))$, $e = (1, \ldots, 1)$, and $f_0 = n - y^T y$, then the gradient is

$$h = -(1/f_0)y^k + (1/n)B^T D^{-1} e$$

and the Hessian is

$$H = -(2/f_0)I - (4/f_0^2)y^k(y^k)^T + (1/n)B^T D^{-2} B$$

Because it is hard to minimize $Q(y)$ subject to $By \leq c$, minimize $Q(y)$ over an ellipsoidal trust region that is centered at y^k and lies inside the polytope described by $By \leq c$. If $\Delta y = y - y^k$, this problem can be written

$$\begin{aligned} \min \quad & \tfrac{1}{2}(\Delta y)^T H \Delta y + h^T \Delta y \\ \text{s.t.} \quad & (\Delta y)^T B^T D^{-2} B \Delta y \leq r^2 \leq 1 \end{aligned} \quad (3.59)$$

where the constant r can be regarded as the "radius" of the ellipsoid. The solution Δy^* is the new search direction. The original constraint set $By \leq c$

has been dropped, but recall that it is indirectly represented by the barrier term in ϕ.

It is relatively easy to solve (3.59) if Q is positive definite, in which case the minimizing point will lie on the boundary of the ellipsoid. This point is the new iterate y^{k+1} about which a new quadratic approximation is generated. The procedure will eventually converge to a local optimum, which is also global if the solution is integer. If the solution is not integer, then one can re-solve the problem after adding cuts, or branch on one of the variables y_j with a fractional value.

The key to such a trust region approach is to adjust the radius r of the trust region so that it is large enough to allow fast convergence but small enough to stay inside the feasible region. This is cleverly done in a Levenberg-Marquardt method by adjusting the Lagrange multiplier μ in the first-order optimality conditions for (3.59) so that the search direction Δy^* that satisfies the condition is a descent direction. These conditions are

$$\begin{aligned} (H + \mu B^T D^{-2} B)\Delta y + h = 0 \\ \mu((\Delta y)^T B^T D^{-2} B \Delta y - r^2) = 0 \end{aligned} \quad (3.60)$$

The following can be shown.

Theorem 41 ([174]) *If B is full rank and y^k is not a local optimum, then there exists a value $\mu_{\min} > 0$ for which the matrix in (3.60) is positive definite for any $\mu \geq \mu_{\min}$, which means (3.60) has a unique solution Δy^*, and for which Δy^* is a descent direction.*

Each iteration of the Levenberg-Marquardt approach begins with the current iterate y^k. Refer to the value of the left-hand side of the constraint in (3.59) as the *length* of the step Δy^*. The step length is a decreasing function of μ when $\mu \geq \mu_{\min}$. A step that is too long may violate the constraint, whereas short steps slow the solution process. One therefore specifies, on the basis of experience, a range of acceptable lengths. First pick a starting value of μ within the range and solve (3.60) for Δy^*. If the length of Δy^* is below the range, decrease μ and solve (3.60) again—unless the length is very close to zero, in which case the algorithm stops with local minimum y^k. If the length is above the range, increase μ and solve (3.60) again. If the length is within the range, set $y^{k+1} = y^k + \Delta y^*$, which completes the iteration.

3.8.3 A Satisfiability Heuristic

The Levenberg-Marquardt procedure of the previous section is augmented with a rounding mechanism in order to solve a satisfiability problem. After each iteration of the Levenberg-Marquardt procedure, round off one component of the solution at a time until a satisfying solution is found. If no satisfying solution is found this way, another iteration of Levenberg-Marquardt is performed, and the rounding mechanism is used again.

More precisely, each iteration of the Levenberg-Marquardt procedure yields an iterate y^k. Let S be the original set of clauses. Beginning with the component y_j of y^k having the largest absolute value, perform the following rounding step. Add the unit clause x_j to S (i.e., round y_j up to 1) if $y_j > 0$ and add $\neg x_j$ to S (round y_j down to -1) otherwise. Then perform unit resolution on S. If all clauses are eliminated, there is a satisfying solution, and the heuristic stops. If the empty clause results, there is a contradiction; in this case perform another iteration of the Levenberg-Marquardt procedure and repeat the rounding procedure. But if clauses remain in S and the empty clause is not deduced, pick the component of y^k with the second largest absolute value and repeat the rounding step.

If the Levenberg-Marquardt procedure converges to a local optimum without finding a satisfying solution, then the satisfiability problem is unresolved. There may yet be a satisfying solution. At this point one may wish to use a cutting plane or branching technique to continue the search.

3.9 Tautology Checking in Logic Circuits

In some applications satisfiability problems occur in the form of tautology checking problems on logic circuits. This is true, for instance, of logic circuit verification problems, which are important in VLSI design.

When a satisfiability problem is defined on a circuit, its integer programming formulation takes on a structure that is remarkably suited for Benders decomposition [19, 216], a well-known mathematical programming technique. In fact, when the problem is solved by the Benders method, the resulting algorithm is equivalent to a nonnumeric algorithm that is apparently quite unlike any method known to electrical engineering or computer science. Furthermore, the connection between Benders decomposition and this new method is closely related to the connection between cutting planes and resolution developed earlier in this chapter.

3.9.1 The Tautology Checking Problem

Figure 3.2 depicts a small logic circuit. The arcs carry signals (0 or 1), and the nodes (or "gates") represent boolean functions, such as AND ($\land$) or OR ($\lor$). There are three inputs (0 or 1), denoted x_1, x_2, and x_3, respectively. The entire circuit represents a boolean function $f(x_1, x_2, x_3)$ whose value is the circuit's output. The output of each node is obtained by applying

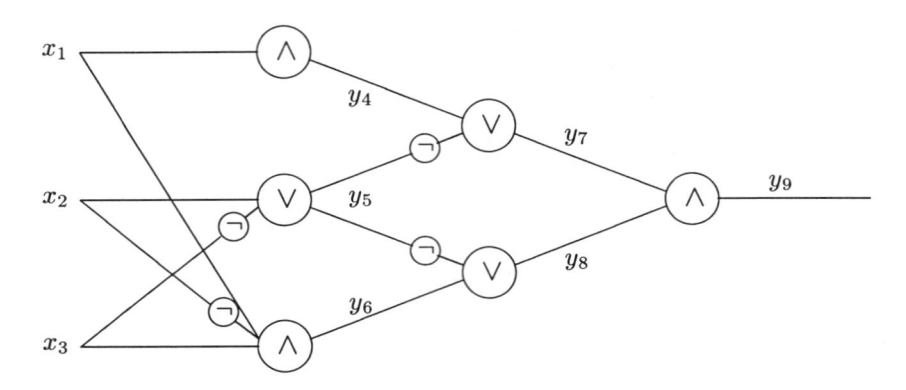

178   CHAPTER 3   PROPOSITIONAL LOGIC: GENERAL CASE

Figure 3.2: A small logic circuit.

its associated boolean function to the node's inputs. In this way the input signals propagate through the network and determine the output signal. Figure 3.2 indicates NOTs as nodes containing ¬.

In the example of Figure 3.2, the inputs x_j feed only into the first "layer" of nodes, y_4, y_5, y_6. But this need not be the case in general. Any directed acyclic graph can be a logic circuit. The nodes without predecessors are the inputs, and those without successors are the outputs. For convenience we focus here on circuits with only one output.

A logic circuit represents a tautology if every input yields a true output (i.e., the boolean function is identically 1). The circuit of Fig 3.2 does not represent a tautology, since input $(x_1, x_2, x_3) = (1, 0, 0)$, for instance, yields output 0.

It is clear that any satisfiability problem can readily be transformed, in a recursive manner, to a tautology checking problem on a network. Figure 3.2, for instance, represents the formula

$$[(x_1 \wedge x_2) \vee \neg(x_2 \vee \neg x_3)] \wedge [\neg(x_2 \vee \neg x_3) \vee (x_1 \wedge \neg x_2 \wedge x_3)]$$

This formula's negation is unsatisfiable if and only if the circuit represents a tautology.

More generally, let F be a formula to be checked for satisfiability. Associate $\neg F$ with the circuit's output node $n(F)$. F is unsatisfiable if and only if the circuit's output is always 1. Now whenever a formula G associated with a node $n(G)$ has the form $\neg H$, associate node $n(H)$ with H and draw an arc from $n(H)$ to $n(G)$, which becomes a NOT node. Whenever G has the form $H_1 \otimes H_2$, where $\otimes$ is some logical operator, draw arcs from $n(H_1)$ and $n(H_2)$ to $n(G)$, which becomes a $\otimes$ node. If G is an atomic

3.9 TAUTOLOGY CHECKING IN LOGIC CIRCUITS

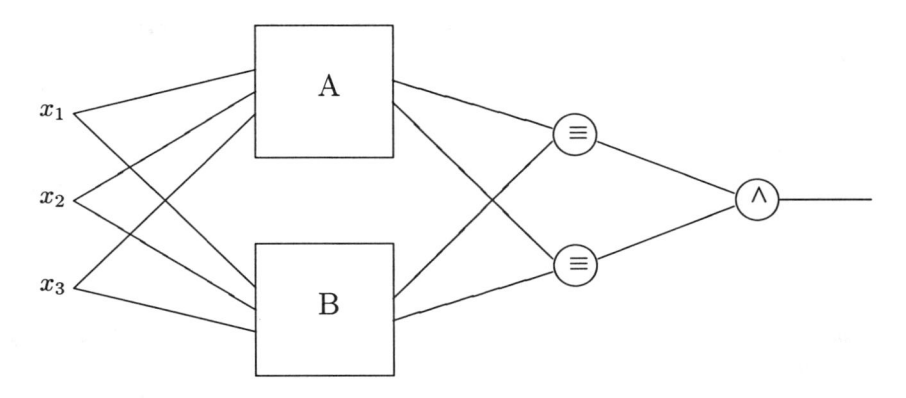

Figure 3.3: Verifying logic circuit A by tautology checking.

proposition, $n(G)$ becomes an input. Nodes representing the same formula can be identified, and inputs representing the same atomic proposition are identified.

A formula in conjunctive normal form, for instance, is readily depicted as a logic circuit with two levels. The first level is the set of inputs, and the second consists of an OR node for every conjunct. The output node is an AND node. The situation is similar for formulas in disjunctive normal form.

The circuit verification problem is typically posed as a tautology checking problem. Suppose one wishes to verify that a circuit A with inputs x_1, x_2, x_3 in Figure 3.2 implements a particular boolean function f. Take another circuit B known to implement f and let x_1, x_2, x_3 be its inputs as well. Each output of A is joined with the corresponding output of B with an equivalence node, and the equivalence nodes are joined with an AND. Obviously A and B are equivalent circuits if and only if the output of this AND node is identically 1.

3.9.2 Solution by Benders Decomposition

The tautology checking problem can be formulated as an integer programming problem and solved by Benders decomposition. Beginning with the example of the previous section, let y_i be the output of each node i, as indicated in Figure 3.3. It will be convenient to let y_1, y_2, y_3 be identical with the inputs x_1, x_2, x_3.

180   CHAPTER 3   PROPOSITIONAL LOGIC: GENERAL CASE

The circuit implements a tautology if and only if the following propositions are unsatisfiable.

$$\begin{aligned}
y_4 &\equiv x_1 \wedge x_2 \\
y_5 &\equiv x_2 \vee \neg x_3 \\
y_6 &\equiv x_1 \wedge \neg x_2 \wedge x_3 \\
y_7 &\equiv y_4 \vee \neg y_5 \\
y_8 &\equiv \neg y_5 \vee y_6 \\
y_9 &\equiv y_7 \wedge y_8, \\
\neg y_9 &
\end{aligned} \qquad (3.61)$$

This satisfiability problem can be converted to an integer programming problem by writing the equivalences in clausal form and converting them to inequalities in the usual way. This yields the problem below, where the x_j have been moved to the right-hand side for reasons that will become evident. The objective function is zero because only feasibility is of interest. Ignore the italic columns on the right for the moment.

$$\begin{array}{lrrrr}
\min & 0 & & RHS & Dual \\
\text{s.t.} & -y_4 & \geq 0 - x_1 & -1 & \\
& -y_4 & \geq 0 \quad -x_2 & 0 & \\
& y_4 & \geq -1 + x_1 + x_2 & 0 & \\
& -y_5 & \geq -1 \quad -x_2 + x_3 & 0 & 2\ 1 \\
& y_5 & \geq 0 \quad +x_2 & 0 & \\
& y_5 & \geq 1 \quad -x_3 & 0 & \\
& -y_6 & \geq 0 - x_1 & -1 & \\
& -y_6 & \geq -1 \quad +x_2 & -1 & \\
& -y_6 & \geq 0 \quad -x_3 & -1 & \\
& y_6 & \geq -1 + x_1 - x_2 + x_3 & 1 & 1 \\
& y_4 - y_5 \quad -y_7 & \geq -1 & -1 & \\
& -y_4 \qquad +y_7 & \geq 0 & 0 & \\
& y_5 \quad +y_7 & \geq 1 & 1 & 1\ 1 \\
& -y_5 + y_6 \quad -y_8 & \geq -1 & -1 & \\
& y_5 \qquad +y_8 & \geq 1 & 1 & 1 \\
& -y_6 \quad +y_8 & \geq 0 & 0 & 1 \\
& y_7 \qquad -y_9 & \geq 0 & 0 & \\
& y_8 - y_9 & \geq 0 & 0 & \\
& -y_7 - y_8 + y_9 & \geq -1 & -1 & 1\ 1 \\
& -y_9 & \geq 0 & -1 & 1\ 1 \\
\end{array} \qquad (3.62)$$

$$x_1, x_2, x_3 \in \{0, 1\}, \ 0 \leq y_4, \ldots, y_9 \leq 1$$

Note that only the x_j are explicitly required to be integer. It will become evident shortly that the y_j will necessarily be integer in any feasible solution.

In general the problem has the form

$$\begin{aligned}
\min \quad & 0 \\
\text{s.t.} \quad & Ax + B\bar{y} \geq a \\
& x_j \in \{0, 1\}, \ j = 1, \ldots, m, \ 0 \leq y_j \leq 1, \text{ all } j
\end{aligned} \qquad (3.63)$$

3.9 TAUTOLOGY CHECKING IN LOGIC CIRCUITS

where $\bar{y}$ is the vector of all y_j except the inputs $y_1, \ldots, y_m$ ($= x_1, \ldots, x_m$). This problem is hard to solve directly, because of the large number of variables (one for every node of the circuit). But it has a peculiar property that can be exploited. Namely, once the values of the inputs $x_1, \ldots, x_m$ are fixed, the values of the remaining variables $y_{m+1}, \ldots, y_n$ are uniquely determined by the circuit. Since it is likely that there are few inputs relative to the total number of nodes, one can use this property to reduce the amount of enumeration necessary to check for satisfiability.

This is done by applying Benders decomposition [19]. The problem is split into a "master problem" containing only the x_j.

$$\begin{aligned} \min \quad & 0 \\ \text{s.t.} \quad & (u^r)^T(a - Ax) - (v^r)^T e \leq 0, \quad r = 1, \ldots, R \\ & x_j \in \{0, 1\}, \quad j = 1, \ldots, m \end{aligned} \quad (3.64)$$

and a "subproblem,"

$$\begin{aligned} \min \quad & 0 \\ \text{s.t.} \quad & B\bar{y} \geq a - Ax \\ & 0 \leq y_j \leq 1, \quad \text{all } j > m \end{aligned} \quad (3.65)$$

where e is a vector of one's and x is regarded as fixed in the subproblem. The inequalities in the master problem are "Benders cuts," to be explained shortly. The motivation for this split is that the subproblem is trivial to solve for the y_j, because one need only propagate the inputs x_j through the circuit to find the y_j that result. The master problem is harder to solve, but it contains relatively few variables—one for each input.

To connect master problem and subproblem, the constraints imposed by the subproblem are incorporated into the master problem by expressing those constraints in terms of the x_j. In other words, the subproblem constraints are *projected* onto the space of the x_j. (Section 3.10 will discuss the concept of projection.) But rather than compute the entire projection, one computes only some of the constraints defining it, as they are needed.

This is done as follows. First solve the master problem for x, which initially is a trivial task, because the master problem initially contains no constraints other than the integrality constraints (i.e., $R = 0$).

Then substitute this value of x in the subproblem and solve it. Because the y_j are not explicitly constrained to be integral, the subproblem is a linear programming problem. What is actually needed is not the solution y of the subproblem, but the solution (u, v) of its *dual* problem (see Appendix):

$$\begin{aligned} \max \quad & u^T(a - Ax) - v^T e \\ \text{s.t.} \quad & u^T B - v^T e \leq 0, \quad u \geq 0 \end{aligned} \quad (3.66)$$

If the dual (3.66) is bounded, then the subproblem (3.65) is feasible, which means that the circuit output is 0, and the procedure stops with the conclusion that the circuit does not represent a tautology. If the subproblem is infeasible, the dual is unbounded, and we let (u, v) be an extreme ray.

Now let $(u^r, v^r) = (u, v)$ for $r = 1$ and add a Benders cut to the master problem. If the dual has more than one extreme ray, one can optionally add more than one cut. Re-solve the master problem for x. If it is infeasible, the output cannot be 0, and the procedure stops with the conclusion that the circuit represents a tautology. Otherwise, let x be a solution and repeat the process. The theory of Benders decomposition states that the algorithm will converge with the correct answer in finitely many steps.

In the example, any binary vector solves the original master problem, for instance $(x_1, x_2, x_3) = (1, 0, 1)$. The subproblem is given by (3.62). When $x = (1, 0, 1)$ is substituted on the right, the resulting right-hand side is shown in the column labeled *RHS*. The resulting problem is infeasible. Two possible extreme rays for the dual problem are shown respectively in the two columns labeled *Dual*. Each dual variable u_i corresponds to the ith constraint of (3.62), and the value of u_i (if it is positive) is shown to the right of the constraint. Each $v_j = 0$ in either extreme ray.

The first dual extreme ray gives rise to the Benders cut $2x_2 - 2x_3 \geq -1$, and the second one generates the cut $-x_1 + 2x_2 - 2x_3 \geq -2$. These are added as constraints in the master problem, which is solved to obtain $(x_1, x_2, x_3) = (1, 0, 0)$. But this input generates a zero circuit output, which means the circuit does not implement a tautology. Indeed, when $x = (1, 0, 0)$ is substituted into the subproblem (3.62), it becomes feasible, and the Benders algorithm terminates.

The strategy of the algorithm, then, is to try a number of inputs x and check the circuit's output each time. If the output is 0, the circuit is not a tautology and the procedure stops. If the output is 1, generate a constraint (Benders cut) and make sure that any future inputs enumerated satisfy this constraint. Ideally the Benders cuts will rule out most inputs, so that it is not necessary to enumerate very many of them. The next section will provide some insight into the logical meaning of the Benders cuts.

3.9.3 Logical Interpretation of the Benders Algorithm

The Benders algorithm of the previous section, applied to logic circuit problems, has an interesting logical interpretation.

To begin with, it is not hard to see by examining (3.62) that when the inputs x_1, x_2, x_3 are fixed, the y_j can be ascertained by unit resolution. This is because each y_j is uniquely determined by the inputs to the corresponding circuit node.

3.9 TAUTOLOGY CHECKING IN LOGIC CIRCUITS

For instance, once y_7 and y_8 are determined, the value of y_9 (see Figure 3.2) is determined by the clauses

$$\begin{aligned} y_7 &\vee \neg y_9 \\ y_8 &\vee \neg y_9 \\ \neg y_7 \vee \neg y_8 &\vee y_9 \end{aligned} \tag{3.67}$$

which represent the equivalence $y_9 \equiv (y_7 \vee y_8)$ defining y_9. These clauses correspond to the last three inequalities (but one) of (3.62). When the inputs are $(x_1, x_2, x_3) = (1, 0, 1)$, for instance, then $(y_7, y_8) = (1, 1)$; that is, y_7 and y_8 are true. One can therefore add the unit clauses y_7 and y_8 to (3.67), whereupon unit resolution yields the unit clause y_9, thus determining $y_9 = 1$. y_7 and y_8 are similarly determined by unit resolution involving their inputs, and so on.

It is not hard to see in general that whenever a logic circuit defines a boolean function $y = f(x_1, \ldots, x_m)$ of its inputs $x_1, \ldots, x_m$, unit resolution determines the value of y. That is, if the network is represented in conjunctive normal form, and an *input premise* x_j is added to this representation when $x_j = 1$ (or $\neg x_j$ when $x_j = 0$), then unit resolution yields the unit clause y if $f(x_1, \ldots, x_m) = 1$ or $\neg y$ if $f(x_1, \ldots, x_m) = 0$. A rigorous proof may be found in [152].

Because unit resolution determines the values of $y_4, \ldots, y_9$, so does the linear relaxation of (3.62). This is because an inequality obtained by unit resolution is simply the sum of the inequalities representing the parents. Thus if unit resolution obtains a unit clause y_j or $\neg y_j$, then a sum of inequalities obtains $y_j \geq 1$ or $-y_j \leq 0$, which indicates that necessarily $y_j = 1$ or $y_j = 0$ in any solution of the linear relaxation. Since all the y_j are determined, it follows that the solution of the linear relaxation is integral if it exists, as claimed earlier.

In particular, if the output y_9 is determined to be 1, there is a unit refutation of (3.62), and the linear relaxation is infeasible. Conversely, whenever there is a unit refutation of (3.62), the circuit output y_9 must be 1. This is because the constraints other than $-y_9 \geq -1$ are always feasible, due to the fact that they merely define the y_j.

The key fact to observe here is that when (3.62) is infeasible, *it may be possible to obtain a unit refutation without using every input premise*. This happens in this example, since merely fixing $x_2 = 0$ and $x_3 = 1$ in (3.62) permits a unit refutation that never uses the input premise x_1. In the refutation below, (P) again marks a premise, and each clause on the right (except premises) is the resolvent of clauses directly above and to the left. Clauses on the left other than premises occur earlier as resolvents on the right.

$$
\begin{array}{llll}
 & & \text{(P)} & x_2 \vee \neg x_3 \vee \neg y_5 \\
\text{(P)} & \neg x_2 & & \neg x_3 \vee \neg y_5 \\
\text{(P)} & x_3 & & \neg y_5 \\
\text{(P)} & y_5 \vee y_8 & & y_8 \\
 & & \text{(P)} & y_5 \vee y_7 \\
 & \neg y_5 & & y_7 \\
\text{(P)} & \neg y_7 \vee \neg y_8 \vee y_9 & & \neg y_8 \vee y_9 \\
 & y_8 & & y_9 \\
\text{(P)} & \neg y_9 & & \emptyset
\end{array}
\qquad (3.68)
$$

This chain of resolutions mirrors the propagation of the input $(x_2, x_3) = (0, 1)$ through the network of Figure 3.2, and it is instructive to trace the reasoning through the network.

The resolution proof (3.68) tells us that every input (x_1, x_2, x_3) with $(x_2, x_3) = (0, 1)$, regardless of the value of x_1, generates an output $y_9 = 1$. It is therefore pointless to try an input in which $(x_2, x_3) = (0, 1)$, because such an input will yield $y_9 = 1$. In other words, any future input should satisfy the constraint $x_2 \vee \neg x_3$, which can be called a *circuit cut*. Thus a circuit cut is a clause with variables in $\{x_1, \ldots, x_m\}$ such that any input violating it yields a circuit output of 1.

It is evident that circuit cuts can play the same role as Benders cuts in a Benders algorithm, because they limit the inputs that one needs to consider. The following will be proved shortly.

Theorem 42 ([152]) *Circuit cuts are equivalent to Benders cuts.*

Here two cuts are equivalent when they are satisfied by the same 0-1 vectors. The circuit cut $x_2 \vee \neg x_3$, for instance, is equivalent to the first Benders cut $2x_2 - 2x_3 \geq -1$ obtained in the previous section. The other Benders cut, $-x_1 + 2x_2 - 2x_3 \geq -2$, is equivalent to $\neg x_1 \vee x_2 \vee \neg x_3$, which is obviously a circuit cut.

The proof of Theorem 42 begins with the fact that since $x_2 \vee \neg x_3$ is a circuit cut, a unit refutation can be obtained when the input premises $\neg x_2$ and x_3 are added to the clauses represented by (3.62). The refutation is shown in (3.68). But note in (3.68) that x_2 always occurs negated and x_3 always occurs posited, except in the unit clauses containing them. This means that one can remove the input premises in (3.68) and obtain a unit

3.9 TAUTOLOGY CHECKING IN LOGIC CIRCUITS

proof of the circuit cut $x_2 \lor \neg x_3$ from the premises in (3.62):

$$
\begin{array}{lll}
 & & \text{(P)} \quad x_2 \lor \neg x_3 \lor \neg y_5 \\
\text{(P)} \quad y_5 \lor y_8 & & \quad\quad\, x_2 \lor \neg x_3 \lor y_8 \\
 & \text{(P)} \quad y_5 \lor y_7 & \\
\quad\quad x_2 \lor \neg x_3 \lor \neg y_5 & & \quad\quad\, x_2 \lor \neg x_3 \lor y_7 \\
\text{(P)} \quad \neg y_7 \lor \neg y_8 \lor y_9 & & \quad\quad\, x_2 \lor \neg x_3 \lor \neg y_8 \lor y_9 \\
\quad\quad x_2 \lor \neg x_3 \lor y_8 & & \quad\quad\, x_2 \lor \neg x_3 \lor y_9 \\
\text{(P)} \quad \neg y_9 & & \quad\quad\, x_2 \lor \neg x_3 \\
\end{array}
\quad (3.69)
$$

Thus in general, any circuit cut can be obtained by a unit proof from the premises in (3.64). This means by Theorem 29 that C is a rank 1 cut for (3.64). In particular, Corollaries 6 and 7 imply that some nonnegative linear combination of the inequalities in (3.64), excluding bounds $0 \leq x_j \leq 1$, yields an inequality $cx \geq \gamma$ that is equivalent to C. Let u be the vector of weights attached to clauses of (3.64) in this linear combination. Then $u^T B = 0$, $u^T Ax = cx$, and $u^T a = \gamma$. Because $u^T Ax \geq u^T a$ is equivalent to C, it remains only to show that $u^T Ax \geq u^T a$ is a Benders cut.

To do this it suffices to show that u is an extreme ray for the dual problem (3.66). Since $u^T B = 0$, the dual constraints are satisfied as equalities. Thus the infeasibility of the primal (3.65) implies that u is an extreme ray for the dual.

In the example, recall that the circuit cut $x_2 \lor \neg x_3$ is a rank 1 cut equivalent to the Benders cut $2x_2 - 2x_3 \geq -1$. To show that $x_2 \lor \neg x_3$ is equivalent to a Benders cut, note that because it can be obtained by unit resolution, it is a rank 1 cut. It is therefore equivalent to a nonnegative linear combination of the inequalities of (3.62) other than $0 \leq x_j \leq 1$. In fact, one can obtain one such linear combination, namely, $x_2 - x_3 \geq -1/2$, by letting the weights be one-half of those in the extreme ray vector in the next-to-last column of (3.62). But this vector of weights is also an extreme ray, since it is a positive multiple of an extreme ray. $x_2 - x_3 \geq -1/2$ is therefore a Benders cut that is equivalent to the circuit cut.

3.9.4 A Nonnumeric Algorithm

The fact that Benders cuts are circuit cuts suggests a nonnumeric version of the Benders algorithm for tautology checking. Rather than compute a Benders cut at each iteration, compute an equivalent circuit cut.

This raises the question as to how to compute a circuit cut. Recall from the previous section that $x_2 \lor \neg x_3$ is a circuit cut in the example of Figure 3.2 because a unit refutation (3.68) uses only $\neg x_2$ and x_3 as input premises. In general, if a unit refutation uses $L_1, \ldots, L_k$ (each of the form

x_j or $\neg x_j$) as input premises, then $\neg L_1 \vee \cdots \vee \neg L_k$ is a circuit cut. So one can find a circuit cut by finding a unit refutation that uses only some of the input premises.

The idea is to reason backward from the circuit's output. Consider again the circuit of Figure 3.2. The previous section found that its output is $y_9 = 1$ when the inputs are $(x_1, x_2, x_3) = (1, 0, 1)$. Note that with these inputs, $y_7 = 1$ and $y_8 = 1$. Since $y_9 \equiv y_7 \wedge y_8$, both $y_7 = 1$ and $y_8 = 1$ are needed to make $y_9 = 1$. Call $\{y_7, y_8\}$ a *determining set* for y_9. Now consider $y_7 \equiv y_4 \vee \neg y_5$. Since $y_4 = y_5 = 0$, the input $y_5 = 0$ is enough to make $y_7 = 1$, so that a determining set is $\{\neg y_5\}$. Consequently, any subset of the inputs $(x_1, x_2, x_3) = (1, 0, 1)$ that makes $y_5 = 0$ and $y_8 = 1$ also makes $y_9 = 1$.

Now consider $y_8 \equiv \neg y_5 \vee y_6$. Since $y_5 = 0$ and $y_6 = 1$, either input suffices to make $y_8 = 1$ as desired. Arbitrarily choose $\{\neg y_5\}$ to be the determining set for y_8. At this point it has been deduced that any subset of the inputs that makes $y_5 = 0$ also makes $y_9 = 1$. Looking at $y_5 \equiv x_2 \vee \neg x_3$, it is clear that $x_2 = 0$ and $x_3 = 1$ are enough to make $y_5 = 0$ and therefore $y_9 = 1$. This means there is a unit refutation using only the input premises $\neg x_2$ and x_3; namely, (3.68). In other words, there is a unit refutation whose input premises are all the literals in determining sets. So $x_2 \vee \neg x_3$ is a circuit cut.

The analysis could have taken the other branch at y_8's node. That is, it could have noted that $y_6 = 1$ suffices to make y_8 equal to 1, because $y_8 \equiv \neg y_5 \vee y_6$. Then since $y_6 \equiv x_1 \wedge \neg x_2 \wedge x_3$, $(x_1, x_2, x_3) = (1, 0, 1)$ are all needed to make $y_6 = 1$, yielding the obvious circuit cut $\neg x_1 \vee x_2 \vee \neg x_3$.

Each iteration of the nonnumeric algorithm therefore runs as follows. The master problem consists of a collection of circuit cuts and is therefore a satisfiability problem in clausal form. Solve it for an input $(x_1, \ldots, x_m)$. If there is no solution, stop with the conclusion that the circuit represents a tautology. Otherwise, if $(x_1, \ldots, x_m)$ results in output 0, stop with the opposite conclusion. If the output is 1, generate one or more new circuit cuts and add them to the master problem. Each circuit cut is the disjunction of the negations of the input premises in the determining sets obtained in a backward pass through the circuit.

Before stating this algorithm precisely, one should consider how to deal with gates other than AND and OR. In general, a gate corresponding to y_i is a boolean function f of certain inputs $y_{i_1}, \ldots, y_{i_p}$. Suppose for the moment that y_i is true ($y_i = 1$) for the current set of inputs. Write f in disjunctive normal form (DNF) $D_1 \vee \ldots \vee D_K$, where each $D_k = \bigwedge_{j \in J_k} L_{kj}$ and each L_{kj} is y_{i_j} or $\neg y_{i_j}$. Then since y_i is true, one or more D_k are true, but only one of these need to be true to make y_i true. Also, D_k is true only if L_{kj} is true for each $j \in J_k$. So the determining set for y_i can be the literals

3.9 TAUTOLOGY CHECKING IN LOGIC CIRCUITS

L_{kj} in any D_k that is true. If y_i is false, one can find the determining set in the same way, except that one begins by writing $\neg f(y_{i_1}, \ldots, y_{i_p})$ rather than $f(y_{i_1}, \ldots, y_{i_p})$ in DNF.

Consider some examples. If y_i corresponds to an AND, then $y_i = \bigwedge_{j \in J} L_j$. If y_i is true, then DNF is $D_1 = \bigwedge_{j \in J} L_j$, and the determining set is $\{L_j \mid j \in J\}$. If y_i is false, then DNF of $\neg y_i$ is $D_1 \vee \cdots \vee D_K = \bigvee_{j \in J} \neg L_j$, and the determining set is any singleton $\{\neg L_j\}$. The situation is analogous for an OR node.

If y_i corresponds to an XOR (exclusive OR), then y_i is true if and only if an odd number of $L_{i_1}, \ldots, L_{i_p}$ are true. Thus f in DNF is a disjunction $D_1 \vee \cdots \vee D_K$, where the D_k are all possible conjunctions of $L_{i_1}, \ldots, L_{i_p}$ in which an odd number of the L_i are negated. If y_i is true, then D_k is true for exactly one k, which means the determining set consists of the literals in D_k. The situation is similar if y_i is false. Thus all the inputs of an XOR gate belong to its determining set, regardless of the gate's output.

The nonnumeric algorithm appears below as Procedure 23. Here each variable y_i (except inputs) is associated with a gate that implements boolean function $f_i(y_{i_1}, \ldots, y_p)$. Recall that $y_1, \ldots, y_m$ are identical with the inputs $x_1, \ldots, x_m$. The procedure CUTS generates all possible circuit cuts in each iteration. If fewer cuts are desired, the main loop that iterates through all $k \in \{1, \ldots, K\}$, thus enumerating all possible determining sets, can be limited to only some of the k.

Procedure 23: *Two-Pass Circuit Verification*

Let $S = \emptyset$.
While S is a satisfiable set of clauses:
    Let x^* be any satisfying solution for S.
    Propagate inputs x^* through the circuit.
    If the output y_t is 0, stop; the circuit does not represent a tautology.
    Call **Cuts**$(\{y_t\}, \emptyset)$ to generate a set of cuts.
    Add the new cuts to S.
Stop; the circuit represents a tautology.

Procedure **Cuts**(P, A).
    If $P = \emptyset$ then generate the cut $\bigvee_{L \in A} \neg L$.
    Else
        Pick $y_i \in P$ and perform **DNF**.
        For each $k \in \{1, \ldots, K\}$: perform **Nextcut**.
End **Cuts**.

Procedure **DNF**.
    If y_i is true then
        Put f_i in DNF $D_1 \vee \ldots \vee D_K$, where each $D_k = \bigwedge_{j \in J_k} L_{kj}$
        and L_{kj} is y_{i_j} or $\neg y_{i_j}$.
    Else put $\neg f_i$ in DNF similarly.
End **DNF**.

Procedure **Nextcut**.
    If D_k is true then
        Let $A' = A$ and $P' = P$.
        For each $j \in J_k$:
            If $j \leq m$ (i.e., y_{i_j} is input) then put L_{kj} into A'.
            Else put y_{i_j} into P'.
        If $J_k \neq \emptyset$ then call **Cuts**$(P' \setminus \{y_i\}, A')$.
End **Nextcut**.

3.9.5 Implementation Issues

In practice [152], the algorithm of the previous section tends to generate a large number of circuit cuts in each iteration, many of which are identical. This either makes the master problem much larger than need be or obliges one to invest time to remove redundant cuts. Also, the mere generation of so many cuts is time consuming.

If only one cut is generated in each iteration, the algorithm can be made much more efficient. As it stands, the algorithm requires two passes through the circuit: a forward pass to evaluate the output, and a backward pass (possibly with many branches) to generate circuit cuts. But if only one cut is generated, a single forward pass is sufficient. One need only associate with each node i an *active set* A_i of inputs x_j, as follows. Each input x_j is associated with the active set $\{x_j\}$ if $x_j = 1$ and $\{\neg x_j\}$ if $x_j = 0$. As soon as the output y_i of a node is determined, consider all possible determining sets for the node. For each determining set compute the union of the active sets associated with the nodes in the determining set. The smallest union becomes the active set A_i associated with y_i. Thus A_i contains circuit inputs that are sufficient to determine the node's output. The active set A_t associated with the circuit's output node (node t) therefore gives rise to a circuit cut $\bigvee_{L \in A_t} \neg L$. Pick the smallest set at each node because this tends to result in a shorter and hence stronger circuit cut.

The algorithm is stated precisely as Procedure 24. Here the nodes are numbered $1, \ldots, t$ so that any node's number is larger than those of its predecessors.

3.10 INFERENCE AS PROJECTION

Procedure 24: *One-Pass Circuit Verification*

Let $S = \emptyset$.
While S is satisfiable:
    Let x^* be a satisfying solution for S.
    For $i = 1, \ldots, m$:
        If $x_i^* = 1$ then let $A_i = \{x_i\}$.
        Else let $A_i = \{\neg x_i\}$.
    For $i = m+1, \ldots, t$:
        Determine whether y_i is true or false.
        Perform **DNF** (see Procedure 23).
        For $k = 1, \ldots, K$ let $A_{ik} = \bigcup_{j \in J_k} A_j$.
        Let $A_i = A_{i_{k*}}$ where $|A_{i_{k*}}| = \min_{k \in K_1} |A_{i_k}|$
            and $K_1 = \{k | D_k \text{ is true}\}$.
    If y_t is false then stop; the circuit does not represent a tautology.
    Add cut $\bigvee_{L \in A_t} \neg L$ to S.
Stop; the circuit represents a tautology.

Since the satisfiability problem is re-solved after each circuit cut is added, it is advantageous to use the incremental satisfiability algorithm of Section 3.1.5. Computational experience suggests that the satisfiability problem can be re-solved, on the average, even more quickly than a cut can be generated [152].

The performance of Algorithm 3.1.5 varies enormously, depending on the structure of the circuit. Nodes whose determining sets necessarily contain all their inputs tend to make a problem hard, because the active sets rapidly grow and yield long, weak cuts. These include XOR nodes and equivalence nodes (negated XORs). But the algorithm appears to run quite rapidly on at least some classes of circuits without such nodes, such as circuits with only AND, OR and NOT nodes [152].

3.10 Inference as Projection

The simplest form of the inference problem asks whether a given proposition can be inferred from a set of premises. A more general problem, and the one more relevant to many applications, is to infer from a set of premises *everything* that is pertinent to a given question.

In an expert system for medical diagnosis, for instance, the clinical observations for a particular patient are added to a set of rules that presumably guide physicians when they diagnose illness. The object is to infer what is wrong with the patient. The simplest kind of inference problem asks, "Can we infer that the patient has appendicitis?" But the more general inference problem asks, "What illnesses can we infer that the patient has?"

A natural way to formulate the more general inference problem is as a *logical projection problem*. Suppose that among the atomic propositions that occur in the medical knowledge base, some are propositions to the effect that the patient has a certain disease, such as "the patient has appendicitis" or "the patient has an intestinal blockage." The diagnostic problem, then, is to draw all inferences that involve only these disease propositions. One might infer, for instance, that "the patient has appendicitis," or "the patient has appendicitis or an intestinal blockage," or "the patient has appendicitis only if he has no intestinal blockage."

If the knowledge base contains atomic propositions $x_1, \ldots, x_n$, let $x_1, \ldots, x_k$ be the disease propositions. The aim is to find all implied propositions that contain no atoms other than $x_1, \ldots, x_k$. This is a problem of projecting the knowledge base onto the variables $x_1, \ldots, x_k$, and it is closely related to the problem of projecting a polyhedron.

The general logical projection problem can be solved by resolution, but there is no known practical procedure for solving it. It can be *partially* solved, however, by polyhedral projection. That is, by projecting onto $x_1, \ldots, x_k$ the polyhedron described by clauses expressed in inequality form, one can obtain *some* of their implications involving only $x_1, \ldots, x_k$. In fact, one obtains precisely those that can be derived by unit proofs (Section 3.5.2). Unsurprisingly, unit proofs completely solve the logical projection problem for Horn clauses.

Although unit proofs are considerably simpler than general resolution, the number of inferences involving $x_1, \ldots, x_k$ they generate can grow exponentially with k. The number is polynomial in n if k is fixed, however. Thus unit resolution can be a practical method of computing some of the inferences that involve $x_1, \ldots, x_k$ when k is small.

H. P. Williams first pointed out that the resolution method of theorem proving is closely related to the Fourier-Motzkin method for polyhedral projection [290, 292, 295]. In fact, one can obtain a resolvent by strengthening the result of a Fourier-Motzkin step in a certain way. But an iteration of the Fourier-Motzkin method projects only onto $\{x_1, \ldots, x_{n-1}\}$. To project a polyhedron onto $\{x_1, \ldots, x_k\}$ in one iteration, one must use a more general procedure. One can then strengthen the result of this procedure in the way that Williams strengthened Fourier-Motzkin. The cutting plane results of Section 3.5.2 can then be used to show that this strengthening yields precisely the inferences that can be obtained by unit proofs. This in turn leads to the above results.

3.10 INFERENCE AS PROJECTION

3.10.1 The Logical Projection Problem

The *projection* of a vector $t = (t_1, \ldots, t_n)$ onto index set $K = \{1, \ldots, k\}$ with $k \leq n$ is $t_K = (t_1, \ldots, t_k)$. Let t be an *extension* of t_K. The projection onto K of a set T of such vectors is $T_K = \{t_K \mid t \in T\}$. That is, T_K is the set of vectors $(t_1, \ldots, t_k)$ that have extensions in T.

If S is a set of logical clauses, let the *satisfaction set* $T(S)$ of S be the (possibly empty) set of vectors of satisfying truth assignments to the atomic propositions $x_1, \ldots, x_n$ in S. A *logical projection* of S onto K is a set $\bar{S}$ of clauses, containing only atomic propositions $x_1, \ldots, x_k$, for which $T(\bar{S}) = T(S)_K$. $\bar{S}$ is not unique, since different sets of clauses may have the same satisfaction set. The *logical projection problem* is to find at least one logical projection of S onto a given K.

A logical projection of S is of interest because, in a sense, it says everything that can be inferred from S about the variables $x_1, \ldots, x_k$. More precisely, we state the following lemma.

Lemma 20 *Any logical projection $\bar{S}$ of S onto K is equivalent to the set S_K of all clauses, containing only $x_1, \ldots, x_k$, that can be inferred from S.*

$\bar{S}$ and S_K are "equivalent" in the sense that $T(\bar{S}) = T(S_K)$. The lemma says, then, that $T(S_K)$ consists precisely of the truth valuations $(t_1, \ldots, t_k)$ that have extensions in $T(S)$. If t has an extension in $T(S)$, then it is consistent with S and nothing implied by S can exclude it, so that $t \in T(S_K)$. Conversely, if t has no extension in $T(S)$, then S implies that $x_1, \ldots, x_k$ cannot have this valuation, so that $t \notin T(S_K)$.

3.10.2 Computing Logical Projections by Resolution

One way to solve the projection problem is by the resolution procedure. In particular, it can be solved with a series of *K-resolutions*, which are resolutions on variables in $\{x_{k+1}, \ldots, x_n\}$. Let the *K-resolution procedure* be the same as the resolution procedure of Section 3.1.1 except that only K-resolutions are used.

Theorem 43 *Let S' be the result of applying the K-resolution procedure to a satisfiable set S of clauses. Then the set $\bar{S}$ obtained by deleting from S' all clauses with at least one variable in $\{x_{k+1}, \ldots, x_n\}$ is a logical projection of S onto K.*

A simple lemma helps explain why this is so. Recall that a literal L containing variable x_j is monotone in a set of clauses if x_j always occurs with the same sign it has in L.

Lemma 21 *If a set S of clauses implies clause C, then some subset of S in which the literals of C are monotone implies C.*

Proof: Let T be the result of removing from S every clause containing a variable whose sign is opposite its sign in C. It suffices to show that T implies C. For this it suffices to show that no truth assignment satisfies T and falsifies C. But any such assignment would satisfy the complement of every literal in C and therefore every clause in $S \setminus T$, which contradicts the assumption that S implies C. □

(This lemma says, incidentally, that regular resolution is complete; see Section 3.2.2.) To understand why Theorem 43 is true, one must understand why $\bar{S}$ implies any clause C with variables in $\{x_1, \ldots, x_k\}$ that S implies. Without loss of generality let $x_1, \ldots, x_l$ be the variables in $\{x_1, \ldots, x_k\}$ that do not occur in C, and let $t = (t_1, \ldots, t_l)$ be any assignment of truth values to them. For any such t define

$$C(t) = C \vee \bigvee_{j=1}^{l} x_j^{t_j}$$

where $x_j^1 = x_j$ and $x_j^0 = \neg x_j$. It suffices to show that $\bar{S}$ implies $C(t)$ for every t, since in this case $\bar{S}$ implies C.

S clearly implies $C(t)$ because it implies C. So by Lemma 21 some $S^+ \subset S$ in which the literals of $C(t)$ are monotone implies $C(t)$. This means, by the completeness of resolution (Theorem 20), that the resolution procedure applied to S^+ generates a clause that absorbs $C(t)$. Since there can be no resolutions on the variables $x_1, \ldots, x_k$ in $C(t)$ (which are monotone), the K-resolution procedure generates a clause that absorbs $C(t)$. Thus $\bar{S}$ implies $C(t)$.

As an example, consider the following set S of clauses:

$$\begin{array}{l} x_1 \vee x_2 \vee x_3 \\ \neg x_1 \vee x_2 \vee \neg x_4 \\ \neg x_1 \vee x_3 \\ \neg x_1 \vee \neg x_3 \vee x_4 \\ x_1 \vee \neg x_3 \end{array} \quad (3.70)$$

The first iteration of $\{1,2\}$-resolution yields resolvents $x_1 \vee x_2$, $\neg x_1 \vee x_4$ and $\neg x_1 \vee x_2 \vee \neg x_3$, while the second yields $\neg x_1 \vee x_2$. So $\bar{S} = \{x_1 \vee x_2, \neg x_1 \vee x_2\}$ is a logical projection onto $\{1,2\}$. Another logical projection is $\{x_2\}$, which has the same satisfaction set.

3.10 INFERENCE AS PROJECTION

3.10.3 Projecting Horn Clauses

Not only does unit resolution check a set of Horn clauses for satisfiability, but unit proofs solve the boolean projection problem for Horn clauses. Let *unit K-resolution* be K-resolution in which at least one of the parents contains only one variable not in $\{x_1, \ldots, x_k\}$. Then if K contains the indices of the variables in a clause C, a unit proof of C is by definition a unit K-resolution proof of C. Let the *unit K-resolution procedure* be the same as the resolution procedure except that all resolutions are unit K-resolutions.

Theorem 44 *Let S' be the result of applying the unit K-resolution procedure to a satisfiable set S of Horn clauses. Then the set $\bar{S}$ obtained by deleting from S' all clauses with at least one variable in $\{x_{k+1}, \ldots, x_n\}$ is a logical projection of S onto K.*

As in the previous section the theorem is true because $\bar{S}$ implies any clause C with variables in $\{x_1, \ldots, x_k\}$ that S implies. To show this it again suffices to show that $\bar{S}$ implies $C(t)$ for any truth assignment $t_1, \ldots, t_l$ to the variables $x_1, \ldots, x_l$ in $\{x_1, \ldots, x_k\}$ that do not occur in C.

Since S implies $C(t)$, by Lemma 21 some $S^+ \subset S$ in which the literals of $C(t)$ are monotone implies $C(t)$. This allows one to construct a unit proof of $C(t)$, as follows. For any clause $D \in S^+$ let D_0 be the portion of D with variables in $\{x_{k+1}, \ldots, x_n\}$. Then $S_0 = \{D_0 \mid D \in S^+\}$ must be unsatisfiable. For if some truth assignment $\{t_{k+1}, \ldots, t_n\}$ satisfied S_0^+, then $(1-t_1, \ldots, 1-t_k, t_{k+1}, \ldots, t_n)$ would satisfy S_0 and falsify $C(t)$, which is contrary to the fact that S^+ implies $C(t)$.

But if S_0 is unsatisfiable, then since it is Horn, there is a series of unit resolutions that begins with the clauses in S_0 and generates the empty clause. Since the variables $x_1, \ldots, x_k$ are monotone in S^+, there is a parallel series of unit K-resolutions that begins with S^+ and generates a clause that absorbs $C(t)$. So the unit K-resolution procedure applied to S generates a clause in $\bar{S}$ that absorbs $C(t)$.

3.10.4 The Polyhedral Projection Problem

Given a polyhedron $P = \{x \mid Ax \geq a\}$, the *polyhedral projection problem* is to find a system $By \geq b$, with $y = (x_1, \ldots, x_k)$, whose feasible set is the projected polyhedron P_K. A *projection cut* is any inequality that is satisfied by every point in P_K.

It is well known that one way to project a polyhedron is to take all nonnegative linear combinations of the inequalities $Ax \geq a$ that cause the coefficients of $x_{k+1}, \ldots, x_n$ to vanish. Let us write the system $Ax \geq a$ as

$$A_1 y + A_2 z \geq a \tag{3.71}$$

194   CHAPTER 3   PROPOSITIONAL LOGIC: GENERAL CASE

where $z = (x_{k+1}, \ldots, x_n)$. Then the projection must satisfy all inequalities $u^T A_1 y \geq u^T a$, where u is any vector in the polyhedral cone $U = \{u \mid u^T A_2 = 0, u \geq 0\}$. These inequalities form an infinite system, but a finite description of P_K can be obtained by using only the vectors u that are extreme rays of U.

Suppose, for instance, one wants to project the system below onto $\{1,2\}$. Note that this system is the linear relaxation of the clauses (3.70).

$$\begin{aligned} x_1 + x_2 + x_3 &\geq 1 \\ -x_1 + x_2 \quad\quad - x_4 &\geq -1 \\ -x_1 \quad\quad + x_3 &\geq 0 \\ -x_1 \quad\quad - x_3 + x_4 &\geq -1 \\ x_1 \quad\quad - x_3 &\geq 0 \\ 0 \leq x_j \leq 1, \ j = 1, \ldots, 4 & \end{aligned} \quad (3.72)$$

The polyhedral cone is

$$U = \{u \mid u_1 + u_3 - u_4 - u_5 + u_8 - u_{12} = -u_2 + u_4 + u_9 - u_{13} = 0\} \quad (3.73)$$

where $u_6, \ldots, u_9$ correspond to the constraints $x_j \geq 0$ and $u_{10}, \ldots, u_{13}$ to $-x_j \geq -1$. U has 14 extreme rays, which give rise to only two inequalities that are not already implied by the bounds $0 \leq x_j \leq 1$, namely, $2x_1 + x_2 \geq 1$ and $-3x_1 + 2x_2 \geq -2$. These inequalities and the bounds describe the projection of (3.72) onto $\{1,2\}$. It is in general very difficult to find the extreme rays of a polyhedral cone, however, since it essentially requires enumeration of the basic solutions of $u^T A_2 = 0$.

Another approach to computing the polyhedral projection is *Fourier-Motzkin elimination*, which is first applied to $Ax \geq a$ to "eliminate" x_n. The result of this elimination is a projection onto $\{x_1, \ldots, x_{n-1}\}$. The procedure is applied to this new system to obtain a projection onto $\{x_1, \ldots, x_{n-2}\}$, and so on.

One can, for instance, eliminate x_4 from (3.72) by writing the two inequalities containing x_4 in the form

$$1 - x_1 + x_2 \geq x_4 \quad (3.74)$$

$$x_4 \geq -1 + x_1 + x_3 \quad (3.75)$$

Pairing the expression on the left with that on the right yields an inequality without x_4:

$$1 - x_1 + x_2 \geq -1 + x_1 + x_3 \quad (3.76)$$

or

$$-2x_1 + x_2 - x_3 \geq -2 \quad (3.77)$$

3.10 INFERENCE AS PROJECTION

Thus (3.77) and the inequalities of (3.72) containing no x_4 comprise a projection system for $\{1, 2, 3\}$.

Now, to eliminate x_3, write (3.77) and the inequalities in (3.72) containing x_3 as

$$\begin{aligned} x_3 &\geq 1 - x_1 - x_2 \\ x_3 &\geq x_1 \\ x_1 &\geq x_3 \\ 2 - 2x_1 + x_2 &\geq x_3 \end{aligned}$$

Pairing obtains four inequalities (after simplifying):

$$\begin{aligned} 2x_1 + x_2 &\geq 1 \\ 0 &\geq 0 \\ -x_1 + 2x_2 &\geq -1 \\ -3x_1 + x_2 &\geq -2 \end{aligned}$$

the second and third of which are redundant of the bounds $0 \leq x_j \leq 1$. This results in the same projection set for $\{1, 2\}$ as before.

In fact, it is easy to see that Fourier-Motzkin elimination is a special case of the projection method described earlier. For instance, combining (3.74) and (3.75) to obtain (3.76) in effect takes a linear combination of (3.74) and (3.75) that causes x_4 to vanish (in this case with unit weights). The computational problem is again difficult, however, because the number of inequalities tends to explode after a few iterations.

3.10.5 Inference by Polyhedral Projection

An inequality can be regarded as a logical formula whose satisfaction set is the set of 0-1 points that satisfy the inequality (Section 3.6). If inequalities are viewed this way, projection cuts are valid inferences.

To make this more precise, let P be the polyhedron described by the linear relaxation of a set S of clauses in inequality form. By definition, a projection cut is satisfied by every point in P (since it is satisfied by every point in P_K and contains only variables in $\{x_1, \ldots, x_k\}$). It is therefore satisfied by every 0-1 point in P, which is to say every point satisfying S, and is therefore implied by S.

In the example of the previous section, the inequalities $2x_1 + x_2 \geq 1$ and $-3x_1 + x_2 \geq -2$ that describe the projected polyhedron are valid logical inferences as well. This can be checked by noting that all points satisfying the logical projection $\{x_1 \vee x_2, \neg x_1 \vee x_2\}$, namely, all 0-1 points of the form $(0, 1, t_3, t_4)$ or $(1, 1, t_3, t_4)$, also satisfy these inequalities.

Since projection cuts are logical inferences involving only variables in $\{x_1, \ldots, x_k\}$, they can help solve the logical projection problem, which is to find all such inferences. But they do not always solve it completely. In the above example they do, since the two projection cuts cut off all points outside of $T(S)_K = \{(0,1), (1,1)\}$. But in general projection cuts allow some points that are not in $T(S)_K$.

To put it differently, it is always true that $T(S)_K \subset P_K$, simply because $T(S) \subset P$. But P_K may contain 0-1 points not in $T(S)_K$, which means that the complete set of projection cuts are not logically equivalent to a logical projection.

As an example, consider the clauses

$$\begin{array}{ccc} x_1 \vee & x_2 \vee & x_3 \\ x_1 \vee & x_2 \vee & \neg x_3 \\ x_1 \vee & \neg x_2 \vee & x_3 \\ x_1 \vee & \neg x_2 \vee & \neg x_3 \end{array}$$

The single clause x_1 is a logical projection onto $K = \{1\}$, so that $T(S)_K$ contains the single point $x_1 = 1$. But the polyhedral projection P_K is the entire interval $[0,1]$, which contains $x_1 = 0$ as well. The polyhedral projection therefore "loses information." It produces a valid inference but does not infer everything that can be inferred. In this case it infers the trivial fact that x_1 is either true or false.

3.10.6 Resolution and Fourier-Motzkin Elimination

It has been seen that polyhedral projection can "lose information" but that resolution does not. H. P. Williams [290] observed that the Fourier-Motzkin method of polyhedral projection can be strengthened with a certain reduction and rounding operation so that it does not lose information. The reason is simply that, when so strengthened, a Fourier-Motzkin step becomes a resolution step.

Consider, for instance, the inequality (3.77) that a Fourier-Motzkin step obtained from (3.74) and (3.75). Add to it the bounds $x_2 \geq 0$ and $-x_3 \leq -1$ (this is an instance of reduction) and divide the sum by 2 to obtain $-x_1 + x_2 - x_3 \geq -3/2$. Round the $-3/2$ up to -1 and obtain the clausal inequality $-x_1 + x_2 - x_3 \geq -1$, which is the resolvent of the clauses represented by (3.74) and (3.75). This is of course the same Chvátal cutting plane operation shown in Section 3.5.1 to yield the resolvent.

3.10.7 Unit Resolution and Polyhedral Projection

By applying reduction and rounding to an inequality that results from a Fourier-Motzkin step, one obtains a resolvent. But the Fourier-Motzkin method is rather limited, since in one iteration it projects only onto $\{1, ..., n-1\}$. This raises the question as to what inference procedure one might obtain by applying reduction and rounding to an inequality resulting from the more general method that projects onto $K = \{1, ..., k\}$ in one iteration. The answer is that one obtains the unit K-resolution procedure.

Theorem 45 ([145]) *Let a set S of clauses be written in the form (3.71), and let $u \in \{u | u^T A_2 = 0, u \geq 0\}$. Then any clause obtained by applying reduction and rounding to $u^T A_1 y \geq u^T a$ is implied by clauses that can be obtained from S by a series of unit K-resolutions.*

Proof: Let C be the result of applying reduction and rounding to $u^T A_1 y \geq u^T a$, and let $C(t)$ be defined as in Section 3.10.2. It suffices to show that, for each t, $C(t)$ is absorbed by a clause obtained by unit K-resolution.

Recall that a unit proof of $C(t)$ uses unit K-resolution, where K is the set of indices of variables occurring in $C(t)$. Also, it is clear from the definition of a rank 1 cut that if C is obtained by applying reduction and rounding to an inequality I that is a nonnegative linear combination of the inequalities in (3.71), then C is a rank 1 cut for (3.71). The same is true of $C(t)$, since one can add multiples of bounds $0 \leq x_j \leq 1$ to I to obtain an inequality that, after reduction and rounding, yields $C(t)$ rather than C. Then by Theorem 28 a series of K-resolutions yields a clause that absorbs $C(t)$. □

It can now be shown that unit K-resolution has the same deductive power as polyhedral projection. It will be convenient to use the following lemma, which says that unit K-resolution does not cut off any integer points in the polyhedral projection.

Lemma 22 *Let S be a set of clauses, P the polyhedron described by the linear relaxation of the inequality representation of S, and $\bar{S}$ the result of applying the unit K-resolution procedure to S and deleting all clauses with variables in $\{x_{k+1}, \ldots, x_n\}$. Then all 0-1 points in P_K lie in $T(\bar{S})$.*

Proof: It suffices to show that any point $(t_1, \ldots, t_n)$ that satisfies the inequalities representing parents of a unit K-resolvent, where $t_1, \ldots, t_k \in \{0, 1\}$, also satisfies the inequality representing the resolvent. Without loss of generality suppose that the parents have the form $ay + c'z + x_n \geq \alpha$ (with $c'_n = 0$), $by - x_n \geq \beta$ and that the resolvent is $cy + c'z \geq \gamma$, where

$c \geq 0$. There are two cases. (a) t has the property that $t_j = 0$ when $c_j = 1$. Then $(t_{k+1}, \ldots, t_n)$ must satisfy $c'z + x_n \geq \alpha$ and $-x_n \geq \beta$ and hence their sum $c'z \geq \gamma$. t therefore satisfies the resolvent. (b) t does not have this property, in which case it obviously satisfies the resolvent. □

Theorem 46 ([145]) *Let S be a satisfiable set of clauses, and P the polyhedron described by their linear relaxation. Let $\bar{S}$ be the result of applying the unit K-resolution procedure to S and deleting all clauses with variables in $\{x_{k+1}, \ldots, x_n\}$. Then the satisfaction set of $\bar{S}$ consists precisely of the 0-1 points in P_K.*

Proof: Let $\text{int}(P_K)$ be the set of 0-1 points in P_K. Lemma 22 implies that $\text{int}(P_K) \subset T(\bar{S})$. It remains to show that $T(\bar{S}) \subset \text{int}(P_K)$.

Take a point $t \notin \text{int}(P_K)$ and show that $t \notin T(\bar{S})$. Thus t is cut off by some projection cut $u^T A_2 y \geq u^T a$. This cut must therefore logically imply a clause C that t falsifies. We will exhibit another projection cut $v^T A_2 y \geq v^T a$ that yields C after reduction and rounding. It then follows from Theorem 45 that C is implied by clauses obtainable by unit K-resolution, which means by Lemma 22 that $t \notin T(\bar{S})$.

For convenience write $u^T A_2 y \geq u^T a$ as $by \geq \beta$. Suppose without loss of generality that $b_j \geq 0$ for $j \in \{1, \ldots, k\}$, since if $b_j < 0$, one can replace x_j with $1 - x_j$. Let $C = \bigvee_{j \in J} x_j$, where $J \subset \{1, \ldots, k\}$. Then since $bx \geq \beta$ implies C, $\sum_{j \notin J} b_j < \beta$. Thus if one subtracts, from $bx \geq \beta$, b_j times the bound $x_j \geq 0$ for each $j \notin J$, one obtains an inequality $cx \geq \gamma$ with $\gamma > 0$. From this it is clear how to alter u to obtain a vector v so that $v^T A_2 \geq v^T a$ is $(c/c_{\max})y \geq \gamma/c_{\max}$, where $c_{\max} = \max_j\{c_j\}$. But rounding and reduction applied to the latter inequality yield C. □

3.10.8 Complexity of Inference by Polyhedral Projection

Although unit K-resolution is much simpler than general resolution, its complexity can grow exponentially with k.

Theorem 47 ([145]) *The unit K-resolution procedure applied to a clause set S with n variables generates a set of clauses whose size, in the worst case, grows exponentially with $k = |K|$, even if $n = 2k - 1$ and $|S| = 2k$.*

Proof: Let S consist of clauses $x_1^t \vee x_{k+1}$, clauses $x_j^t \vee \neg x_{j+k-1} \vee x_{j+k}$ for $j = 2, \cdots, k-1$, and clauses $x_k^t \vee \neg x_{2k-1}$, where $t = 0, 1$. (Recall that $x_j^1 = x_j$, $x_j^0 = \neg x_j$.) Then for any 0-1 sequence $t_1, \ldots, t_k$, one can perform unit K-resolutions to obtain $x_1^{t_1} \vee \ldots \vee x_k^{t_k}$. That is, one can resolve

$x_1^{t_1} \vee x_{k+1}$ with $x_2^{t_2} \vee \neg x_{k+1} \vee x_{k+1}$, their resolvent $x_1^{t_1} \vee x_2^{t_2} \vee x_{k+2}$ with $x_3^{t_3} \vee \neg x_{k+2} \vee x_{k+3}$, and so on. This obtains 2^k distinct clauses in the projection set, none of which absorbs another. □

It may therefore be practical in many situations to use unit K-resolution only for fairly small k. Fortunately, the complexity of unit K-resolution is polynomial when k is fixed.

Theorem 48 ([145]) *The complexity of unit K-resolution is at worst quadratic in the number of literals (i.e., literal occurrences) when k is fixed.*

Proof: Let S be the set of clauses to which unit K-resolution is applied. Also, let the K-*part* of a clause be the portion of the clause with variables in $\{x_1, \ldots, x_k\}$. Label each literal occurring in the non-K-part of a clause in S. When a clause with only one variable in $\{x_{k+1}, \ldots, x_n\}$ is resolved with another clause C, let the literals in the non-K-part of the resolvent inherit their labels in C.

Now take any two clauses D and E (not necessarily distinct) whose variables belong to $\{x_1, \ldots, x_k\}$. Consider the group G_D of all clauses with K-part D that are generated by the unit K-resolution procedure, and the group G_E of all such clauses whose K-part is E. Let L_1 and L_2 be arbitrary labels of literals in S. Since unit K-resolution only checks a pair of clauses for sign alteration when one of them belongs to a clause with only one variable in $\{x_{k+1}, \ldots, x_n\}$, on at most one occasion does it compare a literal with label L_1 in G_D with a literal with label L_2 in G_E. Thus the number of comparisons between literals in G_D with those in G_E is at most quadratic in the number of literals in S. Since the number of pairs D, E is bounded by 2^{2k}, the theorem follows. □

Ordinary unit resolution ($k = 0$) has linear complexity, because a given unit clause occurs only once, and when one resolves on a unit clause, all occurrences of that literal are eliminated. So each literal occurrence is examined only twice: once initially when the literals are counted, and once when the literal is eliminated.

3.11 Other Approaches

There are a few remaining optimization-related approaches to inference in propositional logic that should be mentioned.

G. Patrizi and C. Spera show how to convert a satisfiability problem to an equivalent linear complementarity problem, which they solve heuristically as a parametric linear programming problem [231, 232, 264].

V. Dhar and N. Ranganathan [81] use the pivot and complement heuristic of Balas and Martin [10], followed by branch and bound if necessary, to solve the integer programming formulation of a satisfiability problem. They found that branch and bound always failed to solve a problem in reasonable time whenever the heuristic failed.

K. Truemper uses ideas from combinatorial optimization to solve inference problems by decomposition [276]. Truemper's method is distinguished by the fact that after analyzing a problem, it generates a solution algorithm for that problem and an upper bound on the amount of time necessary to solve it. The same solution algorithm solves any problem instance obtained by deleting rows or columns from the original problem matrix. Truemper and his colleagues implemented the method in a system called Leibniz. They found that many practical problems yield small upper bounds on the running time.

Truemper applies five decomposition methods to determine whether a satifiability problem is made up of 2-SAT, renamable Horn, or balanced components. A MINSAT problem (which minimizes a linear objective function subject to logical clauses) is decomposed into Horn and balanced components. Each component is obtained from the original by removing nonzero entries, taking submatrices, and possibly adjoining additional rows or columns. The problem of finding the best decomposition is itself NP-hard but in practice is solved by heuristic methods. Truemper's method is very much in the spirit of this book because it is based on deep analysis of the inference problem using the methods of combinatorial optimization. Unfortunately Truemper's book appeared at a time too close to the publication of the present one to permit the in-depth treatment it deserves.

P. P. Shenoy [254-256] solves satisfiability problems with local computations that "propagate" valuations through a Markov tree. The algorithm is more effective when the nodes of the Markov tree are associated with small sets of atomic propositions. This approach is closely related to dynamic programming and can in fact be used for discrete optimization [257].

F. Glover and H. J. Greenberg [116] decompose a satisfiability problem into a set of 2-satisfiability problems, which they solve with graph-theoretic techniques.

When inference problems are solved by incorporating logical formulas into a mathematical programming model, it is important to have a "tight" and economical representation by inequalities. A tight formulation is one describing a polyhedron that is in some sense not much larger than the convex hull of the 0-1 points inside it. The usual method of converting a formula to inequalities—conversion to conjunctive normal form and then to clausal inequalities—can result in a loose and lengthy formulation. These issues have been addressed by H. P. Williams [294], as well as by C. Blair,

3.11 OTHER APPROACHES

R. Jeroslow, and J. K. Lowe [23, 168], J. M. Wilson [297], and J. N. Hooker and H. Yan [153]. E. Hadjiconstantinou and G. Mitra [127] show how logical formulas may automatically be converted to inequalities.

Optimization methods can be applied to *inductive* as well as deductive reasoning in propositional logic. Induction is the inference of rules from specific instances. One application of an induction algorithm would be to build a rule base for an expert system, on the basis of a record of past decisions by experts. This problem has been attacked with branch and bound by E. Triantaphyllou, A. L. Soyster, and S. R. T. Kumara [271-274], with interior point methods by A. P. Kamath, N. K. Karmarkar, K. G. Ramakrishnan, and M. G. C. Resende [176], and with boolean methods by E. Boros, I. Crama, P. L. Hammer, J. N. Hooker, and T. Ibaraki [32, 33, 72]. Boros, Hammer, and Hooker give statistical justification for their approach in [33].

Chapter 4

Probabilistic and Related Logics

Probabilistic logic is the result of George Boole's effort to capture uncertainty in logical inference [27, 28]. Its formulas are identical to those of propositional logic, but they are assigned continuous probability values rather than simply truth or falsehood. Today Boole's probabilistic logic is more relevant than ever, because it provides a basis for dealing with uncertainty in knowledge-based systems that is not only well grounded theoretically but has some practical advantages as well.

In a probabilistic knowledge base, each formula is assigned a probability or an interval in which its probability lies. Some conditional probabilities may be specified (or bounded) as well. The inference problem is to determine the probability with which a given conclusion can be drawn. It turns out that a number of different probabilities can be assigned to the conclusion, but they comprise a calculable interval of real numbers.

Because probability intervals as well as point values can be specified, probabilistic logic reasons under incomplete information as well as under uncertainty.

In his careful study of Boole's work [124, 126], T. Hailperin pointed out that the problem of calculating this interval can be captured naturally in a linear programming model, which Boole himself all but formulated. About a decade later N. Nilsson reinvented probabilistic logic and its linear programming formulation [223], and his paper sparked considerable interest [55, 89, 113, 119, 120, 210, 211, 228, 281]. Hailperin provides a historical survey of probabilistic logic in [125]. Due to column generation techniques, the computational problem is well solved for instances having one or two hundred atomic propositions.

The linear model for probabilistic logic can be adapted to perform Bayesian inferences, but it cannot accommodate independence conditions, which give rise to nonlinear constraints. Nonetheless a *nonlinear* programming model can do both. The resulting type of logic might be called *Bayesian logic*.

One way to represent independence conditions is with a "Bayesian network" [234], which is the basis for influence diagrams [226, 250]. In the worst case the number of nonlinear constraints required grows exponentially with the size of the network. However, for a large class of networks the growth is only linear. The column generation methods used for ordinary probabilistic logic can also be adapted to Bayesian logic [2].

Another weakness of probabilistic logic is that it cannot readily be used to *accumulate* evidence for propositions in a knowledge base. Dempster-Shafer theory, which is related to probabilistic logic, was developed for just this purpose. A linear programming formulation of the inference problem in Dempster-Shafer theory clarifies how it differs from probabilistic logic, and a set-covering model can serve as the basis for calculating inferences. The pros and cons of Dempster-Shafer theory vis-à-vis probabilistic logic are briefly discussed below. One of the cons is its strong independence assumption, which can be dropped to yield a type of belief logic in which inferences can be computed with linear programming.

The fact that several logics of uncertainty admit a linear programming model is based on an underlying structural commonality. This idea is developed systematically in [3], which also introduces a linear programming model for second-order probabilistic logic that is not described here.

The most popular device for dealing with uncertainty in practical rule-based systems is a "confidence factor." The confidence factor of a rule's consequent is computed as some reasonable function of the confidence factors of its antecedents. When the rules are "fired" repeatedly, confidence factors may converge to a stable value for each proposition. There are conditions under which convergence occurs after finitely many firings. Confidence factors can be finitely computed even if convergence is infinite, using a mixed integer model proposed by R. Jeroslow. This model relies on a representability theorem of Jeroslow and J. K. Lowe, which, though little known, is of fundamental importance for mixed integer modeling. An entire section is therefore devoted to its statement and proof.

There are a number of other frameworks for reasoning under uncertainty or incomplete information [184]. The latter include default and nonmonotonic logics, surveyed by Reiter [243], which allow inferences to be revised in the light of new evidence. As will be discussed in Chapter 6, Bell et al. [18] indicate how several of these logics can be given integer or linear programming models.

Still other logics of uncertainty have been surveyed by the Léa Sombé group [263]. These include fuzzy logic [204, 275, 300, 303, 304] and related possibilistic logics. A series of volumes on uncertainty in artificial intelligence [136, 177, 193, 194, 251] contains papers on these topics as well as causal reasoning [109], modal logics, [121] and belief nets [135].

4.1 Probabilistic Logic

Probabilistic logic regards each possible assignment of truth values to atomic propositions as a "possible world." Each possible world has some (generally unknown) probability of being the actual world. The probability that a formula is true is the sum of the probabilities of all the possible worlds in which it is true. So, when a probability (or probability range) is assigned to one or more formulas, only certain probability distributions over possible worlds are consistent with this assignment. For each such distribution one can calculate the probability of an inferred conclusion by summing the probabilities of all worlds in which it is true. In this way one finds a range of probabilities for the conclusion that are consistent with the probabilities assigned to the premises.

The theoretical advantage of probabilistic logic, relative to the confidence factors commonly used in expert systems, is that it provides a principled means of computing the probability of an inference rather than an *ad hoc* formula. The main practical advantage is that it allows one to use only as much probabilistic information as is available. A perennial weakness of probability-based reasoning, such as that in decision trees and influence diagrams, is the necessity of supplying a large number of prior and conditional probabilities before any results can be computed. Probabilistic logic makes no such demands.

Because of its linear programming interpretation, probabilistic inference affords a natural application of optimization methods to logic. The linear programming problem that results, however, is impracticably large, because the number of possible worlds and therefore the number of variables tends to grow exponentially with the number of atomic propositions. But mathematical programmers have long used column generation methods for dealing with just this sort of problem. These methods will be applied to probabilistic logic with promising results.

Objections have been raised to the fact that the linear programming model derives intervals rather than point values. Point values can be obtained by computing a maximum entropy solution, but this poses a hard computational problem whose solution may be misleading. The intervals, on the other hand, may be too wide and may reflect excessive conservatism. Both problems can be alleviated within the linear programming model.

4.1.1 A Linear Programming Model

The truth value of a formula in probabilistic logic can be any real number in the interval from 0 to 1 and is interpreted as a probability. Consider, for instance, a knowledge base consisting of three formulas,

$$\begin{array}{c} x_1 \\ x_1 \supset x_2 \\ x_2 \supset x_3 \end{array} \quad (4.1)$$

A *possible world* is an assignment of truth values, true or false, to every atomic proposition. In propositional logic, a *model* is simply a possible world, and x_3 can be inferred from (4.1) because x_3 is true in every model in which (4.1) is true.

Suppose, however, the formulas (4.1) are not known with certainty but have probabilities 0.9, 0.8, and 0.7, respectively. What probabilities can consistently be assigned x_3?

It is important to note that the probability 0.8 assigned to $x_1 \supset x_2$ is not the conditional probability $Pr(x_2 \mid x_1)$ of x_2 given x_1. It is the probability that x_1 is false or x_2 is true, since the material conditional $x_1 \supset x_2$ is equivalent to $\neg x_1 \vee x_2$. The same point applies to $x_2 \supset x_3$. One can, however, specify conditional probabilities as well. Let us also suppose that the conditional probability of x_1, given that x_2 and x_3 are true, is 0.8. That is,

$$Pr(x_1 \mid x_2, x_3) = Pr(x_1, x_2, x_3)/Pr(x_2, x_3) = 0.8 \quad (4.2)$$

The goal is to find a probability range for x_3 that is consistent with the three prior and one conditional probability assignments in (4.1) and (4.2). Probabilistic logic solves this problem by letting a model be, not a possible world, but a distribution of probabilities over all possible worlds. (A model for ordinary propositional logic in effect assigns a probability of one to exactly one possible world and zero to the others. In this sense probabilistic logic is a generalization of propositional logic.)

In this example, there are $2^3 = 8$ possible worlds, corresponding to the 8 truth assignments,

$$\begin{array}{c} (x_1, x_2, x_3) = (0,0,0), (0,0,1), (0,1,0), (0,1,1), \\ (1,0,0), (1,0,1), (1,1,0), (1,1,1) \end{array} \quad (4.3)$$

4.1 PROBABILISTIC LOGIC

Therefore, if $p_1, \ldots, p_8$ are the probabilities assigned to these 8 worlds, one can write the matrix equation

$$\begin{bmatrix} 0 & 0 & 0 & 0 & 1 & 1 & 1 & 1 \\ 1 & 1 & 1 & 1 & 0 & 0 & 1 & 1 \\ 1 & 1 & 0 & 1 & 1 & 1 & 0 & 1 \\ 0 & 0 & 0 & -0.8 & 0 & 0 & 0 & 0.2 \end{bmatrix} \begin{bmatrix} p_1 \\ \vdots \\ p_8 \end{bmatrix} = \begin{bmatrix} 0.9 \\ 0.8 \\ 0.7 \\ 0 \end{bmatrix} \quad (4.4)$$

$$\sum_{i=1}^{8} p_i = 1, \quad p_i \geq 0, \quad i = 1, \ldots, 8$$

The 8 columns of the matrix correspond to the 8 possible worlds (4.3). Since x_1 is true in the last 4 worlds (indicated by the 1's in the first row of the matrix), (4.4) says that x_1's probability 0.9 is the sum of the probabilities $p_5, \ldots, p_8$ of these 4 worlds. The probabilities 0.8 and 0.7 of $x_1 \supset x_2$ and $x_2 \supset x_3$ are similarly computed in rows 2 and 3. The last row of the matrix equation is simply the result of writing (4.2) in the form

$$Pr(x_1, x_2, x_3) - 0.8 Pr(x_2, x_3) = 0$$

which is equivalent to

$$p_8 - 0.8(p_4 + p_8) = 0$$

The constraints in the last line of (4.4) ensure that $(p_1, \ldots, p_8)$ is a probability distribution.

The unknown probability π_0 of x_3 is given by

$$\pi_0 = \begin{bmatrix} 0 & 1 & 0 & 1 & 0 & 1 & 0 & 1 \end{bmatrix} p = c^T p \quad (4.5)$$

It is clear that π_0 can have any value $c^T p$ for which p solves (4.4). Since (4.4) and (4.5) are linear, the possible values of π_0 lie in an interval that can be found by minimizing and maximizing $c^T p$ subject to the constraints in (4.4). This is a linear programming problem. The minimum value of $c^T p$ is 0.5, and the maximum value is 0.7, which means that π_0 can be any probability in the range from 0.5 to 0.7.

In this particular example, all the columns of the linear programming problem (if one counts the objective function coefficients as part of the columns) are distinct. The first two columns, for instance, are $(0, 0, 1, 1, 0)$ and $(1, 0, 1, 1, 0)$, where the first component of each vector is the objective function coefficient. In large problems, however, many columns are likely to be identical, and all but one in each equivalence class can be dropped from the linear model. This coalesces into one world all worlds that give rise to an identical column. In fact, if no conditional probabilities are specified, one can conceive of the set of possible worlds as all consistent assignments of truth values to the formulas in question (each assignment corresponding to a column), rather than as all assignments of truth values to atomic propositions. This is the concept used in Nilsson's work and much of the literature on probabilistic logic.

To write the general linear programming model, suppose that there are n atomic propositions, which give rise to $N = 2^n$ possible worlds. Let the vector p denote the distribution of probabilities over the possible worlds. There are m formulas $F_1, \ldots, F_m$ for whose probabilities one can state lower bounds $\underline{\pi} = (\underline{\pi}_1, \ldots, \underline{\pi}_m)$, respectively, and upper bounds $\overline{\pi} = (\overline{\pi}_1, \ldots, \overline{\pi}_m)$, respectively. (It is of course possible that $\underline{\pi}_i = \overline{\pi}_i$.) In addition there may be lower bounds $\underline{\rho} = (\underline{\rho}_1, \ldots, \underline{\rho}_{m'})$ and upper bounds $\overline{\rho} = (\overline{\rho}_1, \ldots, \overline{\rho}_{m'})$ for conditional probabilities $Pr(G_1 \,|\, H_1)$, ..., $Pr(G_{m'} \,|\, H_{m'})$. Then to find a range for the unknown probability $\pi_0 = c^T p$ of a proposition F_0, the linear model can in general be written

$$\begin{array}{ll} \min/\max & c^T p \\ \text{s.t.} & \underline{\pi} \le Ap \le \overline{\pi} \\ & \overline{B}p \ge 0 \\ & \underline{B}p \le 0 \\ & e^T p = 1, \ p \ge 0, \end{array} \qquad (4.6)$$

where e is a vector of ones. In the example, matrix A consists of the first three rows of the matrix in (4.4), and matrices $\overline{B}$ and $\underline{B}$ (which are identical since $\overline{\rho}_1 = \underline{\rho}_1 = 0.8$) consist of the last row. In general,

$$a_{ij} = \begin{cases} 1 & \text{if } F_i \text{ is true in world } j \\ 0 & \text{otherwise} \end{cases} \qquad (4.7)$$

$$\overline{b}_{ij} = \begin{cases} 1 - \overline{\rho}_i & \text{if } G_i \text{ and } H_i \text{ are both true in world } j \\ -\overline{\rho}_i & \text{if } H_i \text{ but not } G_i \text{ is true in world } j \\ 0 & \text{otherwise.} \end{cases} \qquad (4.8)$$

$$\underline{b}_{ij} = \begin{cases} 1 - \underline{\rho}_i & \text{if } G_i \text{ and } H_i \text{ are both true in world } j \\ -\underline{\rho}_i & \text{if } H_i \text{ but not } G_i \text{ is true in world } j \\ 0 & \text{otherwise} \end{cases} \qquad (4.9)$$

The constraint set of (4.6) may be infeasible. The problem of determining whether it is feasible is the *probabilistic satisfiability problem*, which is easily shown to be NP-complete [113].

Probabilistic logic is easily adapted to accommodate situations in which probability estimates come from uncertain sources [3]. Suppose that proposition G has probability ρ according to a certain expert, and that π is the probability that the expert is right about ρ. Then if H is the proposition stating that the expert is right, the constraints $Pr(G \,|\, H) = \rho$ and $Pr(H) = \pi$ may be added to the linear programming model.

4.1 PROBABILISTIC LOGIC

A model for second-order probabilities is also presented in [3]. On this scheme, one can state the probability π_{ik} that $Pr(F_i) \leq q_{ik}$. A second-order probability distribution for a proposition F_i can be described by specifying π_{ik} for several q_{ik} ranging from 0 to 1. If one is rather sure about the probability $Pr(F_i)$, the second order probability mass is concentrated around that value. Otherwise it is more dispersed.

F_i is a formula of propositional logic containing atomic propositions $x_1, \ldots, x_n$. As usual, let $p_1, \ldots, p_N$ be the (first-order) probabilities of the $N = 2^n$ possible worlds. So $Pr(F_i) = A_i p$, where $A_i = (a_{i1}, \ldots, a_{iN})$ and a_{ij} is given by (4.7). The second-order probability assignments therefore have the form

$$Pr(A_i p \leq q_{ik}) = \pi_{ik}$$

Let H_{ik} be the half-space $\{p \mid A_i p \leq q_{ik}\}$. The intersections of the half-spaces divide R^N into polyhedral regions S_t. If P_t is the second-order probability mass assigned to S_t, then

$$\sum_{S_t \subset H_{ik}} P_t = \pi_{ik} \qquad (4.10)$$

Now for a given proposition F_0, one can compute bounds on the second order probability that $Pr(F_0) \leq \pi_{0k}$ by minimizing and maximizing

$$\sum_{S_t \subset H_{0k}} P_t \qquad (4.11)$$

subject to (4.10) over all given values of π_{ik}, and subject to $\Sigma_t P_t = 1$ and $P_t \geq 0$ for all t.

4.1.2 Sensitivity Analysis

One benefit of using probabilistic logic is that the full sensitivity analysis for linear programming is available. In particular, one can gauge the sensitivity of the result to perturbations in the probabilities assigned propositions in the data base. This can be very useful information, because it alerts us to which assignments are quite critical and therefore perhaps should be more carefully ascertained, and which assignments make little or no difference. The example of the previous section will be used to illustrate sensitivity analysis.

Suppose one wishes to determine the sensitivity of the lower bound 0.5 on $Pr(x_3)$ to the assigned probabilities $Pr(x_1)$, $Pr(x_1 \supset x_2)$, and $Pr(x_2 \supset$

x_3). Perturbing the right-hand side of (4.4),

$$\begin{bmatrix} 0.9 \\ 0.8 \\ 0.7 \\ 0 \end{bmatrix} + \begin{bmatrix} \Delta_1 \\ \Delta_2 \\ \Delta_3 \\ \Delta_4 \end{bmatrix}$$

the resulting change in the lower bound 0.5 is

$$\lambda_1 \Delta_1 + \lambda_2 \Delta_2 + \lambda_3 \Delta_3 + \lambda_4 \Delta_4$$

where $\lambda_1, \ldots, \lambda_4$ are the Lagrange multipliers (dual costs) corresponding to the first four constraints of (4.4)—provided that the perturbation does not result in a change in the optimal basis. These dual costs are always provided with the solution of the linear program, and in the present case they are

$$(\lambda_1, \ldots, \lambda_4) = (1.25, 1.25, 1.25, -1.25)$$

The ranges within which the perturbations can vary without changing the optimal basis, if only one constraint is perturbed at a time, are also provided by most linear program routines:

$$\begin{aligned} -0.4 \leq \Delta_1 \leq 0 \\ -0.4 \leq \Delta_2 \leq 0 \\ -0.4 \leq \Delta_3 \leq 0 \\ 0 \leq \Delta_4 \leq 0.08 \end{aligned}$$

Thus if 0.8 rather than 0.9 is assigned to $Pr(x_1)$, the lower bound will decrease from 0.5 to $0.5 + (1.25)(-0.1) = 0.375$. Here the sensitivity is quite high.

Sensitivity to perturbations in the conditional probabilities is slightly less straightforward. If $Pr(x_1 \mid x_2, x_3)$ changes from 0.8 to $0.8 + \Delta$, the corresponding constraint changes from $-0.8p_4 + 0.2p_8 = 0$ to

$$-(0.8 + \Delta)p_4 + (0.2 - \Delta)p_8 = 0$$

or

$$-0.8p_4 + 0.2p_8 = (p_4 + p_8)\Delta = Pr(x_2, x_3)\Delta$$

This is approximately equivalent to increasing the right-hand side of the constraint by $\Delta_4 = Pr(x_2, x_3)^* \Delta$, where $Pr(x_2, x_3)^* = 0.5$ is the value of $Pr(x_1, x_2)$ obtained by solving the original problem. Thus if the conditional probability rises from 0.8 to 0.9, so that $\Delta = 0.1$, the right-hand side increases by about $\Delta_4 = (0.5)(0.1) = 0.05$. Since this does not change the optimal basis, the lower bound 0.5 changes by about $(0.05)(-1.25) = -0.0625$.

4.1 PROBABILISTIC LOGIC

The exact change in the lower bound is $-0.0555\ldots$. (The change in the lower bound can be determined exactly, without re-solving the problem, if the inverse basis matrix is available. But many linear programming routines do not provide this information, and it is awkward to use.)

In general, a perturbation Δ_i in an assigned probability $Pr(F_i)$ results in a change $\lambda_i \Delta_i$, where λ_i is the corresponding dual cost, provided the optimal basis does not change. A perturbation Δ in an assigned conditional probability $Pr(G_i \mid H_i)$ has approximately the effect of perturbing the right-hand side of the corresponding constraint by $Pr(H_i)\Delta$. The probability $Pr(H_i)$ can be obtained simply by introducing a constraint that defines $Pr(H_i)$. In the example, the constraint is simply

$$p_4 + p_8 = Pr(x_2, x_3)$$

where $Pr(x_2, x_3)$ is treated as an additional variable.

4.1.3 Column Generation Techniques

A difficulty with the linear programming model (4.6) of the probabilistic inference problem is that the number of variables p_j can increase exponentially with the number of atomic propositions. This problem can be alleviated using column generation techniques, which are well known in mathematical programming. Their rationale is that they introduce variables into the problem only as they are needed to improve the solution, so that only a small fraction of the total variable set may eventually be used.

Column generation was suggested for probabilistic logic by Nilsson [223] and by Georgakopolous, Kavvadias, and Papadimitriou [113] in their paper on probabilistic satisfiability. Here three particular column generation methods for probabilistic logic are reviewed—those proposed by Hooker [138], by Jaumard, Hansen, and Aragaö [161], by Brun [35], and by Kavvadias and Papadimitriou [182]. A decomposition method is proposed in [86].

If $a^j = (a_1^j, \ldots, a_m^j)$ is column j of A and similarly for $\overline{b}^j$ and $\underline{b}^j$, the maximization problem in (4.6) can be rewritten as in (4.12) below. Here M is any large number ("Big M"); $s = (s_1, \ldots, s_m)$ and $s' = (s_1, \ldots, s'_{m'})$ are slack variables; $t = (t_1, \ldots, t_m)$ and $t' = (t'_1, \ldots, t'_{m'})$ are surplus variables; and u_0, $u = (u_1, \ldots, u_m)$ and $u' = (u'_1, \ldots, u'_{m'})$ are artificial variables. Associate vectors $\overline{\lambda}$, $\underline{\lambda}$, $\overline{\lambda}'$, and $\underline{\lambda}'$ of dual variables and the dual variable λ_0

with the constraints as shown.

$$\max \sum_{j \in J} p_j c_j - M \sum_{i=0}^{m} u_i - M \sum_{i=1}^{m'} u'_i$$

$$\text{s.t.} \quad \sum_{j \in J} p_j a^j + s = \overline{\pi} \qquad (\overline{\lambda})$$

$$\sum_{j \in J} p_j a^j - t + u = \underline{\pi} \qquad (\underline{\lambda})$$

$$\sum_{j \in J} p_j \overline{b}^j + s' = 0 \qquad (\overline{\lambda'}) \qquad (4.12)$$

$$\sum_{j \in J} p_j \underline{b}^j - t' + u' = 0 \qquad (\underline{\lambda'})$$

$$\sum_{j \in J} p_j + u_0 = 1 \qquad (\lambda_0)$$

$$e^T p = 1, \quad p, s, s', t, t', u, u', u_0 \geq 0$$

Maximization only is treated here, but the minimization problem can be solved by changing the objective function to

$$\max \; -\sum_{j \in J} p_j c^j - M \sum_{j=0}^{m} u_j - M \sum_{j=1}^{m'} u'_j$$

All three methods perform simplex pivots to solve the *master problem* (4.12). Initially the index set J is empty, and the starting basis consists of the slack and artificial columns. On subsequent pivots, a column among those already in the master problem is brought into the basis so as to improve the solution. If there is no such improving column, a *subproblem* is solved in order to generate an improving column $(c_j, a^j, a^j, \overline{b}^j, \underline{b}^j, 1)$. Such a column must satisfy

$$c_j - (\overline{\lambda}^T + \underline{\lambda}^T) a^j - \overline{\lambda}'^T \overline{b}^j - \underline{\lambda}'^T \underline{b}^j > \lambda_0$$

The index j is added to J, and a pivot is performed to bring the new column into the basis. The procedures continue until no more improving columns can be found.

The three methods differ in how they generate an improving column. In the method proposed by Hooker [138], the subproblem is formulated as an integer programming problem. In the example of the previous section, the columns of (4.12) have the form

$$(c_j, a^j, a^j, \overline{b}^j, \underline{b}^j, 1) = (y_0, y_1, y_2, y_3, y_1, y_2, y_3, z_1 - 0.8w_1, z_1 - 0.8w_1, 1)$$

4.1 PROBABILISTIC LOGIC

where

$$\begin{aligned} y_0 &\equiv x_3 \\ y_1 &\equiv x_1 \\ y_2 &\equiv (x_1 \supset x_2) \\ y_3 &\equiv (x_2 \supset x_3) \\ z_1 &\equiv (x_1 \wedge x_2 \wedge x_3) \\ w_1 &\equiv (x_1 \wedge x_2) \end{aligned} \quad (4.13)$$

and (x_1, x_2, x_3) ranges over all possible worlds. The aim is to find a column for which

$$y_0 - (\overline{\lambda}_1 + \underline{\lambda}_1)y_1 - (\overline{\lambda}_2 + \underline{\lambda}_2)y_2 - (\overline{\lambda}_3 + \underline{\lambda}_3)y_3 - (\overline{\lambda}'_1 + \underline{\lambda}'_1)(z_1 - 0.8w_1) \quad (4.14)$$

is greater than λ_0. To find one write (4.13) as a system of linear inequalities in 0-1 variables and maximize (4.14) subject to those inequalities.

In general, the columns of (4.12) are the vectors $(y_0, y, y, \overline{b}, \underline{b}, 1)$ such that $\overline{b}_i = z_i - \overline{\rho}_i w_i$, $\underline{b}_i = z_i - \underline{\rho}_i w_i$, and (x, y_0, y, z, w) is a binary solution of a system

$$D \begin{bmatrix} x \\ y_0 \\ y \\ z \\ w \end{bmatrix} \geq d$$

for some x, where the system encodes a set of equivalences similar to (4.13). The subproblem can therefore be solved by solving the integer model,

$$\begin{aligned} \max \quad & y_0 - \sum_{i=1}^{m}(\overline{\lambda}_i + \underline{\lambda}_i)y_i \\ & - \sum_{i=1}^{m'}[\overline{\lambda}'_i(z_i - \overline{\rho}_i w_i) + \underline{\lambda}'_i(z_i - \underline{\rho}_i w_i)] \\ \text{s.t.} \quad & D \begin{bmatrix} x \\ y_0 \\ y \\ z \\ w \end{bmatrix} \geq d \\ & 0 \leq y_0, y, z, w \leq 1, \quad x \text{ binary} \end{aligned} \quad (4.15)$$

Only x is explicitly required to be binary, because y_0, y, z, and w are binary in any feasible solution in which x is binary. Since it suffices to find a solution for which the value of the objective function is greater than λ_0, it is generally unnecessary to solve (4.15) to optimality. If, for example, a branch-and-bound method is used [222], branching can terminate whenever a sufficiently good solution is found. When no more improving columns exist, however, this must be verified by a complete traversal of the branch-and-bound tree.

In the method proposed by Jaumard et al. [161, 35], the subproblem is formulated as a pseudo-boolean optimization problem. Each variable y_i, z_i and w_i is first written as a numerical function of x. For instance, in the above example y_2 (which is equivalent to the clause $\neg x_1 \vee x_2$) can be written as the function $f_2(x) = 1 - x_1\bar{x}_2$, where $\bar{x}_2$ is an abbreviation for $1 - x_2$. So, the subproblem can be written as the unconstrained pseudo-boolean optimization problem,

$$\max \ f_0(x) - \sum_{i=1}^{3}(\overline{\lambda}_i + \underline{\lambda}_i)f_i(x) - (\overline{\lambda}_1' + \underline{\lambda}_1')(g_i(x) - \rho_i h_i(x))$$
$$= x_3 - (\overline{\lambda}_1 + \underline{\lambda}_1)x_1 - (\overline{\lambda}_2 + \underline{\lambda}_2)(1 - x_1\bar{x}_2)$$
$$- (\overline{\lambda}_3 + \underline{\lambda}_3)(1 - x_2\bar{x}_3) - (\overline{\lambda}_1' + \underline{\lambda}_1')(x_1 x_2 x_3 - 0.8 x_1 x_2),$$

where the x_j are binary.

In general, the subproblem is,

$$\max_x f(x) = f_0(x) - \sum_{i=1}^{k}(\overline{\lambda}_i + \underline{\lambda}_i)f_i(x)$$
$$- \sum_{i=1}^{m'}[\overline{\lambda}_i'(g_i(x) - \overline{p}_i h_i(x)) + \underline{\lambda}_i'(g_i(x) - \underline{p}_i h_i(x))]$$
(4.16)

where f_i, g_i, and h_i are boolean functions represented by the formulas F_i, G_i, and H_i, respectively. If F_i is a clause, then

$$f_i(x) = 1 - \prod_{j \in P_i} \bar{x}_j \prod_{j \in N_i} x_j,$$

where P_i is the index set of positive literals in F_i and N_i is the index set of negative literals, and similarly for G_i and H_i. Jaumard et al. recommend using a tabu search heuristic for solving (4.16) [130, 115]. If it fails to produce a solution value greater than λ_0, they solve (4.16) with an exact method of Crama, Hansen and Jaumard [73] that is based on the method of Hammer, Rosenberg and Rudeanu [129].

Finally, the method of Kavvadias and Papadimitriou [182] proposes what is in effect another heuristic for solving (4.16). For a given value of x, the *potential* of a variable x_j is the change in the objective function when the current value x_j^* of x_j is flipped to $1 - x_j^*$. Thus

$$\phi_j(x) = f(x') - f(x)$$

4.1 PROBABILISTIC LOGIC

where x' is the result of flipping x_j in x. The basic idea of the heuristic is to try flipping the values of one variable after another and stop at the point the objective function is most improved. Each time a variable is flipped, one flips the variable that brings the most improvement. The whole procedure is carried out repeatedly until no further improvement is possible.

More precisely, start with a reasonable initial value of x and proceed as follows. Let x^l be the result of flipping the first l variables and Δ_l the cumulative effect on the objective function, so that $x^0 = x$ and $\Delta_0 = 0$. For each l, beginning with $l = 0$, pick the variable x_{j^*} with the largest potential $\phi_j(x^l)$, let $\Delta_{l+1} = \Delta_l + \phi_{j^*}(x^l)$, and flip the value of $x^l_{j^*}$ in x^l to get x^{l+1}. Stop when $\Delta_l < 0$ or $l = n$, and let Δ_{l^*} be the largest Δ_l found. Let $x = x^{l^*}$ and repeat the entire procedure, unless $\Delta_1 < 0$, in which case the heuristic terminates.

Since the procedure is heuristic, it may fail to find an improving solution even when one exists. Column generation may therefore terminate before an optimal solution is found.

The second and third methods above have been tested computationally, with promising results. Jaumard and coworkers, for instance, solve problems with 70 atomic propositions and 100 clauses in about a minute on a Sun Sparc computer. About 600 columns are generated, out of a possible 2^{70}. These experiments suggest that the inference problem is well solved for a few hundred atomic propositions.

A column generation method can also be applied to second-order probabilistic logic [3]. Recall that the linear programming model is to minimize or maximize (4.11) subject to (4.10) over all given probabilities π_{ik}. Each set S_t in (4.10) is a subset of R^N satisfying some subset of the constraints $A_i p \leq q_{ik}$. Therefore, each S_t can be characterized as the subset of R^N satisfying

$$A_i p \leq q_{ik} + y_{ik}, \text{ all } i, k \tag{4.17}$$

for some 0-1 value of the vector y consisting of y_{ik} for all i, k. So if multiplier λ_{ik} is associated with each constraint (4.10) and λ_0 with the constraint $\Sigma_t P_t = 1$, the column generation subproblem (in the maximization problem) is to maximize

$$y_{0k} - \sum_{ik} \lambda_{ik} y_{ik} - \lambda_0,$$

subject to the constraints (4.17) along with $\Sigma_j p_j = 1$ and $p_j \geq 0$. This is a mixed integer programming problem in which the p_j are the continuous variables and the y_{ik} are the integer variables. Clearly, this approach is practical only if the number 2^n of (first-order) possible worlds is not too large.

4.1.4 Point Values Versus Intervals

Many students of probabilistic logic are troubled by the fact that a linear programming model yields only an interval of probabilities for an inferred proposition. There seems to be a strong desire for a point estimate. Nilsson [223] and Genesereth and Nilsson [111] suggest two methods of obtaining one: a projection method, and a maximum entropy method.

The projection method was originally stated for a problem in which all specified probabilities are point values, and there are no conditional probabilities:

$$\begin{aligned} \min/\max \quad & c^T p \\ \text{s.t.} \quad & Ap = \pi \\ & e^T p = 1, \; p \geq 0 \end{aligned} \qquad (4.18)$$

The row vector c is projected onto the space spanned by the rows of A to obtain a vector $A^T u$. The probability value $u^T \pi$ is then attributed to the inferred proposition F_0. The vector u can be calculated by solving the system $AA^T u = Ac$.

One can imagine various ways to extend this idea to the general probabilistic logic problem (4.6), which has inequality as well as equality constraints. But none seems particularly natural. In fact, it is difficult to see why one would want to compute such a projection in the first place, and what its probabilistic interpretation would be.

A more promising approach is to compute a maximum entropy solution p^* of the constraints in (4.18) and to let the probability estimate be $c^T p^*$. As Nilsson remarks, maximizing entropy is a familiar technique that several authors have applied to probabilistic reasoning [53, 192, 187]. Its rationale is that maximizing entropy minimizes information content and therefore assumes as little as possible.

The *entropy* of a probability distribution p is

$$\sum_{j=1}^{N} -p_j \log p_j \qquad (4.19)$$

A maximum entropy solution of the constraints in (4.18) is obtained simply by maximizing (4.19) subject to these constraints. Cheeseman [53] solves this problem by finding a nonnegative solution of the first-order optimality conditions for the problem (see also [111, 223]).

The general problem (4.6), which involves inequality constraints and conditional probabilities, can similarly be solved by maximizing (4.19) subject to the constraints in (4.6). Cheeseman's approach will no longer work, since it assumes equality constraints, but the problem can be solved by any of several nonlinear programming algorithms [94, 114, 259].

4.1 PROBABILISTIC LOGIC

The maximum entropy solution of the example problem (4.4) is $p = (0, 0, 0, 0.1, 0.1, 0.1, 0.3, 0.4)$, which results in a probability $c^T p = 0.6$ for x_3. This happens to lie in the center of the probability range $[0.5, 0.7]$ obtained earlier.

The exponential explosion of variables p_j afflicts the maximum entropy problem no less than the linear programming model (4.6). The column generation techniques proposed in Section 4.1.3 for the linear problem can in principle be incorporated into a reduced gradient algorithm for the nonlinear problem [259, 299]. But since maximizing entropy tends to equalize the probabilities p_j, a large and possibly exponential number of variables may be positive in an optimal solution, so that a large number of columns must be generated. In the linear programming case, only basic variables are positive in an optimal solution, so that the number of columns generated can be quite small. The maximum entropy problem may therefore be much harder to solve than the linear problem.

Not only are point estimates of inferred probabilities hard to compute, but they must be interpreted with care. In general, an entire range of probabilities are consistent with the probabilities assigned the propositions in the knowledge base. Thus one cannot deduce that the inferred proposition has any particular probability value. A maximum entropy estimate may in some sense be the point estimate that "assumes the least," but it still assumes more than is warranted by the knowledge base. Furthermore, a point estimate can be very misleading when the interval is actually quite wide, because it suggests precision when there is none.

On the other hand, the intervals obtained by a linear programming model can be too conservative. In most applications, probabilities near the ends of a probability interval assigned to a premise are less likely to be the true probability than those near the middle. The probabilities near the end of the interval calculated for the conclusion may therefore be *extremely* unlikely, because their realization may depend on the simultaneous occurrence of a number of given probabilities near the ends of their intervals. If so, the calculated interval is misleadingly wide.

This problem can actually be alleviated within the linear programming model, which offers a flexible framework for a number of devices. Just one will be sketched here. First write the linear programming model (4.6) as follows:

$$\begin{aligned} \min/\max \quad & c^T p \\ \text{s.t.} \quad & \underline{\pi}_i \leq a^i p \leq \overline{\pi}_i, \ i = 1, \ldots, m \\ & \overline{B} p \geq 0 \\ & \underline{B} p \leq 0 \\ & e^T p = 1, \ p \geq 0 \end{aligned} \quad (4.20)$$

Suppose that in addition to specifying an interval $[\underline{\pi}_i, \overline{\pi}_i]$ of probabilities for each proposition i, one specifies a second-order probability distribution over the probabilities in this interval. This will allow one to stipulate that the probabilities near the ends of the interval are less likely. Such a distribution, or more plausibly its logarithm, can be given a piecewise linear approximation as the lower envelope of a set of lines:

$$\log(Pr(\pi_i = t)) = \min_{k=1,\ldots,k_i} \{\alpha_{ik} t + \beta_{ik}\}$$

Here $Pr(\pi_i = t)$ is interpreted as a probability density. A given solution p of (4.20) assigns probability $a^i p$ to proposition i. Assuming independence, the logarithm of the joint probability that the propositions in fact have probabilities $(a^1 p, \ldots, a^m p)$ is

$$\sum_{i=1}^{m} \min_{k=1,\ldots,k_i} \{\alpha_{ik} a^i p + \beta_{ik}\} \qquad (4.21)$$

Thus to eliminate solutions that are extremely unlikely, impose a lower bound L on the quantity in (4.21). (4.20) therefore becomes,

$$\begin{aligned}
\min/\max \quad & c^T p \\
\text{s.t.} \quad & \sum_{i=1}^{m} \ell_i \geq L \\
& \ell_i \leq \alpha_{ik} a^i p + \beta_{ik}, \quad k = 1, \ldots, k_i, \; i = 1, \ldots, m \\
& \underline{\pi}_i \leq a^i p \leq \overline{\pi}_i, \quad i = 1, \\
& \overline{B} p \geq 0 \\
& \underline{B} p \leq 0 \\
& e^T p = 1, \; p \geq 0
\end{aligned}$$

The interval between the minimum and the maximum can now be narrowed by increasing L. A similar device can be used to assign second-order distributions to the conditional probabilities.

4.2 Bayesian Logic

Probabilistic logic has two weaknesses that are imposed by the linearity of its model for reasoning. It cannot accommodate independence assumptions, and it cannot compute Bayesian updates of probabilities—albeit the latter poses a fractional programming problem that can be converted to a linear programming problem.

4.2 BAYESIAN LOGIC

The independence of events may be our most useful source of probabilistic knowledge. We can understand the world only if we assume that most events are significantly influenced by relatively few other events. For instance, we assume (astrology notwithstanding) that the occurrence of an earthquake is unrelated to the position of the distant stars. Weather forecasting, on the other hand, is hard precisely because today's weather may be related to any of millions of minor atmospheric events that occurred around the globe in the last few days.

An adequate knowledge base should therefore incorporate independence assumptions when they can be made. Ordinary probabilistic logic cannot do this, since independence assumptions give rise to nonlinear constraints that destroy the linearity of its linear programming model. But one can accommodate independence assumptions in a nonlinear programming model, provided there are not too many of them.

It is also important to be able to "update" probabilities in a Bayesian fashion. Suppose, for example, that the conditional probability $Pr(A \mid B)$ of A given B is known, as well as prior probabilities $Pr(A)$ and $Pr(B)$ for A and B. Then if one learns that A is true, Bayes's rule computes the probability of B:

$$Pr(B \mid A) = \frac{Pr(A \mid B) Pr(B)}{Pr(A)}$$

The linear model for probabilistic logic can accommodate Bayesian updating if its linear objective function is replaced with the nonlinear expression $Pr(B \mid A) = Pr(A \wedge B)/Pr(B)$. For computational purposes the resulting nonlinear model can be converted to a linear programming problem [52], provided no nonlinear independence constraints are present. But this option will not be considered further, due to the practical importance of using independence constraints as well as Bayesian inference.

This section therefore describes a nonlinear extension of probabilistic logic that permits independence constraints and Bayesian updating. The resulting framework might be called *Bayesian logic*.

A useful device for capturing independence constraints is a Bayesian network [234]. Bayesian networks encode events as nodes and dependence as arcs, and they can capture complex conditional independence relations (i.e., which events are independent, given the occurrence or nonoccurrence of certain other events) [234]. Andersen and Hooker [2] apply the semantics of probabilistic logic to Bayesian networks to obtain a form of Bayesian logic that has both the flexibility of the former and the expressive power of the latter.

When formulated naively, Bayesian networks can give rise to exponentially many nonlinear independence constraints. But for a large class of networks, the number of nonlinear constraints grows only linearly with the number of nodes in the network.

Computational approaches to Bayesian networks have hitherto emphasized local computations that "propagate" values through the network. The easiest class of networks to treat in this manner are "singly connected" networks (as in the work of J. Pearl [233, 234]), but the decomposition technique of F. V. Jensen et al. [164] and S. Lauritzen and D. J. Spiegelhalter [191, 265] is practical for a larger class of networks. Henrion [135] surveys some other approaches. Van der Gaag [283-285] has proposed decomposition and other techniques. These techniques were developed, however, for classical Bayesian network problems in which the given probabilities are point values and every node is associated with an atomic proposition.

To date no computationally feasible propagation technique has been developed for general Bayesian logic, which uses probability intervals and molecular propositions. Spiegelhalter and Lauritzen [265] suggest a statistical approach that attempts to account for uncertainty in the given probabilities, but it cannot deal with intervals. Molecular propositions introduce global connections among the nodes that make decompositions difficult. The Markov trees suggested by P. P. Shenoy [254-256] permits decomposition of some propositional logical problems, but it is unclear whether this technique is helpful in a probabilistic context. (Incidentally, the method of [191] uses a partial k-tree decomposition that is essentially the same as Shenoy's decomposition.) Van der Gaag [283-285] proposes a propagation technique for probabilistic logic, but only for a restricted class of problems.

The global approach of mathematical programming is therefore the focus here. But that is not to say that mathematical programming fails to exploit the structure of the problem. The procedure for encoding independence conditions alluded to above relies on the fact that such conditions need only involve nodes that lie in a certain neighborhood. The resulting mathematical programming problem can be attacked with such solution techniques as column generation, Benders decomposition, and sequential quadratic programming, which in a sense do local computations because they solve the whole problem by solving only parts of it. These techniques can also be specialized to exploit the mathematical form of the constraints.

4.2.1 Possible World Semantics for Bayesian Networks

A *Bayesian network* is a directed network in which each node represents an event and each arc a probabilistic dependence. For present purposes, an event is always the truth or falsehood of a proposition, and nodes are identified with propositions.

An example is depicted in Figure 4.1, in which a symptom (node 1) can

4.2 BAYESIAN LOGIC

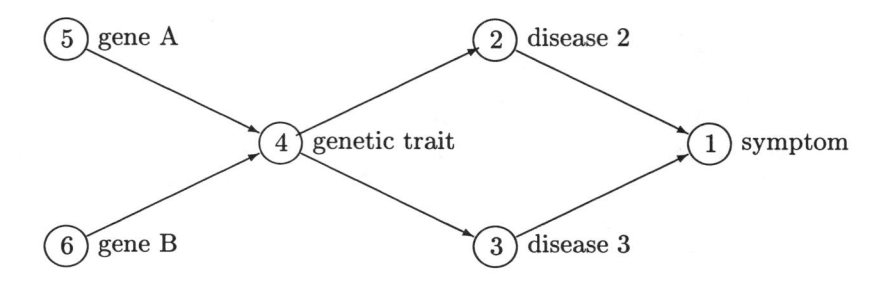

Figure 4.1: A simple Bayesian network.

be evidence for either disease 2 or disease 3 (nodes 2 and 3). The occurrence of the diseases can in turn be influenced by the presence of a certain genetic trait (node 4), to which one is predisposed by the possession of either of two particular genes (nodes 5 and 6).

In classical Bayesian networks, each node j is associated with atomic proposition x_j. (The association will later be extended to molecular propositions.) The probability that x_j is true is $Pr(x_j)$, and the probability that x_j is false is the probability $Pr(\neg x_j) = 1 - Pr(x_j)$ that its denial $\neg x_j$ is true. It will be convenient to let X_j be a variable whose value is either x_j or $\neg x_j$. The *conditional probability* of X_i given X_j is $Pr(X_i | X_j) = Pr(X_i X_j)/Pr(X_j)$, where $Pr(X_i X_j)$ is an abbreviation for $Pr(X_i \wedge X_j)$. Two propositions X_i and X_j are *independent* if $Pr(X_i X_j) = Pr(X_i) Pr(X_j)$. *Conditional independence* is defined as follows: X_i and X_j are independent, given that a set S of propositions are true, if $Pr(X_i X_j | S) = Pr(X_i | S) Pr(X_j | S)$.

The essence of a Bayesian network is that the probability that a node is true, when conditioned on the truth values of all the other nodes, is equal to the probability it is true, conditioned only on the truth values of its immediate predecessors. In Figure 4.1 the probability of observing the symptom depends (directly) only on which diseases the patient has, which is to say $Pr(x_1 | X_2 X_3 X_4 X_5 X_6) = Pr(x_1 | X_2 X_3)$ for all values of $X_2, \ldots, X_6$. (From this it follows that $Pr(x_1 | S) = Pr(x_1 | X_2 X_3)$, where S is any subset of $\{X_2, \ldots, X_6\}$ containing X_2 and X_3.) To put it differently, any influence on the probability of the symptom other than the diseases is mediated by the diseases.

Two nodes i, j in a Bayesian network are (conditionally) independent if X_i, X_j are (conditionally) independent for all values of X_i, X_j. Clearly, two nodes are independent if they have no common ancestor. Two nodes are conditionally independent, given any fixed set of truth values for a set S of nodes, if they have no common ancestor when the nodes in S are removed from the network.

It is straightforward to interpret a Bayesian network with the semantics of probabilistic logic. Prior or conditional probabilities are rendered by constraints of the form (4.6) as in ordinary probabilistic logic. For instance, suppose in the example of Figure 4.1 that the conditional probabilities of observing the symptoms are

$$Pr(x_1 \mid x_2 x_3) = 0.95 \quad Pr(x_1 \mid \neg x_2 x_3) = 0.8$$
$$Pr(x_1 \mid x_2 \neg x_3) = 0.7 \quad Pr(x_1 \mid \neg x_2 \neg x_3) = 0.1$$

So one can specify $\underline{\rho}_1 = \overline{\rho}_1 = 0.95$, $\underline{\rho}_2 = \overline{\rho}_2 = 0.8$, $\underline{\rho}_3 = \overline{\rho}_3 = 0.7$, and $\underline{\rho}_4 = \overline{\rho}_4 = 0.1$ as explained in Section 4.1.1. Suppose also that the predisposition to either disease is captured in the following conditional probabilities:

$$Pr(x_2 \mid x_4) = 0.4 \quad Pr(x_2 \mid \neg x_4) = 0.05$$
$$Pr(x_3 \mid x_4) = 0.2 \quad Pr(x_3 \mid \neg x_4) = 0.1$$

One can specify $\underline{\rho}_5, \ldots, \underline{\rho}_8$ and $\overline{\rho}_5, \ldots, \overline{\rho}_8$ in similar fashion. One can of course specify bounds, rather than exact values, for any of these probabilities.

To illustrate how nodes may be associated with molecular propositions, let us fix the conditional probabilities $Pr(x_4 \mid X_5 X_6)$ of having the genetic trait so that the patient has it precisely when he has one or both of the genes. Thus

$$Pr(x_4 \mid x_5 x_6) = 1 \quad Pr(x_4 \mid \neg x_5 x_6) = 1$$
$$Pr(x_4 \mid x_5 \neg x_6) = 1 \quad Pr(x_4 \mid \neg x_5 \neg x_6) = 0 \quad (4.22)$$

This in effect associates the molecular proposition $x_5 \vee x_6$ with node 4, even within the framework of a conventional Bayesian network. But the network of Figure 4.1 is inappropriate, because it shows nodes 5 and 6 as independent, and there may be no reason to suppose they are independent. The independence assumption might be removed by drawing an arrow from, say, node 5 to node 6. But then the network could not be solved unless the conditional probability $Pr(x_6 \mid x_5)$ were known. In a conventional Bayesian network, then, there is no way to associate a molecular proposition with a node without making possibly unwarranted independence assumptions or presupposing conditional probabilities that may not be available.

Probabilistic logic, however, provides an easy solution to this dilemma. Simply omit nodes 5 and 6 from the Bayesian network, but retain the conditional probability statements (4.22). This in effect associates $x_5 \vee x_6$

4.2 BAYESIAN LOGIC

with node 4 without making additional assumptions. (Thus one specifies $\underline{p}_9 = \overline{p}_9 = \underline{p}_{10} = \overline{p}_{10} = \underline{p}_{11} = \overline{p}_{11} = 1$ and $\underline{p}_{12} = \overline{p}_{12} = 0$.) In general, when there is no knowledge of the conditional probabilities relating a proposition x_j to others, simply omit the proposition from the network. If there is partial knowledge, include x_j in the network, draw arrows from nodes on which x_j depends, and specify appropriate bounds on the conditional probabilities. If there is no information on a particular conditional probability, specify 0 and 1 as bounds.

Finally, suppose that prior probabilities can be estimated for the occurrence of the two genes:

$$Pr(x_5) = 0.25 \quad Pr(x_6) = 0.15$$

Thus one sets $\underline{\pi}_1 = \overline{\pi}_1 = 0.25$ and $\underline{\pi}_2 = \overline{\pi}_2 = 0.15$ in (4.6).

It remains to encode the conditional independence constraints. The most straightforward approach is the following. Using the definition of conditional probability, compute the joint probability distribution of the atomic propositions $x_1, \ldots, x_4$ in Figure 4.1 as follows. (Omit x_5 and x_6, because the corresponding nodes were dropped.)

$$Pr(X_1 X_2 X_3 X_4) = \\ Pr(X_1 \mid X_2 X_3 X_4) Pr(X_2 \mid X_3 X_4) Pr(X_3 \mid X_4) Pr(X_4) \quad (4.23)$$

The computation is valid for any substitution of x_j or $\neg x_j$ for each X_j. Due to the structure of the Bayesian network, two of the conditional probabilities in (4.23) simplify as follows:

$$Pr(X_1 \mid X_2 X_3 X_4) = Pr(X_1 \mid X_2 X_3) \quad (4.24)$$
$$Pr(X_2 \mid X_3 X_4) = Pr(X_2 \mid X_4) \quad (4.25)$$

This means that the joint probability in (4.23) can be computed:

$$Pr(X_1 \ldots X_4) = Pr(X_1 \mid X_2 X_3) Pr(X_2 \mid X_4) Pr(X_3 \mid X_4) Pr(X_4)$$

Thus the independence constraints (4.24) and (4.25) are adequate for calculating the underlying joint distribution and therefore capture the independence properties of the network. Again using the definition of conditional probability, (4.24) and (4.25) can be written as the following nonlinear constraints:

$$Pr(X_1 X_2 X_3 X_4) Pr(X_2 X_3) = Pr(X_1 X_2 X_3) Pr(X_2 X_3 X_4) \quad (4.26)$$
$$Pr(X_2 X_3 X_4) Pr(X_4) = Pr(X_2 X_4) Pr(X_3 X_4) \quad (4.27)$$

Here each variable X_j varies over x_j and $\neg x_j$. To treat (4.26) and (4.27) as constraints, define each probability $Pr(F)$ in terms of the vector p. This is

readily done by treating each "$Pr(F)$" as a variable and adding a constraint of the form
$$Pr(F) = a^T p \qquad (4.28)$$
where
$$a_j = \begin{cases} 1 & \text{if } F \text{ is true in world } j \\ 0 & \text{otherwise} \end{cases}$$

Now suppose bounds are desired on the probability $Pr(x_2 \mid x_1 x_3)$ that a patient with disease 3 and the symptom has disease 2. One can minimize and maximize $Pr(x_1 x_2 x_3)/Pr(x_1 x_3)$ subject to (a) the constraints in (4.6), (b) the independence constraints (4.26) and (4.27), and (c) constraints of the form (4.28) that define all variables $Pr(F)$, including $Pr(x_1 x_2 x_3)$ and $Pr(x_1 x_3)$. In this case obtain the bounds 0.2179 and 0.2836 result.

Unfortunately, one can concoct problem instances with local maxima or minima that are not globally optimal. Thus a nonlinear programming algorithm designed to find local optima, as most are, may find a solution that does not truly maximize or minimize the unknown probability. It is unclear at this writing how often this is likely to happen in practice.

Furthermore, even if globally maximal and minimal probabilities are found, not every probability in between need be realizable. This contrasts with ordinary probabilistic logic, in which the realizable probabilities (if any) always comprise an interval. Again it is unknown whether this is typical of realistic problems.

4.2.2 Using Column Generation with Benders Decomposition

The exponential explosion of variables p_j in probabilistic logic scarcely goes away when nonlinear independence constraints are added. Column generation techniques can only be applied to a linear constraint set, but they can nonetheless be adapted to the nonlinear Bayesian logic problem in at least two ways. One way is to apply Benders decomposition, so that the linear part can be isolated in a subproblem. Another is to use sequential quadratic programming, which computes a linearization of the constraint set at each iteration. We will first describe the use of Benders decomposition.

The general nonlinear programming problem associated with a Bayesian inference problem is as follows. Let π be a vector of variables of the form $Pr(F)$, and let $Pr(F_0 \mid G_0)$ be the probability to be bounded. For convenience, use the notation $Pr(x_i, S_i)$ for $Pr(x_i \wedge y_1 \wedge \cdots \wedge y_k)$ when

4.2 BAYESIAN LOGIC

$S_i = \{y_1, \ldots, y_k\}$. The nonlinear model is

$$\min/\max \quad Pr(F_0 G_0)/Pr(G_0) \tag{4.29}$$
$$\text{s.t.} \quad Pr(x_i, S_i, T_i)Pr(S_i) = Pr(S_i, T_i)Pr(x_i, S_i), \tag{4.30}$$
$$i = 1, \ldots, m \tag{4.31}$$
$$A\pi \geq a \tag{4.32}$$
$$Bp = \pi \tag{4.33}$$
$$Cp \geq c \tag{4.34}$$
$$e^T p = 1 \tag{4.35}$$
$$p \geq 0 \tag{4.36}$$

Here (4.31) are the independence constraints, and (4.32) are other constraints involving the variables $Pr(F)$. Also, (4.33) defines the variables $Pr(F)$ in terms of the variables p_j, while (4.34)–(4.36) represent the constraint set of a conventional probabilistic logic problem (4.6).

As discussed in Section 3.9.2, Benders decomposition allows one to distinguish "complicating" variables and to split the problem into a master problem and a subproblem so that the latter contains no complicating variables. In the present case the complicating variables are those occurring in the nonlinear constraints, namely, the variables $Pr(F)$ in the vector π. Thus (4.29)–(4.32) go into the master problem:

$$\max \quad Pr(F_0 G_0)/Pr(G_0) \tag{4.37}$$
$$\text{s.t.} \quad Pr(x_i, S_i, T_i)Pr(S_i) = Pr(S_i, T_i)Pr(x_i, S_i), \tag{4.38}$$
$$i = 1, \ldots, m \tag{4.39}$$
$$A\pi \geq a \tag{4.40}$$
$$u^r \pi - v^r c + u_0^r \geq 0, \quad r = 1, \ldots, R \tag{4.41}$$
$$0 \leq \pi_i \leq 1, \quad i = 1, \ldots, m \tag{4.42}$$

where the additional constraints (4.41) are "Benders cuts," explained below. The bounds (4.42) make sure that the problem has a finite solution (if any solution). The subproblem is the feasibility problem,

$$\begin{array}{ll} \min & 0 \\ \text{s.t.} & Bp = \pi \quad (u) \\ & Cp \geq c \quad (v) \\ & e^T p = 1 \quad (u_0) \\ & p \geq 0 \end{array} \tag{4.43}$$

where π is regarded as a constant. The master problem is a nonlinear program, but it contains only the variables in π, which are limited in number.

The subproblem contains exponentially many variables p_j, but it is an ordinary probabilistic logic problem (in particular, a probabilistic satisfiability problem) that can be solved by the column generation methods reviewed in Section 4.1.3.

The procedure starts by solving the master problem, without Benders cuts ($R = 0$), for π. It then solves the subproblem, using in (4.43) the value of π just obtained. If the subproblem has a feasible solution, the problem is solved. Otherwise, the dual of the subproblem has an unbounded optimal value along some extreme ray (u^1, v^1, u_0^1), where u and v are row vectors of dual variables corresponding, respectively, to the first two constraints of (4.43), and u_0 corresponds to the third constraint. This ray gives rise to the first Benders cut in (4.41) ($R = 1$). Now re-solve the master problem to obtain a new π and continue in this fashion.

The procedure stops when either the master problem is infeasible or the subproblem is feasible. In the former case, the original problem is infeasible, and in the latter case, the optimal solution of the original problem has been found. The procedure stops in a finite number of steps. (The minimization problem is solved by maximizing the negative of the objective function according to the procedure above.)

If Benders decomposition is used to calculate the lower bound on $Pr(x_2 \mid x_1 x_3)$ derived in the previous section, the solution value is 0.2179 after the addition of 53 Benders cuts.

A difficulty with Benders decomposition is that a nonlinear programming problem must be re-solved each time a new Benders cut is added. Since the number of Benders cuts could be quite large, the resulting computation time could be prohibitive.

This problem can be alleviated by using a sequential quadratic programming approach, which replaces the nonlinear constraints with first-order Taylor expansions about the current solution value. This approach linearizes the constraints without creating a Benders subproblem, so that no additional constraints need be generated. The problem is so structured that columns generated in previous iterations remain usable.

To see how to apply sequential quadratic programming, one can without loss of generality write the nonlinear problem (4.29) in the following form:

$$\begin{aligned} \max/\min \quad & f(\pi) \\ \text{s.t.} \quad & g_i(\pi) = 0, \quad i = 1, \ldots, m \\ & A_1 \pi + A_2 p \geq a \end{aligned}$$

Iteration k provides an estimate (π^k, p^k) of the solution. To perform iter-

4.2 BAYESIAN LOGIC

ation $k+1$, solve the following quadratic subproblem:

$$\begin{aligned} \max/\min \quad & q_k(\Delta\pi) + \sum_{i=1}^{m} u_i^k \Delta\pi^T G_i(\pi^k)\Delta\pi \\ \text{s.t.} \quad & \nabla g_i(\pi^k)^T \Delta\pi + g_i(\pi^k) = 0, \quad i=1,\ldots,m \\ & A_1\Delta\pi + A_2 p \geq a - A_1\pi^k \end{aligned} \quad (4.44)$$

The function $q_k(\Delta\pi)$ in the objective is a quadratic approximation of $f(\pi^k + \Delta\pi)$ about π^k, given by

$$q_k(\Delta\pi) = \tfrac{1}{2}\Delta\pi^T F(\pi^k)\Delta\pi + \nabla f(\pi^k)^T \Delta\pi + f(\pi^k)$$

$F(\pi^k)$ is the Hessian of f at π^k, and similarly for $G_i(\pi^k)$. u_i^k is the Lagrange multiplier for constraint i (i.e., the constraint that linearizes $g_i(\pi) = 0$) in the solution of the subproblem in iteration k. The next iterate is $(\pi^{k+1}, p^{k+1}) = (\pi^k + \Delta\pi^*, p^{k+1})$, where $(\Delta\pi, p) = (\Delta\pi^*, p^{k+1})$ solves (4.44).

The Hessians $G_i(\pi^k)$ are easy to compute because the g_i are quadratic. The coefficients of the p_j do not change from one iteration to the next, so that generated columns can be accumulated across iterations.

A possible variation of this approach is to use the original objective function $f(\pi^k + \Delta\pi)$ rather than its quadratic approximation $q(\Delta\pi)$ in (4.44). Then if the Lagrangian term $\Sigma_{i=1}^m u_i \Delta\pi^T G_i(\pi^k)\Delta\pi$ is omitted, the objective function is the ratio of two linear expressions, $(\pi_i^k + \Delta\pi_i)/(\pi_j^k + \Delta\pi_j)$. This means that (4.44) is a fractional linear programming problem and can be converted to a linear programming problem (provided the constraint $\pi_j^k + \Delta\pi_j \geq \epsilon > 0$ is added) [52]. By removing the Lagrangian convergence is made less certain, so that it may be necessary to use heuristic techniques to bound the variables and guide them toward convergence. Practitioners regularly use such devices in sequential linear programming [17].

4.2.3 Limiting the Number of Independence Constraints

As remarked earlier, one can capture the independence relations in a Bayesian network by imposing the independence constraints that are required to compute the joint probability distribution using the formula

$$\begin{aligned} Pr(X_1 X_2 \ldots X_n) = \\ Pr(X_1 \mid X_2 \ldots X_n) Pr(X_2 \mid X_3 \ldots X_n) \cdots Pr(X_n) \end{aligned} \quad (4.45)$$

The constraints equate some of the factors $Pr(X_k \mid X_{k+1} \ldots X_n)$ with simpler conditional probabilities $Pr(X_k\,|\,S)$, where S is a subset of $\{X_{k+1},\ldots,X_n\}$. Andersen and Hooker [2] exploit this fact to limit the number of equations.

In general, an independence condition has the form

$$Pr(R\,|\,S,T) = Pr(R\,|\,S) \qquad (4.46)$$

where R contains propositional variables $X_1,\ldots,X_r$, S contains s propositional variables, and T contains propositional variables $Y_1,\ldots,Y_t$. R, S, and T may also contain some fixed propositions. Assume the three sets are pairwise disjoint and neither R nor T is empty. Then (4.46) represents 2^{r+s+t} constraints, corresponding to the 2^{r+s+t} possible values of the variables in R, S, T. (Recall that each variable X_i can be either x_i or $\neg x_i$, and similarly for other variables.) Some of the 2^{r+s+t} constraints are redundant, due to the following.

Lemma 23 ([2]) *If $r > 0$ and $t > 0$, the 2^{r+s+t} constraints (4.46) are equivalent to the $(2^r - 1)2^s(2^t - 1)$ constraints corresponding to all values of the variables in R, S, T except those for which $(X_1,\ldots,X_r) = (\neg x_1,\ldots,\neg x_r)$ or $(Y_1,\ldots,Y_t) = (\neg y_1,\ldots,\neg y_t)$. If $r = 0$ and $t > 0$, (4.46) is equivalent to $2^s(2^t - 1)$ constraints. If $r > 0$ and $t = 0$, (4.46) is equivalent to $(2^r - 1)2^s$ constraints. If $r = t = 0$, (4.46) is equivalent to 2^s constraints.*

Proof: Suppose first that $r, t > 0$. There is no need to impose (4.46) with $(X_1,\ldots,X_r) = (\neg x_1,\ldots,\neg x_r)$ because one can sum either side of (4.46) over all $(X_1,\ldots,X_r)$ except $(\neg x_1,\ldots,\neg x_r)$ and take the complement of both sides to obtain

$$Pr(\neg x_1,\ldots,\neg x_r, R_0\,|\,S,T) = Pr(\neg x_1,\ldots,\neg x_r, R_0\,|\,S)$$

where R_0 contains the fixed propositions in R. There is no need to impose (4.46) when $(Y_1,\ldots,Y_t) = (\neg y_1,\ldots,\neg y_t)$ because Bayes's rule allows (4.46) to be rewritten as

$$Pr(T\,|\,R,S)Pr(R\,|\,S)/Pr(T\,|\,S) = Pr(R\,|\,S)$$

which is equivalent to

$$Pr(T\,|\,R,S)Pr(R\,|\,S) = Pr(T\,|\,S)Pr(R\,|\,S) \qquad (4.47)$$

(The equation is not divided by $Pr(R\,|\,S)$ because it could be zero.) Summing each side of (4.47) over all values of $(Y_1,\ldots,Y_t)$ except $(\neg y_1,\ldots,\neg y_t)$, and subtracting $Pr(R\,|\,S)$ from either side, one gets (after reversing the sign)

$$Pr(\neg y_1,\ldots,\neg y_t, T_0\,|\,R,S)Pr(R\,|\,S) = Pr(\neg y_1,\ldots,\neg y_t, T_0\,|\,S)Pr(R\,|\,S)$$

4.2 BAYESIAN LOGIC

where T_0 contains the fixed propositions in T. From this it follows that

$$Pr(R \mid S, \neg y_1, \ldots, \neg y_j) = Pr(R \mid S)$$

as desired. □

Lemma 23 implies that each independence assumption $Pr(X_k \mid X_{k+1} \cdots X_n) = Pr(X_k \mid S)$ requires $2^{n-k} - 2^j$ nonlinear constraints, where S contains j propositions. In the worst case $j = 0$, which means that capturing all the independence conditions can require as many as $2^{n-1} + 2^{n-2} + \cdots + 2^0 - n = 2^n - n - 1$ nonlinear constraints.

Fortunately, this exponential explosion can be avoided in a large class of networks by using only those constraints that are necessary to calculate the desired probability. Consider the Bayesian network of Figure 4.2, and suppose that bounds on $Pr(x_1)$ are desired. It will be convenient to assign each node to one or more *generations* relative to x_1. Node 1 lies in generation 0, and in general all the parents of nodes in generation k that are not themselves in generation k form generation $k + 1$. To determine which independence constraints are necessary, the first step is to observe which ones would be necessary if $Pr(x_1)$ were calculated by conditioning on its immediate predecessors. This continues in a recursive fashion until the computation is complete. Along the way it becomes evident that the independence conditions will never involve nodes in more than two generations. This is the key to an efficient representation of the independence assumptions.

To calculate $Pr(x_1)$, condition on its immediate predecessors:

$$Pr(x_1) = \sum Pr(x_1 \mid X_2 X_{10} X_{11} X_{12} X_{16}) Pr(X_2 X_{10} X_{11} X_{12} X_{16})$$
$$= \sum Pr(x_1 \mid X_2 X_{10} X_{11} X_{12} X_{16}) Pr(X_2) Pr(X_{10} X_{11} X_{12} X_{16})$$

The sum is understood to be taken over all variables on which the probabilities are conditioned, in this case $X_2, X_{10}, X_{11}, X_{12}, X_{16}$. (One does not actually perform this calculation; the aim is merely to see what independence constraints one would need to perform it.) The second equality makes use of the fact that X_2 has no common predecessor with any of the other four nodes, yielding the independence relation,

$$Pr(X_2 \mid X_{10} X_{11} X_{12} X_{16}) = Pr(X_2)$$

which by Lemma 23 is captured in $2^4 - 1 = 15$ nonlinear constraints.

To complete the above calculation, compute $Pr(X_2)$ and $Pr(X_{10} X_{11} X_{12} X_{16})$ by conditioning on nodes in generation 2. First, compute $Pr(X_2)$,

$$Pr(X_2) = \sum Pr(X_2 \mid X_3 X_4) Pr(X_3 X_4)$$

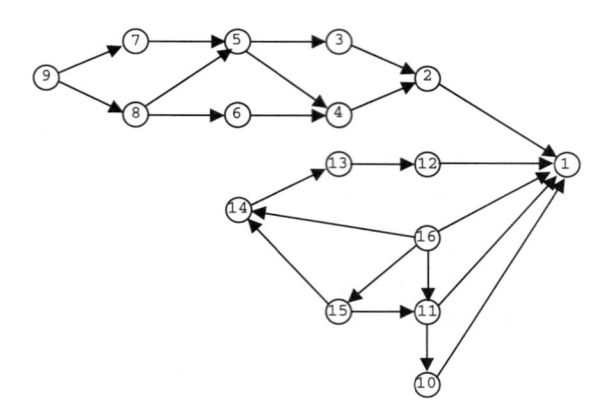

230   CHAPTER 4   PROBABILISTIC AND RELATED LOGICS

Figure 4.2: A larger Bayesian network.

which requires no independence constraints. The joint probability $Pr(X_3 X_4)$ is therefore calculated by conditioning on nodes in generation 3:

$$Pr(X_3 X_4) = \sum Pr(X_3 X_4 \mid X_5 X_6)$$
$$= \sum Pr(X_3 \mid X_4 X_5 X_6) Pr(X_4 \mid X_5 X_6) Pr(X_5 X_6)$$

This calls for the equations $Pr(X_3 \mid X_4 X_5 X_6) = Pr(X_3 \mid X_5)$, which require six nonlinear constraints by Lemma 23. Moving to generation 4,

$$Pr(X_5 X_6) = \sum Pr(X_5 \mid X_6 X_7 X_8) Pr(X_6 \mid X_7 X_8) Pr(X_7 X_8)$$

which requires the conditions $Pr(X_5 \mid X_6 X_7 X_8) = Pr(X_5 \mid X_7 X_8)$ and $Pr(X_6 \mid X_7 X_8) = Pr(X_6 \mid X_8)$, another six constraints. Finally,

$$Pr(X_7 X_8) = \sum Pr(X_7 \mid X_8 X_9) Pr(X_8 \mid X_9) Pr(X_9)$$

which requires two constraints. Note that the independence conditions never involve nodes in more than two consecutive generations.

To complete the other branch of the tree, condition on the nodes X_{13} and X_{15} that are on generation 2 of that branch,

$$Pr(X_{10} X_{11} X_{12} X_{16})$$
$$= \sum Pr(X_{10} \mid X_{11} X_{12} X_{16} X_{13} X_{15}) Pr(X_{11} \mid X_{12} X_{16} X_{13} X_{15})$$
$$= Pr(X_{12} \mid X_{16} X_{13} X_{15}) Pr(X_{16} \mid X_{13} X_{15}) Pr(X_{13} X_{15})$$

4.2 BAYESIAN LOGIC

This requires a total of 51 independence constraints. Next, condition on the generation 3 nodes X_{14} and X_{16}; note that X_{16} also belongs to generation 1.

$$Pr(X_{13}X_{15}) = \sum Pr(X_{13} \,|\, X_{15}X_{14}X_{16})Pr(X_{15} \,|\, X_{14}X_{16})Pr(X_{14}X_{16})$$

and use six constraints to express $Pr(X_{13} \,|\, X_{15}X_{14}X_{16}) = Pr(X_{13} \,|\, X_{14})$, plus two to express $Pr(x_{15} \,|\, X_{14}X_{16}) = Pr(X_{15} \,|\, X_{16})$.

The problem is therefore solved with a total of 89 independence constraints. By contrast, if one used all the constraints necessary to determine the joint distribution in (4.45), there would be (due to Lemma 23) $2^{15} + 2^{14} + \cdots + 2^1 - 70 = 65,465$ nonlinear constraints.

These ideas may be developed formally as follows. Two sets R_i and R_j of nodes in a Bayesian network are independent relative to a set S of nodes if every node in R_i is independent of every node in R_j relative to S. A set R of nodes *splits* into a collection of disjoint sets $R_1, \ldots, R_m$, relative to a set S of nodes, if R is the union of $R_1, \ldots, R_m$ ($m \geq 1$), every pair R_i, R_j ($i \neq j$) are independent relative to S, and no R_i contains subsets that are independent relative to S. One can obtain the following recursive upper bound on the number $N(R \,|\, S)$ of independence constraints needed to compute the joint probabilities $Pr(R \,|\, S)$, where the variables in R range over 2^r possible values, and S is a fixed set of atomic propositions. For a node set U let $T(U)$ be the set of nodes not in U that are immediate predecessors of nodes in U.

Lemma 24 ([2]) *Let a set R of r nodes split into $R_1, \ldots, R_m$ relative to a set S of nodes, where each R_i contains r_i nodes. Then if t_i is the number of nodes in $T(R_i) \setminus S$,*

$$N(R \,|\, S) \leq \alpha 2^{r+1} + \sum_{i=1}^{m} \beta_i (2^{r_i} - 1) 2^{t_i} + N(T(R_i) \setminus S \,|\, S) \qquad (4.48)$$

where

$$\alpha = \begin{cases} 1 & \text{if } m \geq 2 \\ 0 & \text{otherwise} \end{cases} \qquad \beta_i = \begin{cases} 1 & \text{if } S \not\subset T(R_i) \text{ or } r_i > 1 \\ 0 & \text{otherwise} \end{cases}$$

Proof: Since any node in R_i is independent of any node in R_j relative to S, $Pr(R \,|\, S)$ is given by

$$Pr(R \,|\, S) = Pr(R_1 \,|\, R_2 \ldots R_m, S) Pr(R_2 \,|\, R_3 \ldots R_m, S) \ldots Pr(R_m \,|\, S),$$

together with independence constraints (when $m \geq 2$),

$$Pr(R_i \,|\, R_{i+1} \ldots R_m, S) = Pr(R_i \,|\, S), \quad i = 1, \ldots, m-1 \qquad (4.49)$$

and the constraints needed to determine each $Pr(R_i | S)$. Since S is fixed, by Lemma 23 the number of constraints needed to enforce (4.49) for each i (when $m \geq 2$) is $(2^{r_i} - 1)(2^{r_{i+1}+\cdots+r_m} - 1) \leq 2^{r_i+\cdots+r_m}$. Thus, when $m \geq 2$, the total number of constraints in (4.49) for $i = 1, \ldots, m - 1$ is bounded above by 2^{r+1}, and when $m = 1$ it is zero: whence the first term of (4.48).

To find the number of constraints needed to determine each $Pr(R_i | S)$, let $R_i = \{Y_1, \ldots, Y_{r_i}\}$, where the Y_j are numbered so that $Y_i \notin T(Y_j)$ when $i < j$ (this is possible because the network is acyclic). Condition only on the immediate predecessors in $T(R_i)$ that are not in S, since nodes in S are already fixed to true or false. Letting $T_i = T(R_i)$,

$$Pr(R_i | S) = \sum_{T_i \setminus S} Pr(R_i | T_i \setminus S, S) Pr(T_i \setminus S | S)$$
$$= \sum_{T_i \setminus S} Pr(Y_1 | Y_2 \ldots Y_{r_i}, T_i \setminus S, S) Pr(Y_2 | Y_3 \ldots Y_{r_i}, T_i \setminus S, S)$$
$$\ldots Pr(Y_{r_i} | T_i \setminus S, S) Pr(T_i \setminus S | S)$$

The jth factor requires independence constraints,

$$Pr(Y_j | Y_{j+1} \ldots Y_{r_i}, T_i \setminus S, S) = Pr(Y_j | T(Y_j))$$

When $r_i = 1$, T_i contains just the immediate predecessors of the one node in R_i, so that $T_i = T(Y_i)$. Thus when $S \not\subset T_i$ no independence conditions are needed, and otherwise at most 2^{t_i} are needed. Suppose, then, that $r_i \geq 2$. In the worst case $T(Y_i)$ and $T_i \cap S$ are empty and, by Lemma 23, $2^{r_i+t_i-j} - 1$ constraints are needed. So summing over $j = 1, \ldots, r_i$, (4.2.3) requires at most $(2^{r_i} - 1)2^{t_i} - r_i$ constraints. Thus, whether r_i is 1 or greater, the number of constraints is at most $(2^{r_i} - 1)2^{t_i}$, and it is zero when $r_i = 1$ and $S \subset T_i$. Finally, one needs at most $N(T(R_i) \setminus S | S)$ independence constraints to determine $Pr(T_i \setminus S | S)$. The lemma follows. □

Lemma 24 can be used to obtain a closed-form upper bound on the number of constraints needed. It is helpful to use the notion of an *ancestral set*. The ancestral sets of X_1 for the network in Figure 4.2 appear in Figure 4.3. One can recursively define an ancestral set of a node x_j with respect to a set S of nodes other than x_j as follows. $\{x_j\}$ is an ancestral set, and if A is an ancestral set, then so are the sets $A_1, \ldots, A_m$ obtained by splitting $T(A) \setminus S$. $A_1, \ldots, A_m$ are the *parent sets* of A. An *extended ancestral set* of x_j with respect to S is $A \cup T(A) \setminus S$ for any ancestral set A of x_j with respect to S. The extended ancestral sets of X_1 for the network of Figure 4.2 are encircled in Figure 4.4. If $|A|$ is the number of elements in set A, the following is a direct result of Lemma 24.

4.2 BAYESIAN LOGIC

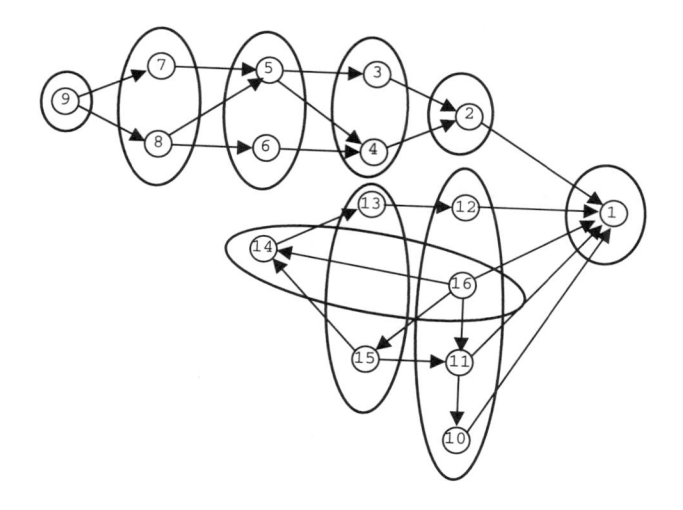

Figure 4.3: Ancestral sets.

Theorem 49 ([2]) *The number $N(x_j \mid S)$ of independence constraints required to determine the probability $Pr(x_j \mid S)$ has the following bound:*

$$N(x_j \mid S) \leq \sum_A \alpha_A 2^{|T(A)\setminus S|+1} + \beta_A(2^{|A|} - 1)2^{|T(A)\setminus S|}$$

where

$$\alpha_A = \begin{cases} 1 & \text{if } m_A \geq 2 \\ 0 & \text{otherwise} \end{cases} \qquad \beta_A = \begin{cases} 1 & \text{if } S \not\subseteq T(A) \text{ or } |A| > 1 \\ 0 & \text{otherwise} \end{cases}$$

and where A ranges over all ancestral sets of X_j relative to S, and A has m_A parent sets.

For instance, to determine $Pr(x_1 \mid x_4)$ in Figure 4.1, note that the ancestral sets of X_1 relative to X_4 are $\{X_1\}$ and its parent sets $\{X_2\}$ and $\{X_3\}$. Thus the number of nonlinear constraints is bounded by $2^{2+1} + 1 \cdot 2^2 + 0 + 0 = 12$ (the actual number is 6). The number of constraints needed to determine $Pr(x_1)$ in Figure 4.2 is bounded by 175 (the actual number is 85).

An immediate corollary of Theorem 49 is that the number of independence constraints grows exponentially with the size of the extended ancestral sets and increases linearly with the number of nodes if this size is bounded.

Corollary 8 ([2]) *Let M bound the number of nodes in any extended ancestral set of X_j relative to S. Then the number of independence constraints*

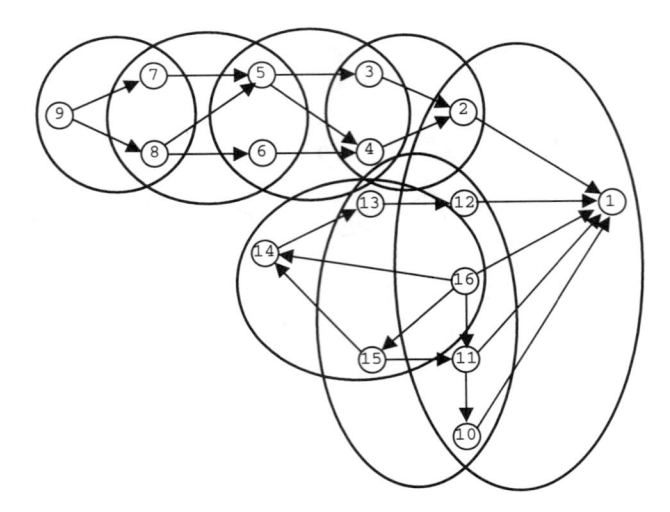

Figure 4.4: Extended ancestral sets.

grows linearly with the number of ancestral sets and therefore with the number of nodes. In particular, if the network contains K ancestral sets, then

$$N(x_j \mid S) \leq K \cdot 2^{M+1}$$

Proof. Since $A \cup T(A) \setminus S$ is an extended ancestral set, $|A| + |T(A) \setminus S|$ is its size, which is at least $|T(A) \setminus S| + 1$. The corollary follows immediately from Theorem 49. □

An important class of Bayesian networks are those that are *singly connected*, since they can be solved relatively easily by a propagation scheme [234]. Singly connected networks are defined as those in which at most one directed path connects any pair of nodes. It is interesting to note that singly connected networks are precisely those in which every (nonempty) ancestral set consists of one node, so that an extended ancestral set consists of a node and its immediate predecessors. It is the maximum in-degree of a node (i.e., the maximum number of immediate predecessors) that determines the difficulty of singly connected networks for Bayesian logic.

Corollary 9 ([2]) *If M is the number of nodes in a singly connected Bayesian network, and D is an upper bound on the in-degree, then*

$$N(X_j \mid S) \leq M \cdot 2^D$$

Proof: The first term of the bound in Theorem 49 vanishes because $m = 1$, and the second term is at most $2^{|T(A) \setminus S|} \leq 2^D$. □

Another important class of graphs for Bayesian networks was defined by Lauritzen and Spiegelhalter [191]. Let the *moral graph* for a Bayesian network be obtained by connecting (i.e., "marrying") all parents of every node with undirected arcs and removing the directions from the original arcs. The graph is *triangulated* by adding arcs so that every cycle of length greater than three contains a chord. (There may be several ways to triangulate the graph.) Lauritzen and Spiegelhalter show that if the size of the largest clique in a triangulation of the moral graph is bounded above by a constant, then the network can be solved in polynomial time. (The complexity of their algorithm is exponential in the size of the largest clique.)

Thus the algorithm for Bayesian logic described here is effective when ancestral sets are small, and the algorithm of Lauritzen and Spiegelhalter is effective when the cliques of the moral graph are small. This raises the question as to which class of networks is larger. It can be shown that if the extended ancestral sets are bounded in size, then the cliques of any moral graph are bounded. On the other hand, it is shown in [2] that the extended ancestral sets can grow without bound even when the cliques of any moral graph are bounded. In this sense the algorithm of Lauritzen and Spiegelhalter is effective on a larger class of networks, but their algorithm solves only classical Bayesian networks.

4.3 Dempster-Shafer Theory

In an effort to develop a more adequate mathematical theory of evidence than Bayesian statistics, G. Shafer [252] extended some statistical ideas of A. P. Dempster [79, 80] to obtain a theory that is closely related to probabilistic logic. The basic ideas of Dempster-Shafer theory are presented here, and a mathematical programming formulation is used to help show how it differs from probabilistic logic. (See [118] for a readable exposition of the theory and how it could be used in expert systems; see also [253].) An integer programming model, in particular a set-covering model, can help solve its inference problems.

Dempster-Shafer theory is designed to combine evidence from possibly conflicting sources, whereas probabilistic logic is not. If probability estimates are inconsistent, the linear programming model is simply infeasible. If evidence is conflicting, however, Dempster's combination rule in Dempster-Shafer theory can nonetheless combine it. In this sense Dempster-Shafer theory is more robust than probabilistic logic, but on the other hand Dempster's combination rule is difficult to justify.

Dempster's rule has two principal features, both of which may be criticized: an assumption that evidence sources are independent in some sense, and a renormalization principle that allows conflicting evidence to be combined. The independence assumption can be dropped without sacrificing the second, more important feature, and this option is explored here. As for the renormalization principle, it can be defended from some of the more serious objections that have been raised against it, but it is hard to support in a positive way.

4.3.1 Basic Ideas of Dempster-Shafer Theory

Dempster-Shafer theory is best introduced with an example, and the following is one of Shafer's own. Sherlock Holmes is investigating a burglary of a shop and determines, by examining the opened safe, that the burglar was left-handed. Holmes also has evidence that the theft was an inside job. One clerk in the store is left-handed, and Holmes must decide with what certainty he can accuse the clerk.

The first step is to identify a *frame of discernment*, which is a set Θ of exhaustive and mutually exclusive possible states of the world. If L means that the thief is left-handed (and $\bar{L}$ that he is not), and if I means that he is an insider (and $\bar{I}$ that he is not), then in this case there are four possible states LI, $L\bar{I}$, $\bar{L}I$, $\bar{L}\bar{I}$. If L and I are interpreted as atomic propositions, then the four states obviously correspond to the four possible truth assignments. Every subset of Θ corresponds to a proposition in the obvious way: $\{LI\}$ to the proposition that the thief was a left-handed insider, $\{LI, L\bar{I}\}$ to the proposition that he was left-handed, and so on. In general, a subset of Θ corresponds to the proposition that is true in precisely the states the subset contains.

The next step is to assign *basic probability numbers* to each subset of Θ, to indicate Holmes's degree of belief in the corresponding proposition. Perhaps the evidence from the safe leads Holmes to assign a probability number of 0.9 to the set $\{LI, L\bar{I}\}$, indicating the "portion of belief" that is "committed to" the hypothesis that the thief is left-handed. Since the basic probabilities numbers must sum to one, Holmes assigns the remaining 0.1 to the entire frame $\Theta = \{LI, L\bar{I}, \bar{L}I, \bar{L}\bar{I}\}$, indicating that this portion of belief is committed to no particular hypothesis. This defines a basic probability function m_1, with $m_1(\{LI, L\bar{I}\}) = 0.9$, $m_1(\Theta) = 0.1$, and $m_1(A) = 0$ for all other subsets A of Θ. Evidence of an inside job is treated similarly to obtain a second basic probability function m_2. In this example, either piece of evidence focuses belief on a single subset of Θ other than Θ itself, but belief can in general be divided among several subsets.

The next task is to combine the left-handed evidence with the insider evidence. For this Shafer uses Dempster's combination rule, which produces a new basic probability function $m = m_1 \oplus m_2$. In this example, the

4.3 DEMPSTER-SHAFER THEORY

Table 4.1: Dempster's Combination Rule

$m_2(\Theta) = 0.2$	$m(\{LI, L\bar{I}\}) = 0.18$	$m(\Theta) = 0.02$
$m_2(\{LI, \bar{L}I\}) = 0.8$	$m(\{LI\}) = 0.72$	$m(\{LI, \bar{L}I\}) = 0.08$
	$m_1(\{LI, L\bar{I}\}) = 0.9$	$m_1(\Theta) = 0.1$

combined basic probability $m(C)$ of a set C is the sum of $m_1(A)m_2(B)$ over all sets A and B whose intersection is C. If m_2 is given by $m_2(\{LI, \bar{L}I\}) = 0.8$ and $m_2(\Theta) = 0.2$, the results are displayed in Table 4.1. For instance, Holmes should commit $(0.9)(0.8) = 0.72$ of his belief to the proposition that the thief is a left-handed insider (i.e., the left-handed clerk). A portion 0.18 of his belief is committed to the more general proposition that the burglar was left-handed, and 0.08 to the general proposition that he was an insider. A small portion 0.02 remains uncommitted.

In this example, every cell of the table corresponds to a distinct intersection, but in general the same intersection could occur several times. If so, one sums the basic probability numbers of all occurrences of the intersection.

Dempster's rule is slightly more complicated when some of the intersections are empty. Suppose, for instance, that Sherlock determines that all the employees are right-handed, so that LI no longer belongs to the frame of discernment. The lower-left square of Table 4.1 now corresponds to the intersection $\{L\bar{I}\} \cap \{\bar{L}I\}$, which is empty. Shafer does not assign it a probability number, since it contains no state of the world and cannot reasonably be the subject of belief. Instead he normalizes the other basic probabilities so that they sum to one, as in Table 4.2. Thus Dempster's combination rule reconciles the 0.9 evidence for a left-handed insider with the 0.8 evidence for a right-handed outsider by reducing the former to 0.64 and the latter to 0.29.

In general, if we set

$$\tilde{m}(C) = \sum_{\substack{A,B \subset \Theta \\ A \cap B = C}} m_1(A)m_2(B) \qquad (4.50)$$

Table 4.2: Normalization in Dempster's Combination Rule

$m_2(\Theta) = 0.2$	$m(\{L\bar{I}\}) = 0.64$	$m(\Theta) = 0.07$
$m_2(\{\bar{L}I\}) = 0.8$	$m(\emptyset) = 0$	$m(\{\bar{L}I\}) = 0.29$

$$m_1(\{L\bar{I}\}) = 0.9 \quad m_1(\Theta) = 0.1$$

then when $C \neq \emptyset$,

$$m(C) = m_1 \oplus m_2(C) = \frac{\tilde{m}(C)}{\sum_{\substack{D \subseteq \Theta \\ D \neq \emptyset}} \tilde{m}(D)} = \frac{\tilde{m}(C)}{1 - \tilde{m}(\emptyset)} \quad (4.51)$$

and $m(\emptyset) = 0$. If there are three basic probability functions, then $m = (m_1 \oplus m_2) \oplus m_3 = m_1 \oplus (m_2 \oplus m_3)$, and similarly for larger numbers of functions.

Now that the composite basic probability function m has been computed, it remains to determine how much credence should be placed in each subset of Θ. To do this Shafer defines a *belief function Bel*, where $Bel(A)$ is the total credence that should be given a subset A of Θ. He takes $Bel(A)$ to be the sum of the basic probability numbers of all subsets of A:

$$Bel(A) = \sum_{B \subset A} m(B) \quad (4.52)$$

Thus the belief allotted to a proposition is the sum of the basic probability numbers of all the propositions that entail it. In the original example (Table 4.1), the belief Holmes should allocate to the guilt of the clerk is $Bel(\{LI\}) = 0.72$, since the only subset of $\{LI\}$ with a positive basic probability is $\{LI\}$ itself.

Not only does the basic probability function m uniquely determine the belief function Bel, but the reverse is true as well. The inclusion-exclusion principle of combinatorics yields

$$m(A) = \sum_{B \subset A} (-1)^{|A \setminus B|} Bel(B) \quad (4.53)$$

Dempster-Shafer theory differs from probabilistic logic in three fundamental ways. First, it has a different convention for distributing probability

4.3 DEMPSTER-SHAFER THEORY

mass over sets of states. Probability mass assigned to a set of states *contributes to* evidence that one of the states is actual, whereas in probabilistic logic it *fixes* the probability that one of the states is actual. In the example, a mass of $m(\{LI, \bar{L}I\}) = 0.08$ was assigned to the set of states in which the thief is an insider. This means only that the evidence that one of these states is actual is at least 0.08. There may be additional evidence, and in fact there is evidence 0.72 that the thief is a left-handed insider ($\{LI\}$). So the total evidence for $\{LI, \bar{L}I\}$ is at least $0.08 + 0.72 = 0.8$ and possibly as high as 1. In probabilistic logic, however, assigning probability 0.08 to $\{LI, \bar{L}I\}$ constrains the probability that one of these states is actual to be exactly 0.08. The next section will make this point more formally.

Second, in probabilistic logic the probabilities of contradictory propositions sum to unity: $Pr(F) + Pr(\neg F) = 1$. This is because $Pr(F)$ is just the probability mass of the set S of states in which F is true, $Pr(\neg F)$ is the mass of the complement $\bar{S}$, and the total mass of one is split between S and $\bar{S}$. But the same is not true of belief values, since possibly $Bel(S) + Bel(\bar{S}) < 1$, reflecting the fact that some of the evidence may be neither for nor against F. Probability mass is still split between S and $\bar{S}$, but $Bel(S)$ reflects only the evidence that *necessarily* supports F because it is assigned to subsets of S. It does not include any probability mass assigned to sets that overlap S. For example, $Bel(\{LI\}) + Bel(\{L\bar{I}, \bar{L}I, \bar{L}\bar{I}\}) = 0.72 + 0 = 0.72$. The remaining mass 0.28 is forced to be neither in $\{LI\}$ nor in its complement.

Third, Dempster's combination rule allows one to combine evidence from conflicting sources, whereas there is no such provision in probabilistic logic. This is discussed further in Section 4.3.5.

4.3.2 A Linear Model of Belief Functions

Linear programming models can formally distinguish probabilistic logic from the idea of a Dempster-Shafer belief function.

Consider again the model for probabilistic logic. Suppose probabilities $\pi_1, \ldots, \pi_k$ are assigned to propositions $F_1, \ldots, F_k$, corresponding to sets $S_1, \ldots, S_k$ of worlds. The goal is to find the probability range of F_0, corresponding to set S_0 of worlds. As in Section 4.1.1, one solves the linear models

$$\begin{aligned} \min / \max \quad & c^T p \\ \text{s.t.} \quad & a^i p = \pi_i, \quad i = 1, \ldots, k \\ & e^T p = 1, \quad p \geq 0 \end{aligned} \qquad (4.54)$$

where

$$a^i_j = \begin{cases} 1 & \text{if } F_i \text{ is true in world } j \\ 0 & \text{otherwise} \end{cases}$$

and similarly for c. Thus the probability mass $a^i p$ of the worlds in which F_i is true must be exactly π_i.

In Dempster-Shafer theory the vector p of probabilities is replaced by a series of vectors p^i, one for each basic probability number $m(S_i)$ assigned. $Bel(S_0)$ and $Bel(\bar{S}_0)$ are equal, respectively, to the minimum and maximum below.

$$\min/\max \quad \sum_{i=1}^{k} c^T p^i$$
$$\text{s.t.} \quad a^i p^i = m(S_i), \quad i = 1, \ldots, k \qquad (4.55)$$
$$\sum_{i=1}^{k} e^T p^i = 1, \quad p^i \geq 0, \ i = 1, \ldots, k$$

Thus the probability mass $\Sigma_{j=1}^{k} a^i p^j$ of states in which F_i is true is *at least* $m(S_i) = a^i p^i$.

One would not actually use the model (4.55) to compute $Bel(S_0)$ and $Bel(\bar{S}_0)$, but it helps to clarify the difference between $Pr(S_0)$ and $Bel(S_0)$. Recall, however, that Dempster-Shafer theory also provides for the combination of basic probability functions before they are used to compute $Bel(S_0)$.

Theorem 50 *The minimum value of (4.55) is $Bel(S_0)$, and the maximum value is $Bel(\bar{S}_0)$.*

Proof: It will be shown that the minimum value is $Bel(S_0)$; the other half of the theorem is similarly proved. For a vector a let $J(a) = \{j \mid a_j = 1\}$, and recall that $Bel(S_0) = \Sigma_{S \subset S_0} m(S)$. Aside from the normalization constraint $\Sigma_i e^T p = 1$, (4.55) is separable into subproblems of the form $\min c^T p^i$ s.t. $a^i p^i = m(S_i)$. If $J(a^i) \subset J(c)$, the minimum value of the subproblem is clearly $m(S_i)$, where p_j^i is set to zero whenever $j \notin J(a^i)$. If $J(a^i) \not\subset J(c)$, the minimum value of the subproblem is zero, which we obtain by setting $p_j^i = m(S_i)$ for some $j \in J(a^i) \setminus J(c)$ and $p_{j'}^i = 0$ for all other j'. Clearly, if every subproblem is solved in this manner, then $\Sigma_{i=1}^{k} e^T p^i = 1$ (since $\Sigma_{i=1}^{k} m(S_i) = 1$), and an optimal solution of (4.55) results. Since $J(a^i) = S_i$, the solution has the desired value $Bel(S_0)$. □

4.3.3 A Set-Covering Model for Dempster's Combination Rule

The difficulty of combining basic probability functions with formula (4.50) increases exponentially with the number of functions. To see this, take the simplest case in which the basic probability functions $m_1, \ldots, m_k$ are what

4.3 DEMPSTER-SHAFER THEORY

Shafer calls *simple support functions*; that is, each m_i assigns a positive value to only one set S_i other than the entire frame Θ. To compute $\tilde{m}(C)$ using formula (4.50), enumerate all intersections of the S_i that are equal to C. Equivalently, if $T_j = S_j \setminus C$, enumerate all intersections of the T_j that are empty. If $S_1, \ldots, S_{k'}$ contain C and $S_{k'+1}, \ldots, S_k$ do not, the most straightforward way to do this is to check all $2^{k'}$ subsets of $\{T_1, \ldots, T_{k'}\}$ to find those whose intersection is empty.

There are various ways to make the enumeration more efficient (albeit still exponential in the worst case). Barnett [15] describes a method that applies when each simple support function assigns probability to a singleton. A method will be presented that is valid for all simple support functions that is based on a set-covering model. Dugat and Sandri describe a related approach in [91].

Let us say that an intersection of the sets in a subset S of $\{T_1, \ldots, T_{k'}\}$ is *minimally empty* if the intersection of the sets in no proper subset of S is empty. It seems intuitively reasonable that one need only enumerate minimally empty intersections when computing $\tilde{m}(C)$ with formula (4.50). This is true, but some care must be taken in doing the computation.

A *set-covering problem* has the form

$$Ax \geq e \\ x_j \in \{0, 1\}, \quad j = 1, \ldots, N \tag{4.56}$$

where A is a 0-1 matrix and e is a vector of 1's. The columns of A correspond to sets and the rows to elements the sets collectively contain. Set

$$a_{ij} = \begin{cases} 1 & \text{if set } j \text{ contains element } i, \\ 0 & \text{otherwise.} \end{cases}$$

Thus a vector x solves (4.56) if the union of the sets j for which $x_j = 1$ contains all the elements. Such an x is called a *cover*. x is a *prime cover* if it properly contains no cover; that is, if no cover y satisfies $y \leq x$ with $y_j \neq x_j$ for some j.

Now associate the sets $\bar{T}_1, \ldots, \bar{T}_{k'}$ with the columns of A and the elements in their union with the rows of A, where $\bar{T}$ is the complement of T. It is clear that if x is a cover, then the intersection of the sets T_j such that $x_j = 1$ is empty. Furthermore, if x is a prime cover, then the intersection is minimally empty. It therefore suffices to generate all the prime covers for (4.56) and to use them in an appropriate calculation to obtain $\tilde{m}(C)$. It will first be shown how to generate the prime covers, and then how to do the calculation.

An initial cover can be obtained simply by finding a feasible solution x^1 of (4.56), such as $x^1 = (1, 1, \ldots, 1)$. x^1 is reduced to a prime cover

y^1 by removing sets from the cover, one by one, until no further sets can be removed without producing a noncover. Suppose, then, that distinct prime covers $y^1, \ldots, y^t$ have been generated. To obtain a $(t+1)$st distinct prime cover, add the following constraints to (4.56) and use an integer programming algorithm to find a feasible solution x^{t+1} of the resulting system:

$$\sum_{\substack{j \\ y_j^\tau = 1}} x_j \leq e^T y^\tau, \quad \tau = 1, \ldots, t \tag{4.57}$$

where e is a vector of ones. Note that each constraint in (4.57) excludes any cover that contains a cover already enumerated. Next, reduce x^{t+1} to a prime cover y^{t+1}, which clearly must be distinct from the prime covers already generated. The process continues until there is no feasible solution of (4.56) with the additional constraints (4.57).

Now that the prime covers are generated, $\tilde{m}(C)$ can be calculated. Suppose that sets $S_1, \ldots, S_5$ contain C and that the remaining sets S_6 and S_7 do not. Thus the formula for $\tilde{m}(C)$ is

$$\sum_{\substack{U_1 \in \{S_1, \Theta\} \\ \vdots \\ U_5 \in \{S_5, \Theta\}}} m_1(U_1) m_2(U_2) m_3(U_3) m_4(U_4) m_5(U_5) m_6(\Theta) m_7(\Theta) \tag{4.58}$$

Let us also suppose that there are four prime covers,

$$\begin{array}{c} \{\bar{T}_1, \bar{T}_2\} \\ \{\bar{T}_2, \bar{T}_3\} \\ \{\bar{T}_4\} \\ \{\bar{T}_1, \bar{T}_3\} \end{array} \tag{4.59}$$

Consider first the terms of (4.58) that correspond to the prime cover $\{\bar{T}_1, \bar{T}_2\}$ (i.e., the terms containing both $m_1(S_1)$ and $m_2(S_2)$):

$$\sum_{\substack{U_3 \in \{S_3, \Theta\} \\ U_4 \in \{S_4, \Theta\} \\ U_5 \in \{S_5, \Theta\}}} m_1(S_1) m_2(S_2) m_3(U_3) m_4(U_4) m_5(U_5) m_6(\Theta) m_7(\Theta) \tag{4.60}$$

Since each m_i is a simple support function, $m_i(S_i) + m_i(\Theta) = 1$ for each i, and (4.60) can be simplified to

$$m_1(S_1) m_2(S_2) m_6(\Theta) m_7(\Theta) \tag{4.61}$$

4.3 DEMPSTER-SHAFER THEORY

Consider next the terms of (4.59) that correspond to the prime cover $\{\bar{T}_2, \bar{T}_3\}$.

$$\sum_{\substack{U_1 \in \{S_1, \Theta\} \\ U_4 \in \{S_4, \Theta\} \\ U_5 \in \{S_5, \Theta\}}} m_1(U_1)m_2(S_2)m_3(S_3)m_4(U_4)m_5(U_5)m_6(\Theta)m_7(\Theta)$$

Some of these terms, namely, those containing $m_1(S_1)$, have already been accounted for in (4.61). The sum of the remaining terms simplifies to

$$m_1(\Theta)m_2(S_2)m_3(S_3)m_6(\Theta)m_7(\Theta)$$

Before considering the other two prime covers, it is useful to display schematically what has been done so far. In (4.62)–(4.67) below an integer j represents the quantity $m_j(S_j)$, and $\bar{j}$ represents $m_j(\Theta)$. Also, $\alpha = m_6(\Theta)m_7(\Theta)$. The terms corresponding to $\{\bar{T}_1, \bar{T}_2\}$ sum to the quantity represented in (4.62).

$$12\alpha \tag{4.62}$$
$$23\alpha \;\to\; \bar{1}23\alpha \tag{4.63}$$
$$4\alpha \;\to\; 14\alpha \;\to\; 1\bar{2}4\alpha \tag{4.64}$$
$$\searrow \bar{1}4\alpha \;\to\; \bar{1}24\alpha \;\to\; \bar{1}2\bar{3}4\alpha \tag{4.65}$$
$$\searrow \bar{1}\bar{2}4\alpha \tag{4.66}$$
$$13\alpha \;\to\; 1\bar{2}3\alpha \;\to\; 1\bar{2}3\bar{4}\alpha \tag{4.67}$$

The terms corresponding to $\{\bar{T}_2, \bar{T}_3\}$ sum to the quantity 23α in (4.63), but to remove those containing 1 (i.e., those containing $m_1(S_1)$) include only those that sum to $\bar{1}23\alpha$. Draw a directed arc $\to$ to $\bar{1}23\alpha$ to indicate that $\bar{1}23\alpha$ is replacing 23α.

Consider now the third prime cover $\{\bar{T}_4\}$. The corresponding terms sum to 4α in (4.64). One must identify the terms in this sum that have already been canvassed in either (4.62) or (4.63). First eliminate those covered by (4.62). Then "branch" on 1 to obtain 14α and $\bar{1}4\alpha$. We see that 14α still involves terms redundant of (4.62), namely, those containing 2, and eliminate these by replacing 14α with $1\bar{2}4\alpha$. But $\bar{1}4\alpha$ contains no terms redundant of (4.62), because all of its terms contain $\bar{1}$, whereas all the terms of (4.62) contain 1. Thus $\bar{1}4\alpha$ needs no further qualification to avoid redundancy with (4.62).

The algorithm has so far produced a tree of four nodes in (4.64), whose two leaves are the quantities $1\bar{2}4\alpha$ and $\bar{1}4\alpha$ that involve no redundancies with (4.62). Each leaf must now be checked for redundancies with (4.63). There are none in $1\bar{2}4\alpha$, but there are in $\bar{1}4\alpha$. Branch on 2 to create nodes $\bar{1}2 4\alpha$ and $\bar{1}\bar{2}4\alpha$. The latter has no further redundancies, but the former does, because it and $\bar{1}23\alpha$ both cover terms containing $\bar{1}234$. Thus $\bar{1}2\bar{3}4\alpha$ replaces $\bar{1}24\alpha$.

The fourth prime cover is treated similarly in (4.67). The sum of all six leaves in the four trees (4.62)–(4.67) is $\tilde{m}(C)$.

The algorithm can now be stated in general. It builds trees as in the example, and it associates with each node the set of integers that represent the terms in the corresponding product. Let $\alpha = \Pi^k_{j=k'+1} m_j(\Theta)$.

Procedure 25: *Set-Covering Algorithm for Dempster's Rule*

Set $\tilde{m}(C) = 0$, $t = 0$.
While the combined constraint set (4.56) and (4.57) is soluble:
    Increase t by 1, and let x^t solve (4.56) and (4.57).
    Reduce x^t to a prime cover y^t.
    Start building tree t with its root, $\{j \mid y^t_j = 1\}$.
    For each tree $\tau = 1, \ldots, t - 1$:
        For each leaf L in tree τ:
            For each leaf M of tree t call **Branch**(M).
    Increase $\tilde{m}(C)$ by $\alpha \Sigma_M \Pi_{j \in M} m_j(S_j) \Pi_{\bar{j} \in M} m_j(\Theta)$,
    where M ranges over all leaves in tree t.

Procedure **Branch**(M)
    If $M \cup L$ contains both j and $\bar{j}$ for no j, then
        Pick any $j \in L \setminus M$.
        Add arc $(M, M \cup \{\bar{j}\})$ to tree t.
        Call **Branch**($M \cup \{\bar{j}\}$).
        If $L \setminus M \neq \{j\}$ then
            Add arc $(M, M \cup \{j\})$ to tree t.
            Call **Branch**($M \cup \{j\}$).

4.3.4 Incomplete Belief Functions

Rather than beginning with basic probability numbers that must be combined somehow, one can begin by specifying right away some belief values $Bel(S)$ that reflect the total evidence bearing on several sets S. In other words, one can begin by defining an *incomplete* belief function and then compute what range of belief function values can consistently be attributed to an inferred proposition. Dubois and Prade [88] suggested this approach,

4.3 DEMPSTER-SHAFER THEORY

which admits a linear programming model, although they used an objective function that minimizes the "specificity" of the solution. It has also been discussed by McLeish [211].

The Sherlock Holmes example illustrates the idea. Recall that Sherlock assigned basic probability $m_1(\{LI, L\bar{I}\}) = m_1(S_1) = 0.9$ to the proposition that the thief was left-handed and $m_2(\{LI, \bar{L}I\}) = m_2(S_2) = 0.8$ to the proposition that the thief was an insider. Treat these assignments as belief function values, so that $Bel(S_1) = 0.9$ and $Bel(S_2) = 0.8$. If $Bel(A)$ were fixed for all other subsets A of Θ (e.g., to zero), the underlying basic probability function m would be uniquely determined. But suppose no other belief values are specified, so that many m's are consistent with our information. The goal is to infer a range for $Bel(\{LI\}) = Bel(S_0)$; that is, a range of belief values that can consistently be assigned to the proposition that the thief was a left-handed insider.

Let p_j, $j = 0, 1, 2$, be the probability mass that is distributed to S_j, and let p_3 be the mass distributed to the universe Θ. The problem has the following linear model:

$$\begin{aligned} &\min / \max \quad p_0 \\ &\text{s.t.} \quad p_0 + p_1 = 0.9 \\ &\phantom{\text{s.t.}} \quad p_0 + p_2 = 0.8 \\ &\phantom{\text{s.t.}} \quad p_0 + p_1 + p_2 + p_3 = 1 \\ &\phantom{\text{s.t.}} \quad p_j \geq 0, \ j = 0, \ldots, 3 \end{aligned} \quad (4.68)$$

The objective function contains only p_0 because S_0 contains no other sets S_j. The first constraint distributes belief value 0.9 to all the subsets of S_1, and similarly for the second constraint. The third constraint distributes the belief value of the universe, one, over all the S_j. The resulting range for $p_0 = Bel(S_0)$ is 0.7 to 0.8.

In this example probabilistic logic and the belief function model yield the same result. That is, if probabilities 0.9 and 0.8 are assigned to the propositions represented by S_1 and S_2, the probability range for the proposition represented by S_0 is again 0.7 to 0.8. But this is not true in general. Suppose, for example, that $S_1 = \{LI, L\bar{I}, \bar{L}I\}$ is the proposition that the thief is not a right-handed outsider, and $S_2 = \{LI, L\bar{I}, \bar{L}\bar{I}\}$ the proposition that the thief is not a right-handed insider. The belief function model is again (4.68), and so the range for $Bel(S_0)$ is again 0.7 to 0.8. But the range for S_0 in probabilistic logic is 0 to 0.8. This is because the state $L\bar{I}$, which lies in $S_1 \cap S_2$ but not in S_0, can absorb probability 0.7 and allow the probability of the state LI in S_0 to go to zero. In Dempster-Shafer theory, however, probability is attached to sets rather than the states in them, and there is no set $\{LI\}$ to absorb probability.

In general sets $S_1, \ldots, S_m$ are assigned belief values $Bel(S_1), \ldots, Bel(S_m)$. To derive a range for $Bel(S_0)$, solve

$$\min/\max \sum_{S_j \subset S_0} p_j$$
$$\text{s.t.} \quad \sum_{S_j \subset S_i} p_j = Bel(S_i), \quad i = 1, \ldots, m$$
$$p_j \geq 0, \quad j = 0, \ldots, m$$

If desired, ranges rather than point values can be given for $Bel(S_i)$, and the above equations replaced with inequalities.

Note that adding a set S_{m+1} to the list $S_0, \ldots, S_m$ can change the solution, even if no belief value is given for S_{m+1} (i.e., the range for $Bel(S_{m+1})$ is 0 to 1). In the last example above, for instance, merely adding the set $S_4 = \{L\bar{I}\}$ to the problem changes the solution from $[0.7, 0.8]$ to $[0, 0.8]$. Also, if the S_j include all possible subsets of the universe Θ, the solution is identical with that obtained by probabilistic logic.

4.3.5 Dempster-Shafer Theory vs. Probabilistic Logic

An obvious attraction of Dempster-Shafer theory is its ability to *accumulate* evidence. Whereas probabilistic logic can do nothing with inconsistent probability judgments except declare them inconsistent, Dempster-Shafer theory can reconcile incompatible judgments. If one says in probabilistic logic that both A and not-A have probability of at least 0.8, the linear programming model has no solution. But in Dempster-Shafer theory one can quite legitimately define two basic probability functions, one of which assigns 0.8 to A and one to not-A. Dempster's combination rule deals with this simply by renormalizing, so that $4/9$ is allotted to each of A and not-A. Thus inconsistency is "normalized away."

Probabilistic logic's inability to merge inconsistent probability judgments is often cited as a weakness. It would seem at first only reasonable to prohibit assigning a belief measure of 0.8 to both A and not-A. But this is only if one regards the belief measure as already summing up what is known about A. It certainly makes no sense, after assessing all the evidence, to be pretty sure of both A and not-A. But it makes perfect sense to accumulate a good deal of evidence both for and against A. This criticism of probabilistic logic, then, seems really to be a demand for a logic that allows accumulation of evidence.

4.3 DEMPSTER-SHAFER THEORY

Dempster-Shafer theory, unlike probabilistic logic, is designed for situations in which belief measures are assigned to propositions *before* disparate evidence sources are combined. It therefore needs a device like Dempster's rule to combine them mathematically.

The difficulty is that no combination rule seems to be as self-evident as probabilistic logic. Given a set of probability assignments, the output of the linear programming model for probabilistic logic is practically unassailable. But not so with Dempster's combination rule. In particular, one can question its normalization device for removing inconsistency and its strong independence assumption.

Shafer's defense for "normalizing away inconsistency" is that it yields reasonable results in several examples [252]. It apparently has no theoretical grounding.

To see how Dempster's rule assumes independence, suppose one combines two simple support functions m_1 and m_2 that assign weight to nondisjoint sets S_1 and S_2, respectively, neither of which contains the other. Then the combined function m satisfies $m(S_1 \cap S_2) = m_1(S_1)m_2(S_2)$. Thus $m(S_1 \cap S_2)$ is computed as though it were the joint probability of two independent propositions.

This can lead to anomalous results if the propositions are not really independent. Suppose in the example of previous sections that Sherlock Holmes finds no left-handed employees but cannot rule out the possibility entirely. Since $\{LI\}$ remains in the frame of discernment, Dempster's rule still forces him to assign credence $m_1(S_1)m_2(S_2) = 0.72$ to the conclusion that the thief is a left-handed employee. The example is more striking if $m_1(S_1) = 0.9$ is evidence for a female thief and $m_2(S_2) = 0.8$ is evidence for a bearded thief.

There are at least three ways to deal with this problem. One is to work within the framework of Dempster's rule by encoding prior information about the joint probability distribution of F_1 and F_2 as basic probability functions. One might, for instance, encode strong *a priori* evidence against the existence of a left-handed employee by stipulating $m_3(S_1 \bar{\cap} S_2) = 1 - \epsilon$. After computing $m = m_1 \oplus m_2 \oplus m_3$, the basic probability of a left-handed inside thief is found to be the suitably small number $m(S_1 \cap S_2) = 0.72\epsilon/(0.28 + 0.72\epsilon)$.

This device also answers an objection raised by Wise and Henrion [298]. They in essence point out that if there is the slightest theoretical possibility that F_1 and F_2 are both true, $S_1 \cap S_2$ is nonempty and therefore receives a possibly large basic probability $m(S_1)m(S_2)$ (such as 0.72 in the example), whereas if F_1 and F_2 are truly incompatible, the intersection is empty and receives no basic probability. Thus there is an unnatural discontinuity between assuming an conjunction is remotely possible and assuming it is impossible. The additional probability function m_3 relieves this problem, however, since $0.72\epsilon/(0.28+0.72\epsilon)$ varies smoothly from 0 to 0.72 as ϵ varies from 0 to 1.

Thus one can get Dempster's rule to yield reasonable numbers by adjusting the values of additional probability functions that characterize the joint distributions of dependent propositions. But the question remains as to why the additional probabilities should combine with m_1 and m_2 as though they were independent.

A second possibility is to use the variation of Dempster-Shafer theory discussed in Section 4.3.4. It dispenses with Dempster's rule by partially defining a belief function Bel. This, however, sacrifices the ability to combine conflicting evidence. In assigning a value to $Bel(S_i)$, one indicates the strength of belief in S_i, after all evidence is considered. The resulting linear model only determines what degree of belief in one proposition is compatible with the belief values assigned other propositions.

A third approach modifies Dempster's combination rule so as to drop the independence assumption. It is discussed in the next section.

4.3.6 A Modification of Dempster's Combination Rule

In Dempster's combination rule, the key mathematical device for reconciling contradictory evidence is normalization. The rule can be modified so as to retain this device while dropping the independence assumption. In the resulting model, probability ranges are calculated via linear programming.

Recall the example of preceding sections, in which $m_1(S_1) = 0.9$ reflects the evidence that the thief is left-handed, and $m_2(S_2) = 0.8$ reflects the evidence that the thief is an insider. Thus each evidence source i assigns positive weight to two sets, S_i and Θ. Using the notation $S_3 = \Theta$, let p_{12}, p_{13}, p_{23}, and p_{33} be the respective probability masses of their intersections $S_1 \cap S_2$, $S_1 \cap S_3 = S_1$, $S_2 \cap S_3 = S_2$, and $S_3 \cap S_3 = \Theta$ (Table 4.3).

The task of a combination rule is to distribute the mass $m_1(S_1)$ over p_{12} and p_{13}, to distribute $m_1(S_3)$ over p_{23} and p_{33}, and similarly for $m_2(S_2)$ and $m_2(S_3)$. Dempster's combination rule does this by imposing a strict independence constraint that says $p_{ij} = m_1(S_i)m_2(S_j)$. But there are weaker and more plausible constraints. The weakest merely requires conservation of probability mass:

$$\begin{aligned} p_{12} + p_{13} &= m_1(S_1) \\ p_{23} + p_{33} &= m_1(S_3) \\ p_{12} + p_{23} &= m_2(S_2) \\ p_{13} + p_{33} &= m_2(S_3) \end{aligned} \quad (4.69)$$

where each $p_{ij} \geq 0$. (Any one of the constraints is redundant.)

4.3 DEMPSTER-SHAFER THEORY

Table 4.3: Distributing Basic Probabilities

	S_1	Θ
$m_2(S_3)$	p_{13}	p_{33}
$m_2(S_2)$	$S_1 \cap S_2$ p_{12}	S_2 p_{23}

$\qquad\qquad\qquad m_1(S_1) \quad m_1(S_3)$

A range of values can now be obtained for the evidence $Bel(S_1 \cap S_2)$ that the thief is a left-handed insider by maximizing and minimizing p_{12} subject to (4.69). This yields a range $[0.7, 0.8]$, which contains the value 0.72 given by Dempster's combination rule. The evidence $Bel(S_1) = p_{12} + p_{13}$ that the thief is left-handed is fixed at $m(S_1) = 0.9$ by the first constraint of (4.69).

If it is stipulated that no employees are left-handed, p_{12} corresponds to the empty set

$$S_1 \cap S_2 = \{L\bar{I}\} \cap \{\bar{L}I\}$$

$Bel(S_1)$ is no longer fixed because it is the fraction $p_{13}/(1 - p_{12})$, which can be maximized and minimized subject to (4.69) to obtain bounds. This poses a fractional linear programming problem, which is easily converted to a linear programming problem [52]. The resulting interval for a left-handed thief is $[0.5, 0.67]$, which contains the mass 0.64 given by Dempster's rule. The interval for an outsider is $[0, 0.33]$, which contains Dempster's 0.29.

In fact it is instructive to examine the general case in which two evidence sources are at odds over a single proposition. Suppose that a set S is assigned a high probability $1 - \delta$ and its complement $\bar{S}$ a high probability $1 - \epsilon$. If $\epsilon \geq \delta$, the resulting range is $[1 - \delta/\epsilon, \epsilon/(\delta + \epsilon)]$ for $Bel(S)$ and $[0, \delta/(\delta+\epsilon)]$ for $Bel(\bar{S})$ when $0 < \delta+\epsilon \leq 1$. Thus it is consistent to discount completely a source that is slightly less certain than the other. One must place total credence in a source that is certain if the competing source is almost certain. If both sources are equally but not completely certain, one can place at most 0.5 credence in either, and as little as zero.

In general, one wishes to combine basic probability functions $m_1, \ldots, m_k$. Let us say that each m_i assigns positive weight to a family of sets $\{S_j \mid j \in J_i\}$. Let $p_{j_1,\ldots,j_k}$ be the probability mass associated with $S_{j_1} \cap$

$\ldots \cap S_{j_k}$. Then the constraint set (4.69) becomes

$$\sum_{t \in T_{ij}} p_t = m_i(S_j) \text{ all } j \in J_i, \ i = 1, \ldots, k, \qquad (4.70)$$
$$p_t \geq 0, \text{ all } t$$

where T_{ij} is the Cartesian product $J_1 \times \cdots \times J_{i-1} \times \{j\} \times J_{i+1} \times \cdots \times J_k$. These constraints always have a feasible solution, because the probabilities given by Dempster's combination rule satisfy them.

To obtain bounds on $Bel(S)$, solve the fractional linear programming problems

$$\min / \max \frac{\sum_{S_{t_1} \cap \ldots \cap S_{t_k} \subset S} p_t}{1 - \sum_{S_{t_1} \cap \ldots \cap S_{t_k} = \emptyset} p_t} \qquad (4.71)$$

s.t. (4.70)

For instance, one can encode prior likelihood that all employees are right-handed by specifying, say, $m_3(S_4) = m_3(S_1 \bar{\cap} S_2) = 1 - \epsilon$. Since $S_1 \cap S_2 \cap S_4 = \emptyset$, maximize and minimize the objective function $Bel(S_3) = p_{123}/(1-p_{124})$, which yields a reasonable range $[0, \epsilon/(0.8-\epsilon)]$ of probabilities that the thief is a left-handed employee.

4.4 Confidence Factors in Rule Systems

Dealing with uncertainty can be a bit more *ad hoc* when inference consists entirely of applying rules. One need only assign a *confidence factor* to each antecedent of a rule and devise some reasonable formula, that is, some *confidence function*, that calculates a confidence factor for the consequent [77, 78, 134].

Confidence factors are typically not probabilities but indicate the state of one's evidence or belief. The most popular confidence function simply takes some fraction of the minimum of the confidence factors of the antecedents. In other words, one can have no more confidence in the consequent (on the basis of a particular rule) than in the most questionable antecedent. If the rule itself is uncertain, even this confidence must be reduced by multiplying it by some number less than one.

The same minimum function (generally without the multiplier) is used in fuzzy control, which appeared at about the same time as expert systems with confidence factors. Fuzzy control differs, however, in that it deals with *vagueness* rather than uncertainty. The numbers attached to antecedents

4.4 CONFIDENCE FACTORS IN RULE SYSTEMS

indicate the degree to which the vague terms in the antecedents describe the actual state of affairs. Consider, for example, the following rule:

> If the pressure is high and falling slowly, then reduce the heat slightly.

In fuzzy control the terms "high," "slowly" and "slightly" are vague, and numbers (defined by "membership functions") are used to indicate the extent to which a given pressure level is "high" and so on. However, even if these terms are precisely defined, perhaps by specifying intervals, one may be uncertain as to what the pressure in fact is or just how slowly it is falling. In this case one might use confidence factors to indicate the degree of certainty.

In expert systems, the consequent of rule A may occur as an antecedent of rule B. So the confidence attached to the antecedent of B is that obtained by applying ("firing") rule A. If several rules having this same consequent are fired, the confidence assigned to the consequent is presumably the largest confidence obtained from any of the rules.

Suppose, however, that one fires a rule B whose consequent is an antecedent of a rule A that has already been fired. Then it appears that one has to fire rule A again to take account of the new confidence level of its antecedent. This necessitates firing rule B again, and so forth in a cycle. One may ask under what conditions this process will converge, and if it does converge, under what conditions it will converge finitely.

Rather than debate the merits of various confidence functions, the task undertaken here is to study how this convergence behavior depends on properties of the confidence functions. A network model of a rule system is used to show that under certain conditions, the number of iterations required for convergence is finite and equal to the length of certain paths in the associated graph. Even when convergence requires infinitely many iterations, Jeroslow [167] showed that the confidence factors can be finitely computed by solving a mixed integer programming model, provided the confidence functions are representable in such a model. Jeroslow [165], building on joint work with J. K. Lowe [169, 200], also proved a general theorem that characterizes functions that are so representable. It is worthwhile to digress somewhat to state and prove this theorem, because it is quite interesting in its own right and is not as well known as it should be. The theorem will then be applied to the computation of confidence factors.

4.4.1 Confidence Factors

A *rule* is an if-then statement of the form

$$x_1 \wedge \cdots \wedge x_m \to y_1 \vee \cdots \vee y_n \qquad (4.72)$$

If there are no antecedents ($m = 0$), the rule simply asserts the consequent. In this case it will be convenient to write t as the antecedent, where t represents a proposition that is necessarily true. Thus both $\to y$ and $t \to y$ assert that y is true. To "fire" a rule is to apply the following inference schema:

$$\begin{array}{c} x_1 \wedge \cdots \wedge x_m \to y_1 \vee \cdots \vee y_n \\ t \to x_1 \\ \vdots \\ t \to x_m \\ \hline t \to y_1 \vee \cdots \vee y_n \end{array} \qquad (4.73)$$

Because firing rules is a complete refutation method only for Horn systems, the discussion will be restricted to the Horn case.

The following small rule base illustrates the computation of confidence factors.

$$\begin{array}{rcl} t & \xrightarrow{1} & x_1 \\ t & \xrightarrow{1} & x_2 \\ t & \xrightarrow{1/3} & x_3 \\ x_1 \wedge x_4 & \xrightarrow{1} & x_3 \\ x_3 & \xrightarrow{1/3} & x_6 \\ x_4 \wedge x_5 & \xrightarrow{2/3} & x_6 \\ x_1 \wedge x_2 & \xrightarrow{2/3} & x_4 \\ x_2 \wedge x_3 & \xrightarrow{2/3} & x_5 \end{array} \qquad (4.74)$$

Each atomic proposition x_j is associated with a confidence factor v_j. Initially it is supposed that each $v_j = 0$ but that t has a confidence factor of 1. For convenience, write $t = x_0$ and $v_0 = 1$. The rth rule having consequent x_k is referred to as rule k, r. It can be written

$$\bigwedge_{j \in J_{kr}} x_j \to x_k \qquad (4.75)$$

When rule (4.75) fires, confidence factor $g_{kr}(v)$ is associated with the consequent x_k, where $v = (v_0, v_1, \ldots, v_n)$. Here g_{kr} is the confidence function associated with the rule. Although it is convenient to let all the v_j occur as arguments of g_{kr}, $g_{kr}(v)$ depends only on the confidence factors of the antecedents; that is, the v_j such that $j \in J_{kr}$. It is assumed that $0 \le g_{kr}(v) \le 1$ for all v.

The example will use the most common sort of confidence function,

$$g_{kr}(v) = \alpha_{kr} \min_{j \in J_{kr}} \{v_j\}$$

4.4 CONFIDENCE FACTORS IN RULE SYSTEMS

Table 4.4: Calculation of Confidence Factors

	\multicolumn{5}{c}{Iteration}				
	1	2	3	4	5
v_1	1	1	1	1	1
v_2	1	1	1	1	1
v_3	1/3	1/3	2/3	2/3	2/3
v_4	0	2/3	2/3	2/3	2/3
v_5	0	2/9	2/9	4/9	4/9
v_6	0	1/9	4/27	4/27	8/27

This implies that one's confidence in the consequent depends on one's confidence in the most questionable antecedent. The coefficient α_{kr} is indicated just above the arrow in (4.74). Thus for the sixth rule (i.e., rule 6,2),

$$g_{62}(v) = (2/3) \min\{v_4, v_5\}$$

The computation of confidence factors proceeds as follows. In the first iteration we compute $g_{kr}(v^0)$ for every rule k, r, where v^0 is the initial value of v. After the pth iteration the updated confidence factor v_j^p for a proposition x_j is the largest confidence factor it receives in rules in which it appears as the consequent. In other words, our confidence in a proposition is governed by the strongest argument for it. Thus the update formula is

$$v_k^p = \max_r \{g_{kr}(v^{p-1})\} \qquad (4.76)$$

In the example, for instance,

$$v_6^p = \max\{g_{61}(v^{p-1}), g_{62}(v^{p-1})\}$$

The computation of confidence factors in the example is displayed in Table 4.4. They converge after five iterations to the values in the last column of the table.

Although rule system (4.74) converges after only five iterations, other rule systems can require infinitely many iterations. Consider, for instance, the following rule system:

$$\begin{aligned} t & \rightarrow x_1 \\ x_2 \wedge x_3 & \rightarrow x_1 \\ x_3 & \rightarrow x_2 \\ t & \rightarrow x_3 \\ x_1 \wedge x_2 & \rightarrow x_3 \end{aligned} \qquad (4.77)$$

Table 4.5: Infinite Convergence of Confidence Factors

	\multicolumn{8}{c}{Iteration}
	1	2	3	4	5	6	7	...	∞
v_1	1/4	1/4	1/4	1/3	11/24	11/24	1/2	...	2/3
v_2	0	1/3	1/3	11/24	11/24	1/2	27/48	...	2/3
v_3	1/3	1/3	11/24	11/24	1/2	27/48	27/48	...	2/3

The corresponding confidence functions are

$$\begin{aligned} g_{11}(v) &= 1/4 \\ g_{12}(v) &= \min\{v_2, v_3\} \\ g_{21}(v) &= v_3 \\ g_{31}(v) &= 1/3 \\ g_{32}(v) &= 1/3 + (1/2)\min\{v_1, v_2\} \end{aligned} \quad (4.78)$$

The rule $t \to x_3$ is redundant of $x_1 \wedge x_2 \to x_3$ because both yield $v_3 = 1/3$ after the first iteration. But for technical reasons it will be convenient to assume that a rule $t \to x_j$ occurs in the rule system, with confidence function $g_{jr} = \beta$, whenever $g_{jr'}(0) = \beta > 0$ for some other rule j, r'. A rule system satisfying this assumption is *initialized*. It may also be desirable to include a rule of the form $x_j \to x_j$ for every x_j, with confidence function $g_{jr}(v) = v_j$. Such rules have no effect in the above example, but in other examples they prevent a proposition's confidence level from decreasing when a rule is fired.

The computation of confidence factors for rule system (4.77) proceeds as shown in Table 4.5. Note that all the confidence factors converge to 2/3 after infinitely many iterations.

4.4.2 A Graphical Model of Confidence Factor Calculation

A graph can represent a set of rules in the following way. A node is created for every atomic proposition and for every rule. For every rule, draw a (directed) arc from the node representing each antecedent x_j to a central node (associated with the rule) and from the central node to the node representing each proposition y_j in the consequent. The graph corresponding to the rule base (4.74) appears in Figure 4.5. Since the rules of (4.74) are Horn, only one arc leaves each central node.

4.4 CONFIDENCE FACTORS IN RULE SYSTEMS

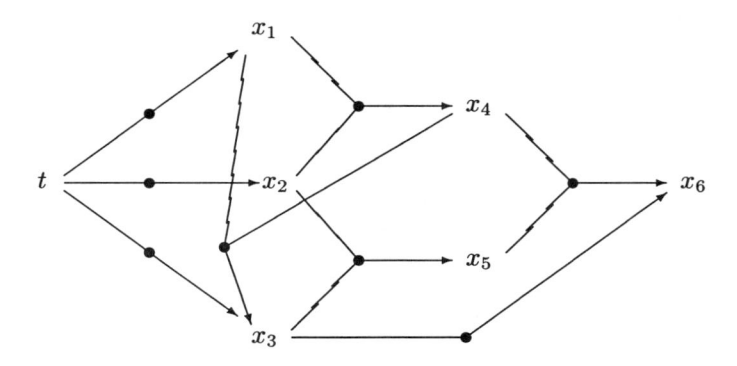

Figure 4.5: A graphical model for a rule system.

An "argument" for a proposition in a rule system can be viewed as a collection of paths in the associated graph from t to the proposition. Such a collection forms a *path tree*, which has the following structure.

Like the graph on which it is based, the nodes of the tree include proposition nodes and central nodes. The immediate predecessor of each proposition node x_j (unless $x_j = t$) is a central node associated with some rule j, r of which x_j is the consequent. The immediate predecessors of the central node are proposition nodes corresponding to each antecedent of rule j, r. Proposition nodes associated with t have no predecessors, and the root of the tree is a proposition node. Figure 4.6 shows a path tree whose root is associated with x_6 in (4.74).

Every path from a leaf to the root of a path tree corresponds to a path in the associated graph. The latter paths need not be simple paths (i.e., they can contain cycles).

Every path tree corresponds to a series of rule firings that results in a certain confidence factor for the proposition at the root. For instance, the tree of Figure 4.6 illustrates how the confidence factor $v_6^5 = 8/27$ of Table 4.4 was calculated for x_6. That is, v_6^5 was derived from v_4^4 and v_5^4 using the formula

$$v_6^5 = (2/3)\min\{v_4^4, v_5^5\},$$

v_4^4 was derived from v_1^3 and v_2^3, and so on. It is clear that if the root is at level 10 and the lowest leaves are at level 0, then the confidence factors associated with level $2p$ of the tree are calculated in the pth iteration of the rule-firing process.

Several different path trees may be rooted at the same proposition. For instance, the path tree of Figure 4.7 is also rooted at x_6. If one calculates

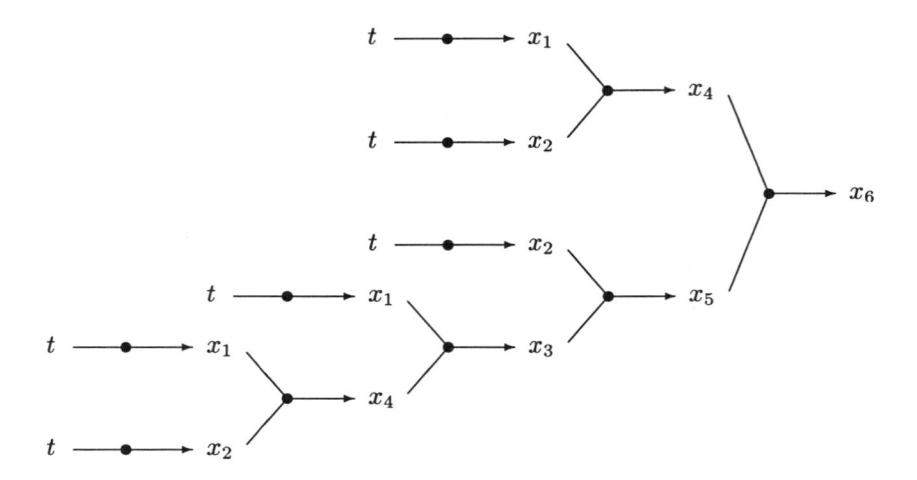

Figure 4.6: A path tree associated with a rule system.

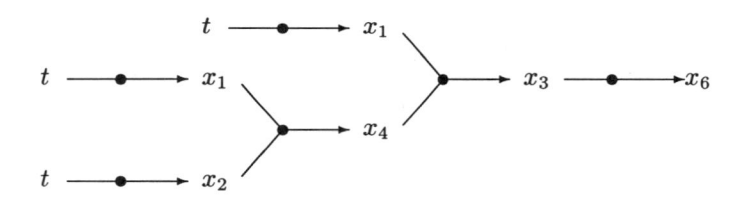

Figure 4.7: A second path tree rooted at x_6.

confidence factors along this tree, one obtains 2/9, which is slightly less than 8/27. Since the confidence factor in fact assigned to x_6 is the one resulting from the "strongest argument" for x_6, v_6^5 is 8/27 rather than 2/9. In fact, it is not hard to see that if the confidence functions are nondecreasing, the confidence level assigned a proposition after the pth iteration is just the maximum of the confidence levels corresponding to the trees of depth at most $2p$ that are rooted at the proposition.

To make this more precise, let $v_j(T)$ be the confidence level associated with a path tree T rooted at x_j. Then $v_j(T)$ is given recursively by,

$$v_j(T) = g_{k(T)r(T)}(v'), \qquad (4.79)$$

where rule $k(T), r(T)$ is the rule associated with the root of T. The component of v' corresponding to each antecedent x_i of rule $k(T), r(T)$ is $v_i(T_i)$,

4.4 CONFIDENCE FACTORS IN RULE SYSTEMS

where T_i is the subtree of T rooted at x_i. If T consists only of its root, then the root is associated with t, and $v_j(T) = v_0(T) = 1$. The claim is stated as follows:

Lemma 25 *In a initialized rule system, the confidence factor v_k^p of x_k after p iterations of rule firing is given by*

$$v_k^p = \max_{T \in T_k^p} \{v_k(T)\} \tag{4.80}$$

where T_k^p is the set of all path trees rooted at x_k of depth at most $2p$.

Proof: The proof is by induction on p. The theorem holds for $p = 0$ because the system is initialized. Suppose therefore that it holds for $p-1$. It will be convenient to use the notation $(a_j)_j$ for the vector $(a_1, \ldots, a_n)$, where n is the number of atomic propositions. From (4.76) and the induction hypothesis, the pth round of rule firing yields

$$v_k^p = \max_r \left\{ g_{kr} \left((\max_{T \in T_j^{p-1}} \{v_j(T)\})_j \right) \right\}$$

If F_r^p is the set of forests consisting of path trees of depth at most $2p$ that are rooted at the antecedents of rule k, r, this can be written

$$v_k^p = \max_r \left\{ g_{kr} \left((\max_{F \in F_r^{p-1}} \{v_j(T_j(F))\})_j \right) \right\}$$

where $T_j(F)$ is the tree in F rooted at x_j. Since each g_{kr} is nondecreasing, this can in turn be written

$$v_k^p = \max_r \left\{ \max_{F \in F_r^{p-1}} \{g_{kr}((v_j(T_j(F)))_j)\} \right\}$$

This implies

$$v_k^p = \max_{T \in T_k^p} \{g_{k,r(T)}((v_j(T_j(F)))_j)\}$$

since every forest in F_r^{p-1} corresponds to a tree in T_k^p obtained by drawing an arc from the root of each tree in the forest to a central node (associated with rule k, r) and from the central node to x_k. The desired result follows from an application of (4.79) to this last expression. □

This connection between confidence factors and path trees provides a tool for understanding when the updated values v^p converge in finitely many iterations. If the confidence factor corresponding to a path tree never decreases when cycles are removed from its paths, then obviously the deepest tree one need consider has depth equal to the longest simple path in the associated graph. It is not hard to see that one can in fact remove cycles, without decreasing confidence factors, if the confidence factor of a rule's consequent is never better than that of any of its antecedents. Such a confidence function is *conservative*, meaning that it satisfies the condition

$$g_{kr}(v) \leq v_j \quad \text{for all } j \in J_{kr}$$

When the confidence functions are conservative, confidence cannot increase as one goes around a cycle.

To make this argument more precise, suppose that a path tree contains a path P such that the corresponding path in the original graph is nonsimple (contains a directed cycle). That is, at least one node occurs twice or more in the path. It is not hard to show that, due to the structure of the graph, some rule k,r is used twice or more in P (i.e., central nodes associated with k,r occur twice or more in P). Let x_k^1 denote the node associated with the consequent of the first occurrence of rule k,r, and x_k^2 denote the consequent node associated with the second occurrence. Let the confidence factor computed for x_k^1 be w_1, and that computed for x_k^2 be w_2. Since x_k^1 is a predecessor of x_k^2 and all the confidence functions used on the path from x_k^1 to x_k^2 are conservative, $w_1 \geq w_2$. Form a new path tree that replaces x_k^2 with x_k^1 and omits all predecessors of x_k^2 except those that are predecessors of x_k^2. Because $w_1 \geq w_2$ and all confidence functions are nondecreasing, the confidence factor calculated for the root does not decrease. Thus one can keep removing cycles, without decreasing the confidence factor that results, until all the paths are simple. This and Lemma 25 prove the following.

Lemma 26 *The confidence factor of a proposition x_k in an initialized rule system with conservative and nondecreasing confidence functions is*

$$\max_T \{v_k(T)\}$$

where T ranges over all path trees that contain no cycles and whose root is associated with x_k.

Because the depth of a path tree is the length of its longest path, the desired result follows.

Theorem 51 *The number of iterations required to calculate the confidence factor of a proposition in an initialized rule system with conservative and nondecreasing confidence functions is at most one-half the length of a longest simple path in the associated graph.*

4.4 CONFIDENCE FACTORS IN RULE SYSTEMS

4.4.3 Jeroslow's Representability Theorem

R. Jeroslow [165], building on joint work with J. K. Lowe [169, 200], characterized subsets of n-space that can be represented as the feasible region of a mixed integer program. They first proved that a set is the feasible region of some mixed integer/linear programming problem (MILP) if and only if it is the union of finitely many polyhedra having the same recession cone (defined below). Jeroslow later proved a more general result to the effect that a set is the feasible region of a certain type of mixed integer/nonlinear programming problem (MINLP) if and only if it is a union of finitely many convex sets having the same set of recession directions. Although these results are not widely known, they might well be regarded as the fundamental theorems of mixed integer modeling.

The MILP representability theorem is proved below, and the MINLP theorem stated. The proof of the latter is more complicated but employs the same strategy. The next section applies the MILP theorem to the computation of confidence factors.

The basic idea of Jeroslow's results is that any set that can be represented in a mixed integer model can be represented in a disjunctive programming problem (i.e., a problem with either/or constraints). This can be illustrated in a classic modeling situation for integer variables. The object is to minimize a fixed charge function $f(x_1)$ subject to $x_1 \in P$, where P is a polyhedron. That is, $f(0) = 0$ but $f(x_1) = c + ax_1$ for $x_1 > 0$, where $c > 0$ is the fixed charge and $a \geq 0$ is the unit variable cost. Thus if $S = \{(x_1, f(x_1)) \mid x_1 \in P\}$, the problem can be stated,

$$\begin{aligned} \min \quad & f(x_1) \\ \text{s.t.} \quad & (x_1, f(x_1)) \in S \end{aligned} \qquad (4.81)$$

This problem can be solved as an MILP if S is represented as the solution set of linear constraints in continuous and integer variables.

This in turn is possible if some constant M bounds x_1, which can be assumed with little sacrifice of generality for sufficiently large M. Here it is assumed for simplicity that $P = \{x_1 \mid 0 \leq x_1 \leq M\}$, so that S is the shaded area in Figure 4.8 plus the line segment from $(0,0)$ to $(0,c)$. The key to representing S is to regard S as the union of two polyhedra: polyhedron P_1, which is the ray extended up from the origin, and polyhedron P_2, which is the infinite strip $\{(x_1, x_2) \mid x_2 \geq c + ax_1, 0 \leq x_1 \leq M\}$. Now write (4.81) as the disjunctive program in (4.82) below. If $\lambda_1 = 1$, any feasible solution of (4.82) is such that x belongs to P_1, and if $\lambda_2 = 1$, any feasible solution is such that x belongs to P_2. Since either λ_1 or λ_2 is 1, the feasible region for (4.82) is the union of P_1 and P_2.

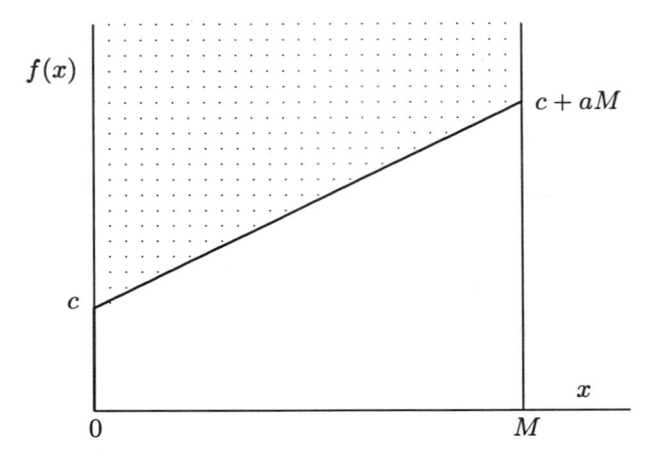

Figure 4.8: A fixed charge problem.

$$\begin{aligned}
\min \quad & x_2 \\
\text{s.t.} \quad & y_1^1 = 0 \cdot \lambda_1 \qquad \lambda_2 \cdot 0 \leq y_1^2 \leq M \lambda_2 \\
& y_2^1 \geq 0 \cdot \lambda_1 \qquad y_2^2 - a y_1^2 \geq \lambda_2 c \\
& x_1 = y_1^1 + y_1^2 \\
& x_2 = y_2^1 + y_2^2 \\
& \lambda_1 + \lambda_2 = 1 \\
& \lambda_1, \lambda_2 \in \{0, 1\}
\end{aligned} \qquad (4.82)$$

The role of the constraint $x_1 \leq M$ is now clear. Without it, y_1^2 (and therefore $x_1 = y_1^1 + y_1^2$) could be arbitrarily large in a feasible solution of (4.82) when $\lambda_1 = 1$. This means that (x_1, x_2) would not belong to P_1. The effect of the constraint $x_1 \leq M$ is to ensure that P_1 and P_2 have the same set "recession directions." A *recession direction* for a set S in n-space is a vector x such that $s + \alpha x \in S$ for all $s \in S$ and all $\alpha \geq 0$. The set of recession directions is denoted $rec(S)$. In the example, both $rec(P_1)$ and $rec(P_2)$ consist of all vectors of the form $(0, x_2)$, where $x_2 \geq 0$, so that $rec(P_1) = rec(P_2) = P_1$.

As often happens, the disjunctive program (4.82) can be simplified. Observing that $y_1^1 = 0$, so that $x_1 = y_1^2$, and setting $\lambda = \lambda_2$, one obtains

$$\begin{aligned}
\min \quad & x_2 \\
\text{s.t.} \quad & 0 \leq x_1 \leq M \lambda \\
& x_2 - a x_1 \geq \lambda c
\end{aligned}$$

which is the traditional model.

4.4 CONFIDENCE FACTORS IN RULE SYSTEMS

It is not hard to generalize the disjunctive construction, but this requires a more rigorous definition of representability. Consider the general mixed integer constraint set below.

$$\begin{aligned}&f(x,y,\lambda) \leq b \\ &x \in \Re^n, \ y \in \Re^p \\ &\lambda = (\lambda_1, \ldots, \lambda_k), \text{ with } \lambda_j \in \{0,1\} \text{ for } j = 1, \ldots, k\end{aligned} \qquad (4.83)$$

Here f is a vector-valued function, so that $f(x,y,\lambda) \leq b$ represents a set of constraints. A set $S \subset \Re^n$ is *represented* by (4.83) if

$$x \in S \text{ if and only if } (x,y,\lambda) \text{ satisfies (4.83) for some } y, \lambda$$

If f is a linear transformation, so that (4.83) is a MILP constraint set, S is *MILP representable*. (Jeroslow uses the term *b-MIP.r*, which means "bounded-mixed-integer-programming-representable.")

Now suppose $S \subset \Re^n$ is the union of polyhedra $P_1, \ldots, P_k$ that have the same set of recession directions. It can be shown as follows that S is MILP representable. Since each P_i is a polyhedron, it is the set of feasible points for a constraint set $f^i(y^i) \leq b^i$, where f^i is a linear transformation and $y^i \in \Re^n$. S is therefore represented by the following disjunctive constraint set.

$$\begin{aligned}&f^i(y^i) \leq \lambda_i b^i, \quad i = 1, \ldots, k, \\ &x = y^1 + \cdots + y^k \\ &\lambda_1 + \cdots + \lambda_k = 1 \\ &\lambda_i \in \{0,1\}, \quad i = 1, \ldots, k\end{aligned} \qquad (4.84)$$

To see this, suppose first that $x \in S$. Then x belongs to some P_{i^*}, which means that x satisfies (4.84) when $\lambda_{i^*} = 1$, $\lambda_i = 0$ for $i \neq i^*$, $y^{i^*} = x$, and $y^i = 0$ for $i \neq i^*$. The constraint $f^i(y^i) \leq \lambda_i b^i$ is satisfied by definition when $i = i^*$, and it is satisfied for other i because $f^i(0) = 0$ and $\lambda_i = 0$.

Conversely, suppose that x, λ, and $y^1, \ldots, y^k$ satisfy (4.84). To show that $x \in S$, note that exactly one λ_i, say, λ_{i^*}, is equal to 1. Then $f_{i^*}(y^{i^*}) \leq b^{i^*}$ is enforced, which means that $y^{i^*} \in P_{i^*}$. For other i's, $f^i(y^i) \leq 0$. Thus $f^i(\alpha y^i) \leq 0$ for all $\alpha \geq 0$, which implies that $y^i \in rec(P_i)$. Since $rec(P_i) = rec(P_{i^*})$ each y^i ($i \neq i^*$) is a recession direction for P_{i^*}, which means that $x = y^{i^*} + \Sigma_{i \neq i^*} y^i$ belongs to P_{i^*}. This is the source of the condition that all the P_i have the same set of recession directions.

It has been shown that S is represented by (4.84). But (4.84) has the form (4.83) with a linear function f, if $y = (y^1, \ldots, y^k)$. Thus S is MILP representable.

Jeroslow showed not only that a finite union of polyhedra with the same recession directions is MILP representable, but the converse as well. To see how, suppose that f in (4.83) is linear. This implies that the feasible region for (4.83) is a union of polyhedra with the same set of recession directions. Simply let $P_{\bar{\lambda}}$ be the set of all solutions of (4.83) for which $\lambda = \bar{\lambda}$, where $\bar{\lambda} \in \{0,1\}^k$. Obviously $P_{\bar{\lambda}}$ is a polyhedron, and the union of $P_{\bar{\lambda}}$ over all $\bar{\lambda} \in \{0,1\}^k$ is the solution set of (4.83). To show that the $P_{\bar{\lambda}}$ have the same recession directions, use the following trick. First define a linear function,

$$f^*(x, y, \lambda) = (f(x, y, \lambda), \lambda, -\lambda)$$

For each $\bar{\lambda}$ let

$$b^{\bar{\lambda}} = (b, \bar{\lambda}, -\bar{\lambda}^i)$$

Note that

$$P_{\bar{\lambda}} = \{x \mid f^*(x, y, \lambda) \leq b^{\bar{\lambda}} \text{ for some } y\}$$

Thus $x \in rec(P_{\bar{\lambda}})$ if and only if $f^*(x, y, \lambda) \leq 0$, which means that $rec(P_{\bar{\lambda}})$ is independent of $\bar{\lambda}$.

This proves the following MILP representability theorem.

Theorem 52 ([165, 169, 200]) *A set in n-space is MILP representable if and only if it is the union of finitely many polyhedra having the same set of recession directions.*

Jeroslow generalized this result to nonlinear constraint sets. He defined a set $S \subset \Re^n$ to be *bounded convex representable (b.c.r.)* if it is represented by a constraint set (4.83) for which f is a positively homogeneous convex function, having no $-\infty$ values, such that $(y, \lambda) = (0,0)$ whenever $f(0, y, \lambda) \leq 0$. (A function $g(x)$ is *positively homogeneous* when $g(\alpha x) = \alpha g(x)$ for all $\alpha \geq 0$.) He proved the following:

Theorem 53 ([165]) *A set in n-space is bounded convex representable if and only if it is the union of finitely many closed, convex sets with the same set of recession directions.*

4.4.4 A Mixed Integer Model for Confidence Factors

Section 4.4.2 showed that, after p iterations of rule firing, the confidence factor of a proposition x_j is the largest confidence factor associated with a path tree rooted at x_j with depth at most $2p$. This means that if the graph associated with the rule system contains no directed cycles, then there is a bound on the length of any path tree, so that the confidence factors converge after finitely many iterations. Even when there are directed cycles, the number of iterations remains finite if the confidence functions are conservative. In fact, it is at most one-half the length of the longest simple path in the graph.

4.4 CONFIDENCE FACTORS IN RULE SYSTEMS

If the graph contains directed cycles and the confidence functions are not all conservative, infinitely many iterations may be needed to achieve convergence. This is true, for instance, of rule system (4.77). Although rule systems with directed cycles are probably common, it may seem unlikely that confidence functions would be nonconservative, so that infinite convergence should rarely occur. But nonconservative rules can play a useful role in a practical rule system. Suppose, for instance, that a rule $x_1 \wedge x_2 \to y$ is given, and that either x_1 or x_2 is by itself conclusive evidence for y. Then confidence in y should be confidence that *either* x_1 or x_2 is true, which should be greater than confidence in x_1 and greater than confidence in x_2. More generally, there are two kinds of rules. One kind shows how to combine pieces of evidence to support a conclusion that is not supported by any piece individually, and this sort of rule would ordinarily have a conservative confidence function. But another kind of rule shows, as in the example, how the case for a conclusion becomes stronger when there are several pieces of evidence that individually support the conclusion. Rules of this type would tend to have nonconservative confidence functions.

In any case, if there is no finite convergence of confidence factors, it may be advantageous to calculate them by solving a mixed integer programming model. The procedure will be illustrated for rule system (4.77) and then stated in general.

The confidence functions for (4.77) are displayed in (4.78). It is intuitively plausible (and is proved below) that the confidence factors v_1, v_2, v_3 can be obtained by solving the mathematical programming problem,

$$
\begin{aligned}
\min \quad & v_1 + v_2 + v_3 \\
\text{s.t.} \quad & v_1 \geq 1/4 \\
& v_1 \geq \min\{v_2, v_3\} \\
& v_2 \geq v_3 \\
& v_3 \geq 1/3 \\
& v_3 \geq 1/3 + (1/2)\min\{v_1, v_2\} \\
& v_1, v_2, v_3 \geq 0
\end{aligned}
\tag{4.85}
$$

The goal is to write this as a mixed integer/linear programming problem (MILP), for which solution algorithms are well developed. Only the second and fifth constraints are not in the right form. If the set of points satisfying either constraint is MILP representable, then it is the feasible region for a set of MILP constraints. This would allow one to replace the second and fifth constraints with MILP constraints.

Begin with the second constraint. It is harmless to suppose that the feasible set S for this constraint satisfies $0 \leq v_j \leq 1$ for $j = 2, 3$. Then S can be viewed as the union of two polyhedra, $P_1 = \{v \mid 0 \leq v_2 \leq 1, \ 0 \leq$

$v_3 \leq 1$, $v_1 \geq v_2$} and $P_2 = \{v \mid 0 \leq v_2 \leq 1, \ 0 \leq v_3 \leq 1, \ v_1 \geq v_3\}$. Both polyhedra have the same recession directions, since $rec(P_1) = rec(P_2) = \{(v_1, 0, 0) \mid v_1 \geq 0\}$. Thus by Theorem 52 one can represent S as the solution set of the following disjunctive constraint set:

$$\begin{array}{ll}
0 \leq v_2^{1,1} \leq \lambda_1 & 0 \leq v_2^{1,2} \leq 1 - \lambda_1 \\
0 \leq v_3^{1,1} \leq \lambda_1 & 0 \leq v_3^{1,2} \leq 1 - \lambda_1 \\
v_2^{1,1} - v_1^{1,1} \leq 0 & v_3^{1,2} - v_1^{1,2} \leq 0 \\
\multicolumn{2}{c}{v_1 = v_1^{1,1} + v_1^{1,2}} \\
\multicolumn{2}{c}{v_2 = v_2^{1,1} + v_2^{1,2}} \\
\multicolumn{2}{c}{v_3 = v_3^{1,1} + v_3^{1,2}} \\
\multicolumn{2}{c}{\lambda_1 \in \{0, 1\}}
\end{array} \quad (4.86)$$

The subscript 1 of λ_1 and the first superscript 1 of $v_j^{1,k}$ simply indicate that the first set is represented disjunctively.

A similar construction defines the feasible set for the fifth constraint. It is the union of the two polyhedra $P_1 = \{v \mid 0 \leq v_1 \leq 1, 0 \leq v_2 \leq 1, v_3 \geq (1/2)v_1\}$ and $P_2 = \{v \mid 0 \leq v_1 \leq 1, 0 \leq v_2 \leq 1, v_3 \geq (1/2)v_2\}$, which have the same recession directions, because $rec(P_1) = rec(P_2) = \{(0, 0, v_3) \mid v_3 \geq 0\}$. The disjunctive constraint set is

$$\begin{array}{ll}
0 \leq v_1^{2,1} \leq \lambda_2 & 0 \leq v_1^{2,2} \leq 1 - \lambda_2 \\
0 \leq v_2^{2,1} \leq \lambda_2 & 0 \leq v_2^{2,2} \leq 1 - \lambda_2 \\
(1/2)v_1^{2,1} - v_3^{2,1} \leq -(1/3)\lambda_2 & (1/2)v_2^{2,2} - v_3^{2,2} \leq -(1/3)(1 - \lambda_2) \\
\multicolumn{2}{c}{v_1 = v_1^{2,1} + v_1^{2,2}} \\
\multicolumn{2}{c}{v_2 = v_2^{2,1} + v_2^{2,2}} \\
\multicolumn{2}{c}{v_3 = v_3^{2,1} + v_3^{2,2}} \\
\multicolumn{2}{c}{\lambda_2 \in \{0, 1\}}
\end{array}$$

$$(4.87)$$

When (4.86) and (4.87), respectively, replace the second and fifth constraints of (4.85), the resulting problem can be solved as an MILP to obtain $v = (2/3, 2/3, 2/3)$, which agrees with Table 4.5.

In general, confidence factors are obtained by solving the mathematical programming problem

$$\begin{aligned}
\min \quad & \sum_{j=1}^{n} v_j \\
\text{s.t.} \quad & v_j \geq g_{jr}(v), \quad \text{for all rules } j, r \\
& v_j \geq v_j^0, \quad j = 1, \ldots, n
\end{aligned} \quad (4.88)$$

4.4 CONFIDENCE FACTORS IN RULE SYSTEMS

If each set $\{v \mid v_j \geq g_{jr}(v)\}$ is MILP representable (perhaps after placing harmless bounds on some v_j), then (4.88) can be written as an MILP. To state the claim precisely, let the limiting values of the confidence factors be

$$\bar{v}_j = \lim_{p \to \infty} v_j^p$$

with $\bar{v} = (\bar{v}_1, \ldots, \bar{v}_n)$.

Theorem 54 ([167]) *Assume that all confidence functions in a rule system are nondecreasing and MILP representable, and that $\bar{v}$ exists and $\bar{v} \geq v^0$. Let v^* be the vector of values of $v_1, \ldots, v_n$ in an optimal solution of an MILP representation of (4.88). Then v^* exists and $\bar{v} = v^*$.*

Proof. Denote by (4.88') the MILP representation of (4.88). It suffices to show that (a) any feasible solution v of (4.88') satisfies $v \geq v^p$ for all p, and (b) $\bar{v}$ is feasible in (4.88').

(a) The reasoning is by induction on p. Clearly, a feasible v satisfies $v \geq v^0$, since this is a constraint of (4.88'). Suppose now that $v \geq v^p$. Then for any j, $v_j \geq g_{jr}(v) \geq g_{jr}(v^p)$ for all rules r with consequent x_j, where the second inequality is due to the fact that g_{jr} is nondecreasing. Thus by the update formula (4.76), $v_j \geq v_j^{p+1}$, so that $v \geq v^{p+1}$.

(b) It is known that $v_j^{p+1} \geq g_{jr}(v^p)$ for all rules j, r and all p. Since the region satisfying $v_j \geq g_{jr}(v)$ is MILP representable, it is a union of polyhedra and therefore a closed set. Thus we can take limits as $p \to \infty$ and obtain $\bar{v}_j \geq g_{jr}(\bar{v})$.

Since by hypothesis $\bar{v}$ satisfies $\bar{v} \geq v^0$, it is a feasible solution of (4.88'). □

The requirement that $\bar{v} \geq v^0$ is innocuous. It is automatically satisfied, for instance, if one includes rules of the form $v_j \to v_j$ to ensure that firing rules does not reduce the confidence factor (or measure of belief) assigned a proposition.

Chapter 5

Predicate Logic

Predicate logic is a refinement of propositional logic that brings substantially more expressive power. This comes at the price of a much more difficult inference problem. However, methods suggested by optimization techniques offer promise not only for improving the state of the art in inference methods, but for obtaining new structural insights as well.

The modeling power of predicate logic derives from two features. (a) It further analyzes atomic formulas of propositional logic by viewing them as attributing properties to individual objects, or as stating that individual objects stand in certain relations to each other. (b) It uses variables that range over objects, as well as constants that refer to particular objects. In "first-order" logic, which alone is considered here, the objects cannot be predicates; they can only be individuals. When all the variables in a formula are given specific values ("instantiated"), the formula can be regarded as a formula of propositional logic.

This already suggests how optimization-inspired methods might be applied to first-order formulas: reduce them somehow to propositional formulas, which can be checked for satisfiability by the methods of Chapters 2 and 3. Indeed a basic result of first-order logic shows how to write for any first-order formula a propositional formula, called the Herbrand extension, that is satisfiable if and only if the original formula is satisfiable. The Herbrand extension may be infinite in length, but when it is not, it allows one to check the satisfiability of a first-order formula using methods designed for propositional logic.

Unfortunately, the propositional formula, even when finite, becomes rapidly more complex for larger problems. This difficulty can be alleviated somewhat by using a "partial instantiation" technique introduced by R. Jeroslow [166]. Two general approaches to partial instantiation, which we call a "primal" and a "dual" approach, are described here. The primal approach begins with a short propositional formula that is too weak and adds clauses to it until it is unsatisfiable or the original first-order formula is proved satisfiable. This is analogous to mathematical row generation methods in optimization, such as Dantzing-Wolfe decomposition, that add constraints only as they are needed. The dual approach begins with a propositional formula that is too strong and weakens it until it is satisfiable or the original first-order formula is proved unsatisfiable.

Probably the best-known method for checking satisfiability of first-order formulas is Robinson's resolution method [244], which comes in several varieties. An introduction to this approach may be found in [198]. Unfortunately, all resolution methods, at least when applied to non-Horn logic, are hindered by the underlying sluggishness of resolution for propositional formulas. The partial instantiation methods presented here permit one to use the fastest available algorithms for propositional satisfiability.

When applied to subsets ("fragments") of first-order logic that have special structure, partial instantiation methods can be adapted to take advantage of the structure. This is illustrated by a hypergraph-based method developed by Gallo and Rago [105] for "datalog" formulas. These are universally quantified Horn formulas, which are widely used in logic programming.

Because full first-order logic is undecidable, no finite algorithm can always settle the satisfiability issue. First-order logic is *semidecidable*, however, meaning that it is always possible to identify an unsatisfiable formula in finite time. The algorithms presented here always terminate if the given formula is unsatisfiable, but termination is guaranteed for a satisfiable formula only if it belongs to a decidable fragment of first-order logic, such as the Löwenheim-Skolem or $\exists\forall$ fragment.

Partial instantiation methods are not optimization methods but are related to optimization in two ways: they use an incremental approach that is similar to row generation, and they reduce the problem to one that can be attacked by propositional satisfiability methods related to optimization. It is possible to use the Herbrand extension, however, actually to formulate a first-order satisfiability problem as an integer programming problem with infinitely many constraints.

One benefit of this formulation is that it connects Herbrand's famous compactness result, which in many ways provides the foundation for theorem proving in first order logic, with an analogous compactness property of integer and linear programming problems. Because the latter property is stronger, it may lead to stronger results in theorem proving. The semi-infinite programming framework also permits a generalization of the least

5.1 BASIC CONCEPTS OF PREDICATE LOGIC

element property of propositional Horn clauses to Horn clauses in first-order logic, thus obtaining a well-known result for definite logic programs.

The oldest application of quantitative methods to predicate logic is to the logic of arithmetic, which is logic in which the predicates have specific numerical meanings. In Section 5.5 we show how linear and integer programming can be used to check the validity of quantified formulas in certain types of real and integer arithmetic that do not involve the multiplication of variables.

It is best to begin with a brief and informal introduction to predicate logic. A more complete treatment can be found in [111] or [198].

5.1 Basic Concepts of Predicate Logic

5.1.1 Formulas

First-order predicate logic is more versatile than propositional logic because it includes properties and relations, which are denoted by *predicates*. In the formula

$$P(a) \supset Q(a,b) \tag{5.1}$$

the *constants* a and b denote *individuals*, P is a 1-place predicate, and Q is a 2-place predicate. If $P(a)$ means that a is a father and $Q(a,b)$ means that a is the father of b, then (5.1) means that if a is a father then he is the father of b.

Predicate logic is also called *quantified* logic because it contains the *universal quantifier* $\forall$, which means "for every," and the *existential quantifier* $\exists$, which means "there exists." The formula

$$\forall x(P(x) \supset \exists y Q(x,y)) \tag{5.2}$$

says that for everything x that is a father, there is something y of which it is the father. Here x is a *variable* that is *bound* by the quantifier $\forall$, and y is bound by $\exists$. Unbound variables are said to be *free*. The logic under consideration is *first-order* because only individuals and not predicates can be bound by quantifiers.

A formula without variables, such as (5.1), can be regarded as a formula of propositional logic and is said to be at *ground level*. Thus (5.1) is an *instance* of (5.2) because it is the result of replacing every variable with some possibly different symbol. In fact, (5.1) is a *ground instance* because it is at ground level. A formula F is a *generalization* of G if G is an instance of F. Note that every formula is an instance and a generalization of itself.

First-order logic can be augmented with *functions*, which are indispensable in logic programming [266]. For instance, a function $f(t,a)$ might be interpreted as the result of adding an individual a to a list t. So the list a, b, c, d could be represented $f(f(f(f(0, a), b), c), d)$, where 0 represents the empty list.

There are variations of first-order logic that will not be directly addressed here. First-order logic *with equality* can be obtained by adding the 2-place predicate "is equal to" and the necessary inference rules. *Many-sorted* or *typed* first-order logic can be very useful in practice. In the simplest case each argument of a predicate may be required to refer to only one type of thing. For instance, one might require the first argument of Q above to be a constant or a variable that can refer only to men. It is straightforward to adapt the development below to many-sorted logic.

For more complete treatments of first-order logic, see [198, 213].

5.1.2 Interpretations

The idea of an *interpretation* is useful for defining truth and falsehood of formulas. In predicate logic an interpretation specifies the *domain* (the set of constants that can instantiate variables) as well as which ground level predicates are true. An interpretation that makes a formula true is a *model, in first-order logic* for the formula.

For instance, an interpretation I for (5.2) might say that $\{a, b, c\}$ is the domain, $P(a)$ and $P(b)$ are the only true ground instances of $P(x)$, and $Q(a, b)$ and $Q(b, c)$ are the only true ground instances of $Q(x, y)$. One checks whether (5.2) is true in I in the obvious way: by checking whether for each element x of the domain $\{a, b, c\}$, $P(x) \supset Q(x, y)$ is true in I for some element y of $\{a, b, c\}$. So one must check whether

$$P(a) \supset Q(a, y) \text{ is true in } I \text{ for some } y, \text{ and}$$
$$P(b) \supset Q(b, y) \text{ is true in } I \text{ for some } y, \text{ and}$$
$$P(c) \supset Q(c, y) \text{ is true in } I \text{ for some } y$$

Because all three are true, I is a model for (5.2). Also, (5.2) is *satisfiable* because it has at least one model. It is not *valid*, however, because it is not true in every interpretation. If I' is identical to I except that the true ground instances of $P(x)$ are $P(a)$, $P(b)$, and $P(c)$, then (5.2) is false in I', since there is no individual of whom c is the father.

The versatility of predicate logic exacts a price, part of which is that the problem of determining satisfiability (or validity) is not only hard but *insoluble* in general. In propositional logic there is a finite procedure for checking whether a formula is satisfiable: just try all assignments of truth values to the atomic propositions. But Church [57] proved in 1938 that first-order predicate logic is *undecidable*, meaning that there is no such

5.1 BASIC CONCEPTS OF PREDICATE LOGIC

procedure that works for every formula. This is true despite the fact that first-order logic can be given a *complete* axiomitization, as proved by Gödel in 1930 [117]; that is, there is a finite set of axioms from which every valid formula can be deduced. The difficulty is that there is no finite algorithm that will always determine whether such a deduction exists.

Although first-order logic is undecidable, it is *semidecidable*, meaning that there is a procedure that will always terminate with a proof of unsatifiability if a formula is unsatisfiable. The partial instantiation procedures given below have this property. If a formula is satisfiable, however, the procedures may never terminate with an indication whether the formula is satisfiable.

5.1.3 Skolem Normal Form

A technical device that was historically useful for proving the semidecidability of first-order logic, and that remains useful for algorithmic purposes, is to express formulas in *Skolem normal form*; that is, as a conjunction of universally quantified clauses. Such a formula can be written

$$\forall x_1 C_1 \wedge \cdots \wedge \forall x_m C_m$$

where each C_i is a quantifier-free clause, $x_i = (x_{i1}, \ldots, x_{in_i})$ is a vector of the variables occurring in C_i, and $\forall x_i$ is short for $\forall x_{i1} \cdots \forall x_{in_i}$. Furthermore, the clauses are *standardized apart*, meaning that they have no variables in common.

Any first-order formula can converted to Skolem normal form, in the sense that the original formula is satisfiable if and only if the converted formula is satisfiable. The first step is to put it into *prenex form*, which means that all of its quantifiers (if any) precede everything else. The part of the formula following the quantifiers is the *matrix* (in the sense of a template). For instance, prenex form for formula (5.2) is

$$\forall x \exists y (P(x) \supset Q(x,y)) \qquad (5.3)$$

The following transformations, together with rules (2.3)-(2.5) in Section 2.1.2, suffice to convert any formula to prenex form.

$$\begin{aligned} \neg \forall x A &\equiv \exists x \neg A \\ \neg \exists x A &\equiv \forall x \neg A \\ (\forall x A \vee B) &\equiv \forall x (A \vee B) \\ (\exists x A \vee B) &\equiv \exists x (A \vee B) \\ (\exists x A \wedge B) &\equiv \exists x (A \wedge B) \end{aligned} \qquad (5.4)$$

where x does not occur free in B, plus

$$(\forall x A \wedge \forall x B) \equiv \forall x(A \wedge B)$$
$$(\exists x A \vee \exists x B) \equiv \exists x(A \vee B)$$

and finally,
$$\forall x A \equiv \forall y A' \tag{5.5}$$
where y does not occur free in A, and A' is A with every occurrence of x replaced by y. Transformations (5.5) are used to standardize variables apart.

To convert (5.2) into (5.3), first write (5.2) as $\forall x(\neg P(x) \vee \exists y Q(x,y))$. Then apply the fourth rule in (5.4) to obtain $\forall x \exists y(\neg P(x) \vee Q(x,y))$, which is equivalent to (5.3).

Once a formula is rendered in prenex form, it is *Skolemized* in order to eliminate the existential quantifiers. This is simplest when all the quantifiers are existential. An example would be the formula

$$\exists x \exists y \exists z (P(x,y) \wedge \neg P(x,a) \wedge Q(z)) \tag{5.6}$$

Every variable is replaced by a distinct constant that does not occur anywhere else in the formula. Therefore (5.6) becomes,

$$P(b,c) \wedge \neg P(b,a) \wedge Q(d)$$

The process is similar when all the existential quantifiers precede all the universal quantifiers. Such formulas comprise the $\exists \forall$ *fragment* of first-order logic, also known as the *Schönfinkel-Bernays* fragment. Consider, for example, the formula

$$\neg \{\forall z \forall w \exists x [\exists y (\neg P(x,y) \wedge \neg P(z,x))] \vee \exists y P(y,w)]\}$$

The formula must first be converted to prenex form. Bringing the first $\neg$ inside, one obtains

$$\exists z \exists w \forall x [\forall y (P(x,y) \vee P(z,x)) \wedge \forall y \neg P(y,w)]$$

One can now use rules (5.4) to bring the quantifiers to the front,

$$\exists z \exists w \forall x \forall y [(P(x,y) \vee P(z,x)) \wedge \neg P(y,w)]$$

to obtain prenex form. This reveals that the formula is in the $\exists \forall$ fragment. Skolemization yields

$$\forall x \forall y [(P(x,y) \vee P(a,x)) \wedge \neg P(y,b)]$$

5.1 BASIC CONCEPTS OF PREDICATE LOGIC

Next, the matrix of the Skolemized formula is put into clausal (disjunctive normal) form, which in this case it already exhibits. Each clause is given its own universal quantifiers,

$$\forall x \forall y[P(x,y) \vee P(a,x)] \wedge \forall y \neg P(y,b)$$

and the variables are standardized apart, resulting in Skolem normal form:

$$\forall x \forall y(P(x,y) \vee P(a,x)) \wedge \forall z \neg P(z,b) \tag{5.7}$$

The $\exists\forall$ fragment for first-order logic is important because it is decidable. Within this fragment the partial instantiation procedures described in the next section always terminate with an indication of whether a formula is satisfiable.

When universal quantifiers precede existential quantifiers, it is necessary to introduce *Skolem functions*. As an example consider the formula

$$\exists x \forall y \exists z (P(x,y) \wedge \neg P(x,z)) \tag{5.8}$$

One can instantiate x with a Skolem constant as usual, but not so z, because it is in the scope of $\forall y$. It is true that if (5.8) is satisfiable, it should remain satisfiable after a constant is substituted for z, but this constant may depend on y. The remedy is to replace z with a Skolem function $g(y)$, which one can regard as a constant that depends on y:

$$\forall y(P(a,y) \wedge \neg P(a,g(y)))$$

Skolem normal form is

$$\forall y P(a,y) \wedge \forall z \neg P(a,g(z)) \tag{5.9}$$

The complications introduced by Skolem functions can be seen in a famous example of Bernays and Schönfinkel that is cited by Church [58]. Let $P(x,y)$ be interpreted as saying that x precedes y. Then the formula

$$\forall x \forall y \forall z [(P(x,y) \wedge P(y,z)) \rightarrow P(x,z)] \wedge \forall x \neg P(x,x) \wedge \forall x \exists y P(x,y)$$

says that precedence is transitive (if x precedes y and y precedes z, then x precedes z), and furthermore, nothing precedes itself, and everything precedes something. The formula in Skolem normal form is

$$\forall x \forall y \forall z [((P(x,y) \wedge P(y,z)) \rightarrow P(x,z)) \wedge \neg P(x,x) \wedge P(x,s(x))] \tag{5.10}$$

Formula (5.10) is satisfiable, and one model for it has domain $\{a, s(a), s(s(a)), \ldots\}$, where $s(t)$ is interpreted as a successor of t. The model regards any $P(t, s(t))$ as true, where t is any element of the domain. The difficulty is that this model has an infinite domain, and it is not hard to see that any model for (5.10) must have an infinite domain. So an algorithm that tries to prove satisfiability by constructing a model will run forever.

5.1.4 Herbrand's Theorem

It seems intuitively reasonable that a formula in Skolem normal form can be checked for satisfiability by trying all possible instantiations of the variables and checking whether the resulting ground instances can all be true at once. But what set of constants should one use to instantiate the variables? Herbrand's theorem answers this question. It is the basis for the partial instantiation methods discussed in the next section.

Consider formula (5.7), for example. It is satisfiable if it is true in some interpretation. Because it is in Skolem normal form, it contains only universal quantifiers, which simplifies the task of checking whether it is true. One need only conjoin all possible ground instances of each clause and check whether the resulting ground level formula is true. If the domain is $\{a,b\}$, for instance, the conjunction of the ground instances for (5.7) is

$$
\begin{array}{r}
(P(b,a) \vee P(a,a)) \wedge \\
(P(b,b) \vee P(a,a)) \wedge \\
(P(b,a) \vee P(a,b)) \wedge \\
(P(b,b) \vee P(a,b)) \wedge \\
\neg P(a,a) \wedge \\
\neg P(b,a)
\end{array}
\tag{5.11}
$$

If there is an interpretation that makes this conjunction true—that is, if the conjunction is satisfiable—then it makes (5.7) true as well, showing that (5.7) is satisfiable. So perhaps one can check (5.7) for satisfiability by checking the ground level formula (5.11) for the same. But one must consider the converse: Does the unsatisfiability of (5.11) imply the unsatisfiability of the original formula (5.7)? As it happens (5.11) is unsatisfiable, which shows, at least on the face of it, only that there is no model *with domain* $\{a,b\}$ for (5.7).

Could there be a model with some other domain? J. Herbrand [137] showed that the answer is no. It is enough to consider a minimal domain $\{a,b\}$ that contains only constants that occur in the formula and, if there are functions, terms that can be built up from constants and functions.

The treatment of functions can be illustrated by formula (5.9). Because any term can be substituted for z to obtain another term, a minimal domain must contain $\{a, g(a), g(g(a)), \ldots\}$. An interpretation makes (5.9) true only if it makes true every possible ground instance obtained by instantiating from the domain. In the present case there is no such interpretation, because there is in fact a single ground instance that is already unsatisfiable:

$$P(a, g(a)) \wedge \neg P(a, g(a))$$

In fact, Herbrand proved a *compactness* theorem: if a formula is unsatisfiable, then some conjunction of finitely many ground instances of the formula is unsatisfiable. (In the example, just one instance was enough to show unsatisfiability.) This fact lies behind the semidecidability of first-order logic.

To state Herbrand's results more precisely, let a *term* be an expression that can be built up from constants, variables, and function symbols. That is, every constant and variable is a term, and for any n-place function f and any vector $t = (t_1, \ldots, t_n)$ of terms, $f(t)$ is a term. A *constant term* is a term that contains no variables. Let the *Herband universe* for a formula F consist of all possible constant terms built up from constants and functions in F; if F contains no constants, let the constant a be a term. The *Herbrand base* consists of all ground instances of predicates that appear in F. That is, if P is an n-place predicate that appears in F, then $P(t_1, , \ldots, t_n)$ belongs to the Herbrand base for any vector $(t_1, \ldots, t_n)$ of terms in the Herbrand universe. A *Herbrand interpretation* assigns a truth value to each element of the Herbrand base. A *Herbrand ground instance* of F is a ground instance obtained by instantiating from the Herbrand universe.

Theorem 55 ([137]) *A formula in Skolem normal form is unsatisfiable if and only if it is false in all of its Herbrand interpretations. Furthermore, it is unsatisfiable if and only if some finite conjunction of Herbrand ground instances of its clauses is unsatisfiable (compactness).*

The conjunction of all Herbrand ground instances is known as the *Herbrand extension*, which can be an infinitely long formula.

5.2 Partial Instantiation Methods

Herbrand's theorem states that one can check the satisfiability of a formula in Skolem normal form by checking the satisfiability of a ground level formula: the Herbrand extension, or the conjunction of all Herbrand ground instances. Unfortunately, there are likely to be a very large number of such instances—even when there are finitely many of them, as in the $\exists\forall$ fragment of first-order logic.

It is often possible to check satisfiability, however, by means of *partial instantiations*; that is, by examining certain instances in which not all of the variables are replaced by constants. They can be treated as formulas of propositional logic, so that their satisfiability can be tested using the efficient methods of Chapters 2 and 3.

R. Jeroslow, who did pathbreaking work in partial instantiation methods, proposed an approach that is essentially the primal approach we describe here [166]. Kagan, Nerode, and Subrahmanian [173] described an algorithm similar to Jeroslow's in which his mechanism of direct and indirect covers is implmented in a tree of partial instantiations.

The primal algorithm presented here, which is based on [150], differs from Jeroslow's in three ways: (a) it first convert formulas to clausal form (i.e., conjunctive normal form), which simplifies the algorithm; (b) it solves the propositional satisfiability problem by giving truth values to only some literals ("satisfiers"), a device that accelerates the generation of clauses; and (c) it generalizes the method to deal with functions, which are indispensable in logic programming. The dual algorithm, also based on [150], is a generalization of the hypergraph-based method in [103].

Key to the efficiency of partial instantiation methods is the availability of a fast satisfiability algorithm for propositional logic and, in particular, a fast algorithm for re-solving a problem after a new clause has been added (the *incremental* satisfiability problem). In Section 3.1.5 we showed how to adapt the Davis-Putnam-Loveland algorithm to exploit the information gained in the solution of the original problem to accelerate the solution of the incremented problem. This algorithm can often re-solve a satisfiability problem, after adding a clause, in only a small fraction of the time it would take to solve the incremented problem from scratch.

5.2.1 Partial Instantiation

To see how partial instantiations can provide a satisfiability test, consider the formula
$$\forall x P(a,x) \wedge \forall y \neg P(b,y) \tag{5.12}$$
Because the Herbrand universe is $\{a,b\}$, the formula is satisfiable if and only if the following is satisfiable:
$$P(a,a) \wedge P(a,b) \wedge \neg P(b,a) \wedge \neg P(b,b) \tag{5.13}$$
Rather than instantiate both x and y in this fashion, one can begin by instantiating neither:
$$\begin{array}{cc} P(a,x) \wedge \neg P(b,y) \\ T \qquad\quad F \end{array} \tag{5.14}$$
One can treat each predicate in (5.14) as an atomic proposition and try to solve the resulting propositional satisfiability problem. In this case there is a solution, indicated by the "T" and "F" below the atoms.

Although (5.14) is satisfiable, does this mean that the original formula (5.12) is satisfiable? It does, provided the solution of (5.14) can be used to construct an interpretation in which (5.13) is true. A natural way of doing so is the following: let a ground level predicate be true when it instantiates

5.2 PARTIAL INSTANTIATION METHODS 277

a true predicate that occurs posited in (5.14) and false when it instantiates a false predicate that occurs negated. Thus $P(a,a)$ and $P(a,b)$ are true, since they instantiate the true predicate $P(a,x)$, and $P(b,a)$ and $P(b,b)$ are false, since they instantiate the false predicate $P(b,y)$. Each of the four conjuncts in (5.13) is satisfied because the predicates in them inherit the same truth values that satisfy (5.14). So (5.12) is satisfiable.

If this plan is to work, predicates that are the same but for a renaming of variables (i.e., *variants*) must receive the same truth value in the valuation. There are no variants in the above example, but if there were a $P(a,x)$ and a $P(a,z)$, assigning them different truth values would present an ambiguity as to which truth value is assigned to $P(a,a)$ and $P(a,b)$. It is therefore assumed that any valuation is *variant-independent*, meaning that variants receive the same truth value.

In the scheme just enunciated, one need not specify the truth value of a ground level predicate when it instantiates only a negated true predicate or a posited false predicate. This is because the formula is given in clausal form. For instance, if the matrix is

$$P(a,x) \vee \neg P(y,b) \vee Q(c)$$
$$\quad F \qquad\quad T \qquad\, T$$

with the truth valuation shown, the falsehood of $P(a,x)$ and the truth of $P(y,b)$ are not needed to satisfy the formula. Thus the ground level equivalent is satisfied regardless of the truth value given $P(a,b)$.

The strategy, then, is to *extend* a variant-independent truth valuation for partially instantiated predicates to a valuation for ground level predicates. This approach can break down, however, if a ground level predicate instantiates both a true posited predicate and a false negated predicate.

This happens in the case of formula (5.7). The matrix, whose clauses can be denoted C_1 and C_2, has the satisfying solution shown below:

$$C_1 \wedge C_2 \;=\; (P(x_1,y_1) \vee P(a,x_1)) \wedge \neg P(z_2,b) \qquad (5.15)$$
$$\qquad\qquad\qquad T \qquad\quad T \qquad\qquad F$$

Note that the variables have been renamed (for reasons that will become evident in the next section). The ground level equivalent of (5.15) is

$$(P(a,a) \vee P(a,a)) \wedge$$
$$(P(a,b) \vee P(a,a)) \wedge$$
$$(P(b,a) \vee P(a,b)) \wedge$$
$$(P(b,b) \vee P(a,b)) \wedge$$
$$\neg P(a,b) \wedge$$
$$\neg P(b,b)$$

Each of the conjuncts would be satisfied if its predicates inherited their truth values from those shown in (5.15). But this cannot happen, because it would make $P(a,b)$ and $P(b,b)$ both true and false. Each instantiates a true posited predicate and a false negated predicate. In R. Jeroslow's terminology, the valuation in (5.15) is *blocked*. To show (5.7) is satisfiable, then, one must find an *unblocked* satisfying valuation of the partially instantiated formula.

The task of checking for blockage can be simplified by using the idea of a *satisfier* introduced by Gallo and Rago [103]. To satisfy (5.15) it is enough to have one true literal in each clause, as, for example,

$$C_1 \wedge C_2 \;=\; \underset{T}{(P(x_1,y_1) \vee P(a,x_1))} \wedge \underset{F}{\neg P(z_2,b)} \tag{5.16}$$

The literal chosen to satisfy a clause is its satisfier. Only those ground level predicates that instantiate satisfiers inherit truth values. If a ground predicate is not an instance of a satisfier, as in the case of $P(b,a)$ above, it receives no truth value. This causes no difficulty (barring blockage, of course) because there will still be at least one true literal in every ground clause. It is now easier to check for blockages because only the satisfiers need be examined.

Formally, satisfiers are defined by a mapping that associates each clause C of a given formula with an atom $S(C)$ of the clause. $S(C)$ is a *positive satisfier* of C if $S(C)$ is posited in C and is a *negative satisfier* otherwise. S is a *satisfier mapping* if no positive satisfier is a variant of any negative satisfier.

5.2.2 A Primal Approach to Avoiding Blockage

Two ways to find unblocked valuations are the "primal" and "dual" approaches mentioned earlier. The primal approach will be considered first.

Consider again the example (5.16) of the previous section. The satisfier mapping shown in (5.16) results in blockage because a positive satisfier and a negative satisfier, namely, $P(a,x_1)$ and $P(z_2,b)$, can be instantiated to yield the same predicate $P(a,b)$. All of its instantiations (in this case, only $P(a,b)$) inherit contradictory truth values. The substitution $(x_1,z_2) = (b,a)$ that makes them the same is a *unifier* of $P(a,x_1)$ and $P(z_2,b)$. The strategy is to avoid the contradiction by letting all atoms that instantiate both $P(a,x_1)$ and $P(z_2,b)$ inherit $P(a,b)$'s truth value.

This is done as follows. Conjoin $C_1 \wedge C_2$ with a formula C_3 obtained by making the substitution $(x_1,z_2) = (b,a)$ in C_1, and with a formula C_4

5.2 PARTIAL INSTANTIATION METHODS

obtained by making the same substitution in C_2:

$$C_3 = \underset{T}{P(b,y_3) \vee P(a,b)} \qquad (5.17)$$

$$C_4 = \underset{F}{\neg P(a,b)}$$

The variable in C_3 is renamed y_3 in order to standardize apart. $C_1 \wedge C_2 \wedge C_3 \wedge C_4$ has satisfier mappings, one of which is indicated by the T's and F's in (5.16) and (5.17).

Now each ground level predicate receives its truth value from the most specific (i.e., most fully instantiated) satisfier of which it is an instance. So $P(a,b)$, for instance, inherits its truth value (false) from the unifier $P(a,b)$ and not from $P(a,x_1)$, because the former is more specific.

The satisfier mapping in (5.16) and (5.17) must now be checked for blockage. Again the differently valued satisfiers $P(a,x_1)$ in C_1 and $P(z_2,b)$ in C_2 are generalizations of $P(z_2,b)$. But this is no longer a blockage, because resolving the apparent blockage would create generalizations (in fact, variants) of clauses that already appear in $C_1 \wedge C_2 \wedge C_3 \wedge C_4$ and so would not resolve any blockages that are not already resolved. There is a blockage elsewhere, however, because the negative satisfier $P(z_2,b)$ in C_2 and the positive satisfier $P(b,y_3)$ in C_3 are generalizations of $P(b,b)$. It is resolved by applying the unifying substitution $(x,y_3) = (b,b)$ to C_3 to generate a new clause C_5, and to C_2 to obtain C_6. Because C_5 (as well as C_6) is not a generalization of a clause that already occurs in $C_1 \wedge C_2 \wedge C_3 \wedge C_4$, the blockage is genuine and must indeed be resolved.

$$C_5 = P(b,b) \vee P(a,b)$$
$$C_6 = \neg P(b,b).$$

Because $C_1 \wedge \cdots \wedge C_6$ is unsatisfiable, so is (5.7).

To state the primal method in general, it is necessary to lay some groundwork. Let a *substitution* σ be a mapping that replaces each variable with some term. $F\sigma$ is the result of performing substitution σ in formula F. A *renaming* substitution replaces all variables with variables. σ is a *unifier* of $P(t)$ and $P(t')$ if $P(t)\sigma = P(t')\tau$. It is a *most general unifier* (mgu) if for every unifier τ of $P(t)$ and $P(t')$, $P(t)\tau$ is an instance of $P(t)\sigma$.

A linear-time algorithm for finding an mgu was introduced by Robinson [244]. First standardize $P(t)$ and $P(t')$ apart so that they contain no variables in common. Define the *disagreement set* D of $P(t), P(t')$ to consist of the first pair of corresponding terms in $P(t), P(t')$ that differ. For instance, the disagreement set of $P(x, f(g(a)), y)$ and $P(x, f(z), b)$ is $\{g(a), z\}$. Then the unification algorithm may be stated as follows.

Procedure 26: *Unification*

Step 1. Let $k = 0$ and let σ_0 be the identity substitution (i.e., it has no effect).

Step 2. If $P(t)\sigma_k = P(t')\sigma_k$ then stop, because σ_k is an mgu. Otherwise find the disagreement set D_k of $P(t)\sigma_k$ and $P(t')\sigma_k$.

Step 3. If one element of D_k is a variable v that does not occur in the other element t of D_k, then do the following: let σ_{k+1} be the substitution that applies σ_k and subsequently replaces all occurrences of v with t; increment k by 1; and go to step 2. Otherwise stop, because $P(t)$ and $P(t')$ are not unifiable.

The algorithm has exponential complexity if implemented naively but can also be implemented so that its running time grows linearly with the length of $P(t), P(t')$ [198].

The operation in step 3 of checking whether v occurs in t is known as the "occur check." The occur check is often omitted from PROLOG implementations on the ground that it is computationally expensive, although it seems more likely that a properly implemented occur check would make a relatively minor addition to computation time.

Next the idea of blockage must be defined precisely. Given a satisfier mapping S for a quantifier-free formula F, a pair of satisfiers $P(t), P(t')$ in F are *blocked* if:

(a) $P(t)$ is a positive satisfier and $P(t')$ a negative satisfier.

(b) $P(t)$ and $P(t')$ have a most general unifier σ.

(c) There are clauses C and C' in F of which $P(t)$ and $P(t')$ are, respectively, satisfiers, and for which either $C\sigma$ is a generalization of no clause in F or $C'\sigma$ is a generalization of no clause in F (or both).

The primal method may now be stated in general. Let $F = \bigwedge_{i=1}^{m} \forall x_i C_i$ be a first-order formula in Skolem normal form.

Procedure 27: *Primal Partial Instantiation*

Step 1. *(Initialization.)* Set $F_0 = C_1 \wedge \cdots \wedge C_m$ and $k = 0$.

Step 2. *(Ground satisfiability.)* Try to find a satisfier mapping S for F_k.

Step 3. *(Termination check.)* If S does not exist, stop; F is unsatisfiable. Otherwise if S is unblocked, stop; F is satisfiable.

Step 4. *(Refinement—S is blocked.)* Let C_h and C_i be two clauses in F_k whose satisfiers are blocked, and let σ be a most general unifier of

5.2 PARTIAL INSTANTIATION METHODS

$S(C_h)$ and $S(C_i)$. Set $F_{k+1} = F_k \wedge C_h\sigma \wedge C_k\sigma$ after standardizing apart, set $k = k+1$, and go to step 2.

The algorithm applies to formulas with functions. For instance, consider the formula $C_1 \wedge \forall x_2 C_2 \wedge \forall x_3 C_3 \wedge C_4$, where

$$\begin{aligned} C_1 &= \underset{T}{P(s(a))} \\ C_2 &= \underset{F}{\neg P(x_2) \vee Q(s(x_2))} \\ C_3 &= \underset{F}{\neg Q(s(x_3)) \vee R(s(x_3))} \\ C_4 &= \underset{F}{\neg R(s(a))} \end{aligned} \qquad (5.18)$$

Thus $F_0 = C_1 \wedge C_2 \wedge C_3 \wedge C_4$. A satisfier mapping for F_0 is shown. $S(C_1)$ and $S(C_2)$—that is, $P(s(a))$ and $P(x_2)$—are blocked because they have an mgu that substitutes $x_2 = s(a)$, yielding $P(s(a))\sigma = P(x_2)\sigma = P(s(a))$. Step 4 creates $F_1 = F_0 \wedge C_5$, where $C_5 = C_2\sigma$, or

$$C_5 = \underset{T}{\neg P(s(a)) \vee Q(s(s(a)))}$$

It is not necessary to add $C_1\sigma$ to the formula because σ has no effect on C_1. The previous satisfier mapping can be extended as shown. $Q(s(x_3))$ and $Q(s(s(a)))$ are blocked, and step 4 creates $F_2 = F_1 \wedge C_6$, where

$$C_6 = \underset{T}{\neg Q(s(s(a))) \vee R(s(s(a)))}$$

The satisfier mapping shown is unblocked, and the algorithm terminates with satisfiability.

To illustrate proof of unsatisfiability, let $F = C_1 \wedge \forall x_2 C_2 \wedge \forall x_3 C_3$, where

$$\begin{aligned} C_1 &= \underset{T}{P(s(a))} \\ C_2 &= \underset{T}{\neg P(x_2) \vee Q(s(x_2))} \\ C_3 &= \underset{F}{\neg P(s(x_3)) \vee \neg Q(s(s(a)))} \end{aligned}$$

The blockage between $P(s(a))$ and $P(s(x_3))$ creates $F_1 = F_0 \wedge C_4$, with

$$C_4 = \underset{F}{\neg P(s(a)) \vee \neg Q(s(s(a)))}$$

The satisfier mapping can be extended as shown. The blockage between $Q(s(x_2))$ and $Q(s(s(a)))$ creates $F_2 = F_1 \wedge C_5$, with

$$C_5 = \neg P(s(a)) \vee Q(s(s(a)))$$

Because F_2 is an unsatisfiable as a propositional formula, the algorithm terminates.

The algorithm always terminates in the $\exists\forall$ fragment because each clause has finitely many instances that are not generalizations of other instances. The algorithm can run forever on a full first-order formula, however, as can be verified by applying it to the Schönfinkel-Bernays formula (5.10).

The correctness of the algorithm is based on the following lemma. The lemma (or a lemma similar to it) was first proved by Jeroslow [166], but the proof given here is that of [150].

Lemma 27 *Given $F = \bigwedge_{i=1}^{m} \forall x_i C_i$, let S be a satisfier mapping for the quantifier-free formula $F' = C_1 \wedge \cdots \wedge C_m$. Then F is satisfiable if S is unblocked.*

Proof. It suffices to show that if S is unblocked, one can define a Herbrand interpretation I in which every ground instance of F is true.

Let a generalization $P(t)$ of $P(d)$ be a *most specific generalization* of $P(d)$ with respect to a class of formulas if every generalization of $P(d)$ in the class is a generalization of $P(t)$. Now I can be defined as follows. For any atom $P(d)$ in the Herbrand base of F,

- $P(d)$ is true if F' contains a positive satisfier $P(t)$ that is a most specific generalization of $P(d)$ with respect to the satisfiers of F', but no negative satisfier that is a most specific generalization of $P(d)$.

- $P(d)$ is false if F' contains a negative satisfier $P(t)$ that is a most specific generalization of $P(d)$ with respect to the satisfiers of F', but no positive satisfier that is a most specific generalization of $P(d)$.

- $P(d)$ is arbitrarily true or false otherwise.

Note that no ground predicate can assume both a true and a false value.

It suffices to show that any ground instance of any clause of F is true in I. Let

$$C = \neg P_1(t_1) \vee \cdots \vee \neg P_q(t_q) \vee P_{q+1}(t_{q+1}) \vee \cdots \vee P_m(t_m)$$

5.2 PARTIAL INSTANTIATION METHODS

be any clause in F and

$$C_0 = \neg P_1(d_1) \vee \cdots \vee \neg P_q(d_q) \vee P_{q+1}(d_{q+1}) \vee \cdots \vee P_m(d_m)$$

be any ground instance of C. (Not all literals of C_0 need be distinct.)

Of the clauses in F' that are most specific generalizations of C_0, at least one of them C' is an instantiation of C (perhaps C itself). Let

$$C' = \neg P_1(u_1) \vee \cdots \vee \neg P_q(u_q) \vee P_{q+1}(u_{q+1}) \vee \cdots \vee P_m(u_m)$$

be one such clause. C' has a satisfier $P_i(u_i)$. The aim is to show that $P_i(d_i)$ inherits $P_i(u_i)$'s truth value and therefore makes C_0 true. For this it suffices to show that (a) no most specific satisfier that generalizes $P_i(d_i)$ has a truth value opposite that of $P_i(u_i)$, and (b) some such satisfier has the same truth value of $P_i(u_i)$.

(a) Suppose there is a most specific satisfier $P_i(v_i)$ that generalizes $P_i(d_i)$ and that has the opposite truth value. Since $P_i(d_i)$ instantiates both $P_i(u_i)$ and $P_i(v_i)$, the latter two have an mgu τ such that $P_i(u_i') = P_i(u_i)\tau$ generalizes $P_i(d_i)$. For some substitution τ_2, $C'\tau_1\tau_2 = C_0$. Note first that $P(u_i')$ is more specific than $P(u_i)$. This is because $P_i(u_i)$ and $P_i(v_i)$ have different truth values and are therefore not variants of each other. So if $P(u_i)$ and $P(u_i')$ were variants, $P_i(v_i)$ would not be a most specific satisfier that generalizes $P_i(d_i)$. Next, note that because the pair $P_i(u_i)$ and $P_i(v_i)$ are by assumption unblocked, $C'\tau_1$ must be a generalization of some clause in F'. $C'\tau_1$ is more specific than C', since as just shown $P(u_i') = P(u_i)\tau_1$ is more specific than $P(u_i)$. Also, $C'\tau_1$ generalizes C_0 because $C_0 = C'\tau_1\tau_2$. But this contradicts the assumption that C' is a most specific generalization of C_0 in F' that is an instantiation of C. So any most specific satisfier that generalizes $P_i(d_i)$ must have the same truth value as $P_i(u_i)$.

(b) One can now show that there in fact exists a most specific satisfier that generalizes $P_i(d_i)$, which by (a) must have the same truth value as $P_i(u_i)$. But this is clear, because some instantiation of $P_i(u_i)$, perhaps $P_i(u_i)$ itself, is such a satisfier.

This completes the proof. □

Two further lemmas are needed. The first is straightforward to show.

Lemma 28 *A formula F is satisfiable if and only if $F \wedge \bigwedge_{k=1}^{K} F\sigma_k$ is satisfiable, where σ_k is any substitution.*

Lemma 29 *Let $F = \bigvee_{i=1}^{m} \forall x_i C_i$. If $F' = C_1 \wedge \cdots \wedge C_m$ has no satisfier mapping, then F is unsatisfiable.*

Proof: It suffices to construct a ground instance $F'\sigma$ of F' such that a pair of atoms in F' are variants of each other if and only if their instances in $F'\sigma$ are identical. Then $F'\sigma$ is unsatisfiable, and by Theorem 55 F is unsatisfiable.

To construct $F'\sigma$, define the following equivalence relation among variables. When two atoms $P(t), P(t')$ are variants of each other, so that $P(t) = P(t')\tau$ for a renaming substitution τ, let any variable x_j be equivalent to $x_j\tau$. Associate each equivalence class of variables with a distinct constant that does not already occur in F'. Now let σ replace each variable with the constant associated with its equivalence class. □

Correctness of the algorithm can now be proved.

Theorem 56 *Given a formula F in Skolem normal form, the primal algorithm indicates satisfiability only if F is satisfiable and unsatisfiability only if F is unsatisfiable.*

If the algorithm indicates satisfiability, it obtains for a formula F_k an unblocked satisfier mapping. So F is satisfiable by Lemmas 27 and 28. If the algorithm indicates unsatisfiability, then in the last step it generates a formula F_k that has no satisfier mapping. Then F is unsatisfiable by Lemmas 28 and 29.

A partial completeness result is proved in [150].

Theorem 57 ([150]) *If F is unsatisfiable, then the primal algorithm terminates with an indication of satisfiability.*

The proof is based on a mechanism that keeps track of the nesting depth of functions in unifiers. It is shown that due to Herbrand's compactness theorem, the nesting depth and therefore the number of iterations can be bounded.

5.2.3 A Dual Approach to Avoiding Blockage

The primal approach proceeds in a reactive manner. It begins with a weak, partially instantiated version of the problem and instantiates predicates only to the extent necessary to avoid blockages in the current satisfying truth assignment. The dual approach tries to anticipate blockages in advance and adds clauses that prevent them from happening. This results in an overly strong version of the problem that is very likely to be unsatisfiable. The problem is gradually weakened by instantiating predicates, until the original problem is proved satisfiable or proved unsatisfiable.

Since blockages occur when unifiable predicates receive different truth values, one obvious way to avoid blockage is to add clauses that equate the

5.2 PARTIAL INSTANTIATION METHODS

truth values of all pairs of unifiable predicates. More precisely, blockage can be avoided as follows: whenever a posited predicate $P(t)$ and a negated predicate $P(t')$ can be unified, add the clause $\neg P(t) \vee P(t')$ to force $P(t')$ to be true whenever $P(t)$ is.

Formula (5.7) can again illustrate the idea. As before $C_1 \wedge C_2$ is the matrix shown in (5.16). If $C_1 \wedge C_2$ were unsatisfiable, one could as before conclude that (5.7) is unsatisfiable. Since there is a variant-independent solution, one can ignore the solution and ward off blockage by conjoining with $C_1 \wedge C_2$ the conjunction B of the following clauses:

$$\neg P(x_1', y_1') \vee P(z_1', b)$$
$$\neg P(a, x_2') \vee P(z_2', b)$$

The variables are renamed to standardize apart.

Any variant-independent solution of $C_1 \wedge C_2 \wedge B$ provides an unblocked valuation for $C_1 \wedge C_2$. So the next task is to check whether such a solution exists. In practice this can be done by first checking $C_1 \wedge C_2$ for satisfiability and, if it is satisfiable, using an incremental satisfiability algorithm (Section 3.1.5) to recheck satisfiability after adding the clauses of B one at a time. The process stops when unsatisfiability is detected, with the conclusion that $C_1 \wedge \bar{B}$ is unsatisfiable, where $\bar{B}$ is a conjunction of some clauses in B. In the present case, $B = \bar{B}$.

Next, arbitrarily select a clause in $\bar{B}$, such as the last clause $\neg P(a, x_2') \vee P(z_2', b)$ added to $\bar{B}$, and carry out the substitutions in C_1 and C_2 that unify $P(a, x_1)$ and $P(z_2, b)$ to obtain C_3 and C_4:

$$C_3 = P(b, y_3) \vee P(a, b)$$
$$C_4 = \neg P(a, b)$$

Since $C_1 \wedge C_2 \wedge C_3 \wedge C_4$ is satisfiable, start generating clauses in B. Upon the generation of the third clause below, $C_1 \wedge C_2 \wedge C_3 \wedge C_4 \wedge \bar{B}$ becomes unsatisfiable:

$$\neg P(x_1', y_1') \vee P(z_1', b)$$
$$\neg P(x_2', y_2') \vee P(a, b) \qquad (5.19)$$
$$\neg P(b, y_3') \vee P(z_3', b)$$

Note that the unifiable pair consisting of $P(a, x_1)$ and $P(z_2, b)$ does not generate a clause this time. This is because the instantiated clauses C_5 and C_6 that would result are generalizations of existing clauses C_3 and C_4, respectively, and so would avoid only blockages that are already avoided.

The substitutions that unify the two atoms in the last clause of (5.19) generate two more clauses:

$$C_5 = P(b, b) \wedge P(a, b)$$
$$C_6 = \neg P(b, b)$$

Because $C_1 \wedge \cdots \wedge C_6$ is unsatisfiable, so is (5.7).

A precise statement of the algorithm follows. Let $F = \bigwedge_{i=1}^{m} \forall x_i C_i$ be a formula in Skolem normal form.

Procedure 28: *Dual Partial Instantiation*

Step 1. (Initialization.) Set $F_0 = C_1 \wedge \cdots \wedge C_m$ and $k = 0$.

Step 2. (Ground satisfiability.) If F_k cannot be satisfied by a variant-independent truth assignment, stop; F is unsatisfiable.

Step 3. (Blockage avoidance.) Find a conjunction $\bar{B}$ of clauses such that $F_k \wedge \bar{B}$ has no satisfying variant-independent truth assignment, and such that each clause in $\bar{B}$ has the form $\neg P(t) \vee P(t')$, where $P(t)$ and $P(t')$ satisfy the following conditions: (a) $P(t)$ occurs posited in some C_i and $P(t')$ occurs negated in C_j (possibly $i = j$); and (b) $P(t)$ and $P(t')$ have an mgu σ, such that either $C_i\sigma$ is a generalization of no clause in F_k or $C_j\sigma$ is a generalization of no clause in F_k. If there is no such $\bar{B}$, stop; the original formula is satisfiable.

Step 4. (Refinement.) Pick a clause $\neg P(t) \vee P(t')$ in $\bar{B}$ (such as the last clause added to $\bar{B}$, as suggested above). Let C_i, C_j, σ be as in the previous step. Set $F_{k+1} = F_k \wedge C_i\sigma \wedge C_j\sigma$. Increment k by one and go to step 2.

Like the primal algorithm, this dual algorithm can be applied to formulas with functions, although termination is not guaranteed. If F_0 is (5.18), for instance, the conjunction B of the following clauses is generated in step 3:

$$\neg P(s(a)) \vee P(x_1')$$
$$\neg R(s(x_2')) \vee R(s(a))$$

Because $C_1 \wedge C_2 \wedge C_3 \wedge C_4 \wedge B$ is satisfiable, the original formula is satisfiable.

It is not hard to see why the algorithm works. If it terminates with a variant-independent satisfying valuation for $F_k \wedge B$, then that valuation is unblocked by construction. So by Lemmas 27 and 28, F is satisfiable. Furthermore, if the algorithm terminates with an unsatisfiable F_k, then F is unsatisfiable by Lemma 29.

Theorem 58 *Given a formula F in Skolem normal form, the dual algorithm indicates satisfiability only if F is satisfiable and unsatisfiability only if F is unsatisfiable.*

The following is shown in [150].

5.3 A METHOD BASED ON HYPERGRAPHS

Theorem 59 *If F is unsatisfiable, then the dual algorithm terminates with an indication of unsatisfiability.*

The practical rationale for the primal algorithm of the last section is that it may find a solution by luck before very many instantiations are generated. This obviously calls for a heuristic that selects which blockages to resolve in such a way as to find a satisfying solution as soon as possible. The rationale for the dual algorithm is that an intelligent selection of which blockage-avoiding clauses to generate first may reveal unsatisfiability early in the process.

Because the primal approach is oriented toward finding a solution and the dual toward finding a refutation, and because one does not know at the outset which will succeed, it is reasonable to combine the two approaches to obtain a primal-dual algorithm. This device is often used successfully in optimization. In the present context a primal-dual algorithm might begin with a dual phase by generating only a predetermined subset of the clauses that avoid blockage. If unsatisfiability results within this subset, instantiation would proceed as in the dual algorithm. Otherwise the algorithm would move to a primal phase by checking the solution for blockage. If blockage is found, instantiation would proceed as in the primal algorithm. Ideally, the dual phase could rule out the most likely kinds of blockage with a relatively small number of clauses, while allowing the primal strategy of finding any early solution by luck to operate as well.

5.3 A Method Based on Hypergraphs

The partial instantiation method can be specialized to universally quantified Horn formulas, also known as *datalog* formulas. This special case is instructive not only because of its practical importance in logic programming, but because it illustrates how a partial instantiated method can be designed to take advantage of special structure.

A hypergraph interpretation of propositional logic suggests an interesting implementation of the dual algorithm. This interpretation was proposed by G. Gallo, G. Longo, S. Nguyen, and S. Pallottino [102, 104], and the inference algorithm for datalog formulas to be presented here was developed by Gallo and Rago [105]. The hypergraph algorithm was only later shown in [146] to be a specialization of a dual approach to partial instantiation.

The first section below describes a hypergraph model for Horn propositional logic. Section 5.3.2 presents a shortest path algorithm for checking satisfiability in such hypergraphs, and Section 5.3.3 extends the hypergraph model to universally quantified logic.

5.3.1 A Hypergraph Model for Propositional Logic

Hypergraph models for propositional logic have been proposed by Gallo et al. [102], as well as by Jeroslow et al. [170] in connection with gain-free Leontief flows. The former treatment is followed here.

Whereas a *graph* can be viewed abstractly as a collection of two-element subsets (arcs) of a finite set of nodes, a *hypergraph* is a collection of subsets of any size (*hyperarcs*) of a finite set of nodes. A *directed* graph can be obtained by regarding the two-element subsets as ordered pairs. One possible generalization of this idea to hypergraphs is to regard each hyperarc as an ordered set of disjoint subsets of the set of nodes, where each subset is called a *layer* of the hyperarc. An arc of an ordinary directed graph has two layers, each consisting of one node.

Recall that any clause of propositional logic can be written as a rule,

$$(x_1 \wedge \cdots \wedge x_m) \to (y_1 \vee \cdots \vee y_n) \qquad (5.20)$$

where the x_j's and y_j's are atomic propositions (and possibly $m = 0$ or $n = 0$). If every atomic proposition is represented by a node of a hypergraph, then every rule (5.20) can be represented as a hyperarc with two (possibly empty) layers; the first layer (the *tail* consists of nodes representing $x_1, \ldots, x_m$, and the second (the *head*) of nodes representing $y_1, \ldots, y_n$. Thus when one establishes that all of the nodes in the tail of a hyperarc are true, one can infer that at least one node in the head is true.

Gallo et al. study such hypergraphs for general propositional logic [102], but here they are restricted to Horn rules. Because a Horn rule has at most one consequent, the head of the corresponding hyperarc contains at most one node. Such a hyperarc is a *B-arc* (short for "backward arc"), and a hypergraph made up of B-arcs is a *B-graph*. Thus when one establishes that all the nodes in the tail of a B-arc are true, one can infer that the node in the head is true.

As an example, consider the following Horn rule base:

Rule 1. $t \to x_1$
Rule 2. $t \to x_2$
Rule 3. $t \to x_3$
Rule 4. $(x_1 \wedge x_2) \to x_4$
Rule 5. $(x_3 \wedge x_4) \to x_5$
Rule 6. $x_5 \to x_1$
Rule 7. $(x_2 \wedge x_4 \wedge x_5) \to x_6$
Rule 8. $(x_3 \wedge x_5) \to f$
Rule 9. $(x_5 \wedge x_6) \to f$

5.3 A METHOD BASED ON HYPERGRAPHS

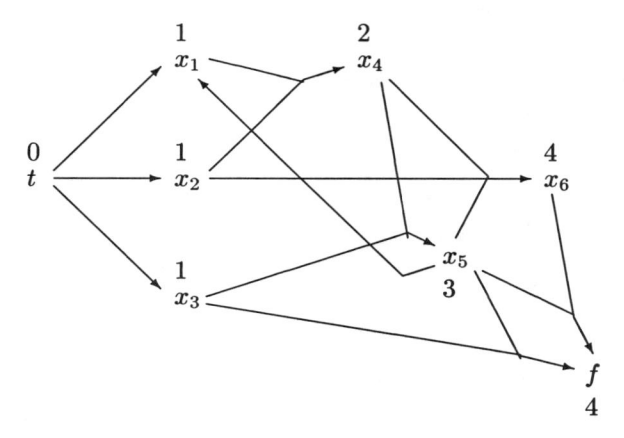

Figure 5.1: A hypergraph model of a Horn rule system.

where t and f, respectively, represent necessarily true and necessarily false propositions; thus $t \to x_1$ asserts x_1, and $x_3 \wedge x_5 \to f$ denies $x_3 \wedge x_5$. The corresponding hypergraph appears in Figure 5.1. Note that B-arcs representing rules with two or three antecedents are drawn as two- or three-pronged arrows (B-arcs).

In a hypergraph like that of Fig 5.1, truth can be regarded as flowing out of t and through portions of the graph. If truth reaches all the nodes in the tail of a given B-arc (i.e., all the antecedents of the corresponding rule), then it flows into the head (the consequent). Recall that a Horn rule system is inconsistent if and only if forward chaining yields a contradiction. Thus it is intuitively clear that it is inconsistent if and only if truth flows all the way to the f node, as it does in Figure 5.1.

To make this more precise, let $H(e)$ be the head of a B-arc e and $T(e)$ its tail. A *path* in a B-graph is a series of B-arcs $e_1, \ldots, e_m$ such that $H(e_i) \subset T(e_{i+1})$ for $i = 1, \ldots, m-1$. A path is a *cycle* if $H(e_m) \subset T(e_1)$. A *B-path* from x to y is a union of paths that allows truth to flow from x to y. This means that a path must run from x to every node in the B-path, one of which is y. The precise definition of a B-path from x to y in B-graph H is that it is a minimal cycle-free B-graph H' whose nodes and B-arcs are nodes and B-arcs of H, such that x and y are nodes of H' and x is connected to every node of H' by some path in H'. For instance, rules 1 to 5 form a B-path from t to x_5 in Figure 5.1, but rules 1 to 3 and 5 do not because they induce no path from t to x_4. Also rules 1 to 6 do not form a B-path from t to x_4 because they induce a cycle joining x_1, x_4, and x_5. Clearly, truth flows to a node if and only if there is a B-path from t to the node.

Lemma 30 ([102]) *A Horn rule system is satisfiable if and only if there is no B-path from t to f in the corresponding B-graph.*

5.3.2 Shortest Paths in B-Graphs

The inference algorithm of Section 5.3.3 will require the computation of shortest B-paths in a B-graph. Let a length $w(e)$ be associated with each B-arc e and define the *length* of a path in a B-graph to be the sum of the lengths of the B-arcs in the path. The *length* of a B-arc from x to y is the length of the longest path from x to y in the B-path. Suppose, for instance, that every B-arc in the B-graph of Fig 5.1 has unit length. Then the B-path induced by rules 1 to 5 has length 3.

The task at hand is to find the shortest B-path from t to all nodes of a B-graph. This can be done with a labeling algorithm similar to Dijkstra's shortest-path algorithm for ordinary graphs [82]. A distance label $D(x)$ is associated with every node x, which is the length of the shortest path to x found so far. $D(t)$ is set to zero to indicate that the shortest path from t to t has length 0, and $D(x)$ is set to infinity for every other node x to indicate that no paths to it have been found. A node x has been *reached* when $D(x) < \infty$. A node is *processed* when one tries to reach other nodes from it. The list L contains nodes that have been reached but have not been processed since they were last reached. Initially $L = \{t\}$.

At each step of the algorithm, pick a node x from L to process, unless L is empty, in which case the algorithm terminates. For each B-arc e whose tail contains x, no unreached nodes, and no nodes belonging to L (other than x), do the following: let $y = H(e)$, update $D(y)$ by setting

$$D(y) = \min\left\{D(y), w(e) + \max_{z \in T(e)} \{D(z)\}\right\}$$

remove x from L, and add y to L if $D(y)$ is changed (unless y already belongs to L).

When the algorithm terminates, $D(x)$ is the shortest distance from t to x. If $D(x) = \infty$, there is no B-path from t to x. The labels $D(x)$ for the B-graph of Fig 5.1 are indicated next to each node, on the assumption that each B-arc has unit length.

5.3.3 Extension to Universally Quantified Logic

The dual partial instantiation method of Section 5.2.3 may now be extended to datalog formulas using a hypergraph model. The shortest B-path algorithm of the previous section provides an intuitively appealing heuristic for generating the conjunction $\bar{B}$ of clauses that avoid blockage.

5.3 A METHOD BASED ON HYPERGRAPHS

Consider the following rule set, which is adapted from an example of Gallo and Rago [105]. Let $P(x,y)$ mean that x is a parent of y, and $A(x,y)$ mean that x is an ancestor of y.

$$\begin{aligned}
& t \to P(a,b) \\
& t \to P(b,c) \\
& (\forall x)(\forall y)(P(x,y) \to A(x,y)) \\
& (\forall x)(\forall y)(\forall z)(P(x,y) \wedge A(y,z) \to A(x,z)) \\
& (\forall x)(A(b,x) \to f)
\end{aligned} \quad (5.21)$$

The first two rules say that a, b, and c form a line of descent. The question is whether it is consistent with this information to say that b is no one's ancestor (clearly it is not).

The problem in Skolem normal form is

$$C_1 \wedge C_2 \wedge \forall(x_3, y_3) C_3 \wedge \forall(x_4, y_4, z_4) C_4 \wedge \forall x_5 C_5$$

with

$$\begin{aligned}
C_1 &= P(a,b) \\
C_2 &= P(b,c) \\
C_3 &= P(x_3, y_3) \to A(x_3, y_3) \\
C_4 &= (P(x_4, y_4) \to A(x_4, y_4)) \to A(x_4, z_4) \\
C_5 &= \neg A(b, x_5)
\end{aligned}$$

To apply the dual partial instantiation algorithm, each F_k is represented in hypergraph form. For instance, $F_0 = C_1 \wedge \cdots \wedge C_5$ is depicted by solid lines in Figure 5.2. The hyperarcs drawn with solid lines are denoted *blue arcs* to distinfguish them as representing the clauses of F_k rather than the clauses of B, which will be denoted by *red arcs* (dashed lines). Gallo and Rago also use some additional blue arcs that connect any head predicate with a tail predicate that is an instance of it. In Figure 5.2, the blue arcs $A(x_3, y_3) \to A(y_4, z_4)$ and $A(x_4, z_4) \to A(b, x_5)$ are added. When new clauses C_i are generated by instantiation, new blue arcs are added to represent them.

Now the satisfiability check of step 2 can be carried out by applying a shortest B-path algorithm. Because the additional blue arcs connect every head predicate to every tail predicate of which it is a variant, the algorithm will find the problem to be satisfiable only if there is a variant-independent satisfying assignment. It is convenient to say that a node is *blue-reachable* if there is a B-path from t to it consisting of blue arcs. So F_k is unsatisfiable if f is blue-reachable.

The blockage-avoiding clauses in B, generated in step 3, are now represented in the hypergraph as red arcs (dashed lines in Figure 5.2). $F_k \wedge B$ is unsatisfiable if there is a B-path from t to f using both blue and red arcs.

Figure 5.2

$P(a,b) \dashrightarrow P(x_3, y_3) \longrightarrow A(x_3, y_3)$

t

$P(b,c)$

$A(y_4, z_4)$

$A(x_4, z_4)$

$P(x_4, y_4)$

$A(b, x_5) \longrightarrow f$

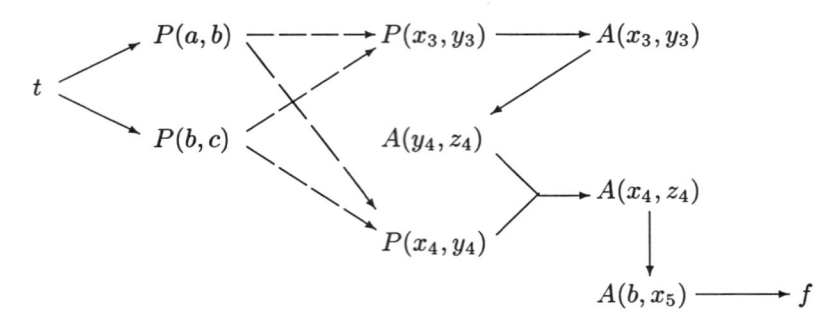

Figure 5.2: A hypergraph for datalog formulas.

The shortest B-path computation provides a heuristic for choosing a subset $\bar{B}$ of the red arcs that result in unsatisfiability, and for selecting the clause from $\bar{B}$ that is to generate the next instantiation. The motivation is as follows. If F is unsatisfiable, the goal is to instantiate enough clauses so that when their blue arcs are added to the hypergraph, f is blue-reachable. Presumably a short blue B-path to f will take fewer iterations to generate than a long one. If so, a reasonable heuristic is to instantiate the rule corresponding to a red arc in a shortest B-path, on the theory that the shortest B-path may evolve into a short blue B-path (if there is one). $\bar{B}$ should therefore consist of the red arcs on a shortest B-path from t to f. In Figure 5.2, one such path contains all the predicates except $P(b,c)$, so that $\bar{B}$ is the conjunction of the two red arcs $P(a,b) \rightarrow P(x_3, y_3)$ and $P(a,b) \rightarrow P(x_4, y_4)$.

It remains to pick a red arc in step 4 to induce an instantiation. There is no reason one cannot pick several, and a reasonable heuristic is to pick the initial red arcs in the shortest B-path; that is, all red arcs whose tail nodes are blue-reachable. This tends to build an advancing frontier of blue arcs. In Figure 5.2, this would be both arcs in $\bar{B}$. So (a,b) is substituted for (x_i, y_i) in C_3 and C_4, resulting in two new clauses. When these clauses are added to the hypergraph, and identical nodes identified, the blue arcs of Figure 5.3 result.

At this point the algorithm reverts to step 2, and the process is repeated with a new set of red arcs and a new shortest B-path. Figure 5.4 depicts the result of the second iteration and Figure 5.5 the result of the third, in which f is blue reachable and the formula is proved to be unsatisfiable.

The algorithm may be stated as follows. Let $F = \bigwedge_{i=1}^{m} \forall x_i C_i$ be a datalog formula (i.e., a Horn formula in Skolem normal form). It is understood that each B-graph is simplified by identifying nodes that represent the same atom.

5.3 A METHOD BASED ON HYPERGRAPHS

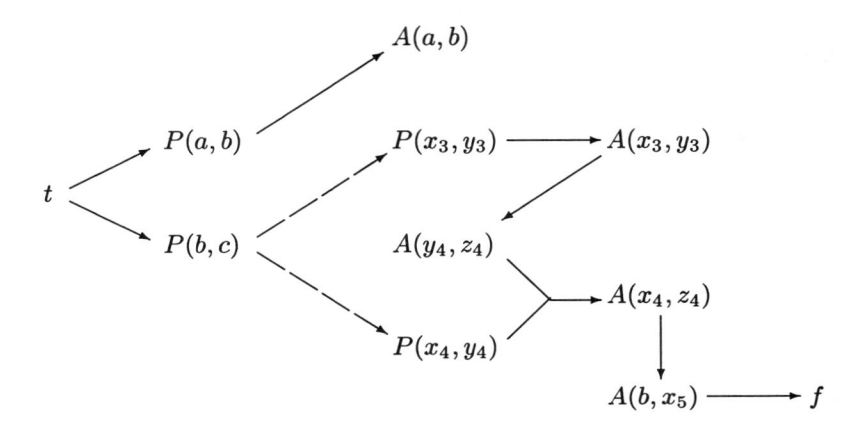

Figure 5.3: The hypergraph after one iteration.

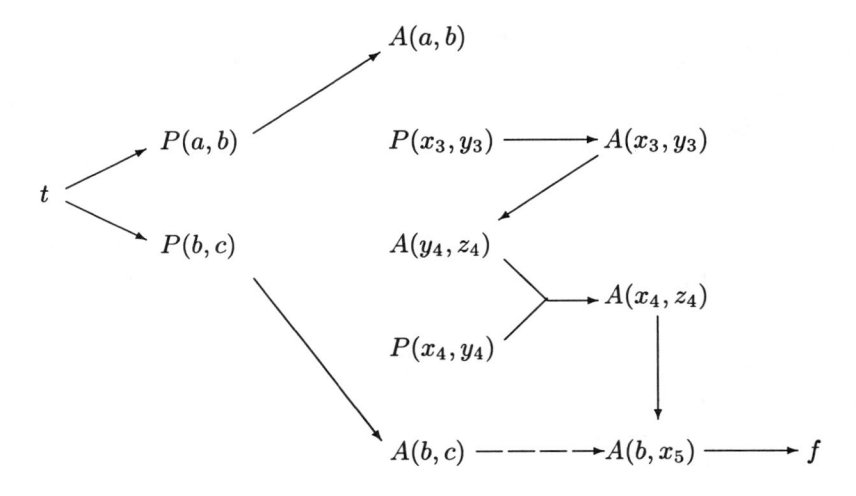

Figure 5.4: The hypergraph after two iterations.

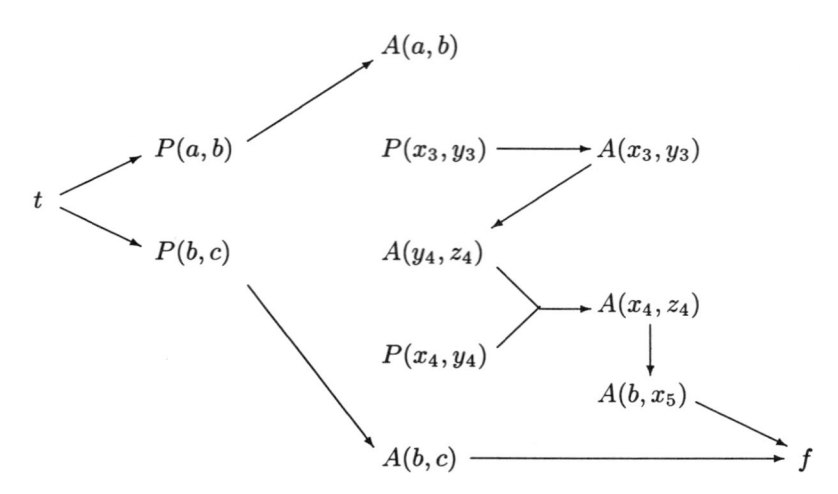

```
                        A(a, b)
                       ↗
         P(a, b)          P(x₃, y₃) ──────→ A(x₃, y₃)
       ↗
  t                                ↙
       ↘ P(b, c)       A(y₄, z₄)
                                              ↘ A(x₄, z₄)
                       P(x₄, y₄) ↗              ↓
                                              A(b, x₅)
                                                     ↘
         ↘  A(b, c) ─────────────────────────────→ f
```

Figure 5.5: The hypergraph after three iterations.

Procedure 28: *Hypergraph-Based Partial Instantiation*

Step 1. (*Initialization.*) Let H_0 be a B-graph representing $C_1 \wedge \cdots \wedge C_m$, and set $k = 0$.

Step 2. (*Ground satisfiability.*) Add B-arcs to H_k representing all rules of the form $P(t) \rightarrow P(t')$, where $P(t)$ is in the head of a B-arc of H_k, $P(t')$ is in the tail of a B-arc, and $P(t')$ is an instance of $P(t)$. If there is a B-path in H_k from t to f, stop; F is unsatisfiable.

Step 3. (*Blockage avoidance.*) Let H' consist of H_k plus a red arc representing the rule $P(t) \rightarrow P(t')$ for every $P(t), P(t')$ such that: (a) $P(t)$ is in the head of a B-arc A of H_k; (b) $P(t')$ is in the tail of a B-arc A' of H_k; (c) there is an mgu $P(t)\sigma = P(t')\tau$, where τ is not a renaming substitution; and (d) if C, C' are the clauses represented by B-arcs A, A', then either $C\sigma$ generalizes no clause represented by a B-arc in H_k, or this is true of $C'\tau$. Compute a shortest B-path in H' from t to f. If there is no such path, stop; F is satisfiable.

Step 4. (*Refinement.*) Let H_{k+1} consist of all the B-arcs of H_k plus each red arc A of H' for which (a) A lies in the shortest B-path computed in the previous step; and (b) the tail nodes of A are reachable in H_k from t. Increase k by one and return to step 2.

5.3.4 Answering Queries

In the process of discovering a contradiction in the example of the previous section, the algorithm found an instantiation of the final rule $A(b, x) \to f$ in (5.21), namely, $A(b, c) \to f$, that completes the blue path from t to f. This reveals that not only is it inconsistent to say that no one is b's ancestor, but that c in particular is b's ancestor.

This means that one can pose a query, "Who are b's descendants?", and obtain an answer by adding $A(b, x) \to f$ to the data set and finding an instance that completes a blue path from t to f. If no such path is found, the query has no answer. To find *all* descendants of b, one can simply remove the blue arc $A(b, c) \to f$ and continue the algorithm. If the query has another answer, another blue B-path will be found. In this case, the algorithm terminates immediately, because c is b's only descendant.

In general, one can pose a query, "What instantiates predicate $P(x)$?", by adding $(\forall x) P(x) \to f$ to the data base. The above algorithm is applied until a blue path from t to f is found; if none is found, it stops. Then if $P(t) \to f$ is the partial instantiation of $P(x) \to f$ that occurs in the blue path, all complete instantiations of $P(t)$ answer the query. The blue arc from $P(t)$ to f is then removed, and the procedure repeated.

5.4 An Infinite 0-1 Programming Model

The approach taken so far has been to reduce an inference problem in first order logic to a series of propositional satisfiability problems involving partial Herbrand extensions. The propositional inference problem can then be embedded in a mathematical programming model, but purely for the sake of solving the problem rather than obtaining structural insight.

In this section, the Herbrand extension is used for structural analysis based on topology. The framework this yields is that of infinite dimensional mathematical programming. Elementary techniques of topology are used to show that Herbrand's theorem is a simple consequence of compactness in *product topologies*. By generalizing the proofs of Section 2.3.2 one can also prove the existence of an integer least element for predicate Horn relaxations. This geometric proof of the unique "minimal" model property of first-order Horn logic provides a mathematical programming foundation for analyzing model theory of logic programming.

While the use of the embedding to analyze the Herbrand extension of a first-order formula is an interesting application, the real challenge would be to "liberate" theorem proving from the clutches of the restrictive (and sometimes unnatural) Herbrand universe and yet maintain the semidecidable complexity of theorem proving. The framework provided here offers a glimmer of hope for accomplishing this objective, as our compactness theorems apply to infinite linear (0-1) programming problems whose constraints and variables need not be denumerable.

A compactness theorem is first proved for infinite dimensional 0-1 programming problems. This result is then applied to Herbrand theory and first-order Horn logic.

5.4.1 Infinite Dimensional 0-1 Programming

Consider a mathematical program of the form

$$\mathcal{D} = \{x \in \{0,1\}^\omega \mid Ax \geq b\} \tag{5.22}$$

where ω denotes (uncountable) infinity. Each row of the matrix A has entries that are $0, \pm 1$, and each entry of the (uncountably) infinite column b is one less than the number of -1's in the corresponding row of A. So this is just an infinite version of the integer programming formulation of satisfiability in propositional logic (Section 2.2). The finite support of the rows of A is the important structural property that permits the compactness theorems based on product topologies to go through in the ensuing development.

A discussion of Horn logic leads one to the continuous (linear programming) relaxation of the infinite mathematical program (5.22):

$$\bar{\mathcal{D}} = \{x \in [0,1]^\omega \mid Ax \geq b\} \tag{5.23}$$

Let $\{A_\alpha x \geq b_\alpha\}_{\alpha \in \mathcal{I}}$ denote a suitable indexing of all finite subfamilies of $\{Ax \geq b\}$. And for each α in the uncountable set $\mathcal{I}$ let

$$\mathcal{D}_\alpha = \{x \in \{0,1\}^\omega \mid A_\alpha x \geq b_\alpha\}$$

$$\bar{\mathcal{D}}_\alpha = \{x \in [0,1]^\omega \mid A_\alpha x \geq b_\alpha\}$$

Thus,

$$\mathcal{D} = \bigcap_{\alpha \in \mathcal{I}} \mathcal{D}_\alpha$$

$$\bar{\mathcal{D}} = \bigcap_{\alpha \in \mathcal{I}} \bar{\mathcal{D}}_\alpha$$

The analysis of finite dimensional mathematical programming problems is based on elementary techniques from combinatorics and polyhedral theory. The situation in the infinite dimensional case is more complicated.

5.4 AN INFINITE 0-1 PROGRAMMING MODEL

Constraint qualification is a sticky issue even for semi-infinite mathematical programs. The standard approach in infinite dimensional mathematical programming is to impose an appropriate (weak) topological framework on the feasible region and then use the power of topology to develop the structural theory.

5.4.2 A Compactness Theorem

A classical result in finite dimensional linear programming states that if a finite system of linear inequalities in $\Re^d$ is infeasible, there is a "small" $(d+1)$ subsystem that is also infeasible. This compactness theorem is a special case of the ubiquitous Helly's theorem. Analogous theorems are also known for linear constraints on integer valued variables (cf. [249]). In the infinite dimensional case, one could hope for the "small" witness of infeasibility to simply be a *finite* witness. This is exactly what will be proved for mathematical programs of the form (5.23) and (5.22).

Let $\mathcal{S}_\gamma$, $\gamma \in \mathcal{G}$, be copies of a Hausdorff space $\mathcal{S}$. Let $\mathcal{S}^{\mathcal{G}} = \prod_{\gamma \in \mathcal{G}} \mathcal{S}_\gamma$. The *product topology* on $\mathcal{S}^{\mathcal{G}}$ is the topology defined by a basis $\prod_\gamma O_\gamma$, where the O_γ are open in $\mathcal{S}_\gamma$ and $O_\gamma = \mathcal{S}_\gamma$ for all but at most finitely many $\gamma \in \mathcal{G}$. A classical theorem on compact sets with the product topology is that of Tychonoff (cf. [220], page 232) which states the following:

Theorem 60 *Arbitrary (uncountable) products of compact sets with the product topology are compact.*

Taking $\{0,1\}$ and $[0,1]$ as compact sets of Hausdorff spaces $\{0,1\}$ and $[0,1]$, respectively, an application of Tychonoff's theorem yields the following:

Corollary 10 $\{0,1\}^\omega$ *and* $[0,1]^\omega$ *(with the product topology) are compact.*

Next to be demonstrated is that $\mathcal{D}_\alpha$ and $\bar{\mathcal{D}}_\alpha$, with product topologies, are also compact for any α in $\mathcal{I}$. This follows from the corollary and the lemma below.

Lemma 31 *The set* $\{x \mid A_\alpha x \geq b_\alpha\}$ $(\alpha \in \mathcal{I})$ *is closed and hence compact.*

Proof: Let y be a point in the complement of $\{x \mid A_\alpha \geq b_\alpha\}$. So there must be at least one violated constraint in the system $A_\alpha x \geq b_\alpha$ of the form

$$\sum_{j \in J_i} A_{ij} y_j < b_i$$

Because $|J_i|$ is finite,

$$B_\epsilon = \{z \mid |z_j - y_j| < \epsilon, \forall j \in J_i\}$$

is an open set. And for sufficiently small ϵ, $B_\epsilon \subset \{x \mid A_\alpha \geq b_\alpha\}^C$, where C designates the complement. Hence $\{x \mid A_\alpha \geq b_\alpha\}^C$ is open and $\{x \mid A_\alpha \geq b_\alpha\}$ is closed. □

The main theorem can now be proved.

Theorem 61 $\mathcal{D}$ *($\bar{\mathcal{D}}$) is empty if and only if $\mathcal{D}_\alpha$ ($\bar{\mathcal{D}}_\alpha$) is empty for some $\alpha \in \mathcal{I}$.*

Proof: Suppose $\mathcal{D}$ is empty and $\mathcal{D}_\alpha$ is nonempty for all $\alpha \in \mathcal{I}$. Then, for every $\mathcal{K} \subset \mathcal{I}$ with $|\mathcal{K}| < \infty$,

$$\bigcap_{\alpha \in \mathcal{K}} \mathcal{D}_\alpha \neq \emptyset$$

So by the finite intersection property (cf. [220] page 171), $\mathcal{D}$ is nonempty—a contradiction. The proof for $\bar{\mathcal{D}}$ is identical. □

5.4.3 Herbrand Theory and Infinite 0-1 Programs

As noted earlier in the chapter, any formula of first-order logic can be converted to Skolem normal form. This formula is in turn equivalent to its Herbrand extension, a (possibly infinite) formula of propositional logic. It is straightforward to formulate the satisfiability problem for this formula in a 0-1 programming problem. Because the Herbrand extension may contain a (countably) infinite number of clauses, the embedding will be a special case of the infinite 0-1 problem (5.22). Herbrand's compactness result (Theorem 55) now follows immediately from Theorem 61.

This result may be viewed as the cornerstone of theorem proving in predicate logic, because it implies that constructing an unsatisfiability proof for a formula known to be unsatisfiable is always a finite task. Simply develop the Herbrand extension incrementally and check the resulting finite propositional formula for satisfiability. Of course there have been many refinements to this scheme since Herbrand, but the basic construct remains the same. The infinite 0-1 embedding that was just used to prove Herbrand's theorem can also be specialized and honed to shed light on these more recent aspects of theorem proving. This is illustrated in the next section.

5.4.4 Minimum Solutions

The minimum element property of Horn formulas discussed in Section 2.3.2 can be extended to quantified Horn formulas in Skolem normal form, to obtain a well-known result regarding minimal models for definite logic programs. The formulas are Horn in the sense that each clause in the matrix contains at most one positive atom. The minimal element property is easily demonstrated by using the relaxed linear embedding (5.23) of the Herbrand extension.

Assuming now that H is a Horn formula in Skolem normal form, consider the following infinite dimensional optimization problem.

$$\inf \sum_j x_j \qquad (5.24)$$

$$\text{s.t.} \quad Ax \geq b$$

$$x \in [0,1]^\omega \qquad (5.25)$$

where the linear inequalities $Ax \geq b$ are simply the clausal inequalities corresponding to the ground clauses of H. The syntactic restriction on Horn clauses translates to the restriction that each row of A has at most one $+1$ entry (all other entries are either 0 or -1's—only finitely many of the latter, though). It will now be proved that if the infinite linear program (5.24) has a feasible solution then it has an integer optimal (0-1) solution. Moreover, this solution will be a least element of the feasible space: that is, it will simultaneously minimize all components over all feasible solutions.

Lemma 32 *If the linear programming problem (5.24) is feasible then it has a minimum solution.*

Proof: Let $\psi_n = \sum_{j=1}^n x_j$ and $\Psi = \sup_n \psi_n$. As the supremum of continuous functions, Ψ is lower semi-continuous. So the optimization problem (5.24) seeks to find the infimum of a lower semi-continuous function over a compact set. Therefore, the minimum is attained. □

Lemma 33 *If x^1 and x^2 are both feasible solutions for (5.24) then so is $\bar{x} = (\bar{x}_j)_j = (\min\{x_j^1, x_j^2\})_j$.*

Proof: Let x^i be partitioned into (y^i, z^i) ($i = 1, 2$) such that the components of y^1 are no larger than the components of y^2 and the components of z^2 are no larger than the components of z^1. Now if an inequality in the constraints of (5.24) has a $+1$ coefficient on a y variable (or if the inequality has no $+1$ coefficient at all), note that (y^1, z^1) satisfies the inequality and therefore so does (y^1, z^2) since the z-coefficients are all nonpositive. Similarly, if an inequality in the constraints of (5.24) has a +1 coefficient on a z variable we note that (y^2, z^2) satisfies the inequality and therefore so does (y^1, z^2) since the y-coefficients are all nonpositive. Therefore, in all cases, (y^1, z^2) is feasible. □

Theorem 62 *If the linear program (5.24) is feasible, then it has a unique 0-1 optimal solution which is the least element of the feasible set.*

Proof: If the feasible region of (5.24) is nonempty, an optimal solution exists. Let x^* be such an optimal solution. If x^* has all 0-1 components there is nothing to prove. Else let $\tilde{A}\tilde{x} \geq \tilde{b}$ be obtained from $Ax \geq b$ by fixing all components $x_j = x_j^*$ for all 0-1 valued x_j^* and clearing the constants to the right-hand-side of the inequalities to obtain $\tilde{b}$. Note that $\tilde{b}$ is integer valued and therefore nonpositive. If an inequality had $b_i > 0$, then $b_i \geq 1$, which is impossible to satisfy with each $\tilde{x}_j$ fractional and in $[0,1]$. Hence one can set $\tilde{x}$ to 0 and maintain feasibility. This contradicts the optimality of x^* in (5.24). □

The interpretation of this theorem in the logic setting is that if a Horn formula H has a model then it has a least model (a unique minimal model). This is an important result in model theory (semantics) of so-called definite logic programs.

Neither of the applications described here (Herbrand's theorem and Horn model theory) uses the full power of the embedding and the compactness theorem. This is because the Herbrand extension grounds a predicate formula into a countably infinite propositional form while the embedding (and the compactness theorems) admit uncountable structures. One focus of future work should be to see if this latitude could be exploited to work out a decision theory for first-order logic that goes beyond the Herbrand setting.

Another important issue related to the uncountable nature of the embedding is whether the proof of the compactness theorem can be made constructive even for this case. An obvious idea is to identify a countable subsystem to which to restrict the search. This is in effect what is done in logic, because the Herbrand universe and the Herbrand extension provide just such a substructure.

These embedding results can be usefully applied to better our understanding of constraint logic programming (CLP). In CLP there has been an attempt to mix pure first-order logic predicates with constraint predicates that can be of the type considered in mathematical programming formulations. The embedding now permits a unified setting for carrying out this integration. This theme will be elaborated upon in Section 6.4.

5.5 The Logic of Arithmetic

5.5.1 Decision Methods for Arithmetic

Presburger [237] proved in 1929 that when quantifiers and logical connectives are added to the arithmetic of natural numbers (resulting in Presburger arithmetic), a decidable system results. More precisely, Presburger arithmetic is built up from natural numbers (plus negative integers if desired), integer variables, addition (and subtraction if desired), arithmetical relations ($<, \leq, >, \geq, =$), quantifiers ($\forall$ and $\exists$), and the usual propositional connectives. The formulas must be *closed* (i.e., contain no free variables). Since a sum $x + x$ can be written $2x$, multiplication by an integer is implicitly present, but the product of two variables cannot appear.

Although Presburger arithmetic is decidable, the most efficient known decision procedure is Cooper's [69], which has worst-case complexity on the order of $2^{2^{2^n}}$, where n is the length of the formula in question. Work by Oppen [227] suggests that this is essentially the best one can do in the worst case.

The $\forall$ fragment of Presburger arithmetic, however, is more tractable, since the decision complexity is bounded above by 2^n. Bledsoe [25, 26] developed a *SUP-INF method* for the $\forall$ fragment that was later improved by Shostack [261]. Shostack [262] also extended the method to the $\forall$ fragment with uninterpreted predicate letters and function symbols added.

There are also decision methods for the arithmetic of real numbers without multiplication; see [36] for a general discussion. If Presburger arithmetic is extended so as to contain real constants and variables that range over real numbers, the result is *Presburger real arithmetic*. The $\forall$ fragment of this arithmetic is sometimes called *Bledsoe real arithmetic*. The SUP and INF algorithms within the SUP-INF method provide a decision method for Bledsoe real arithmetic, and Shostack's method for accommodating function symbols (and therefore Skolem functions) provides a decision procedure for full Presburger real arithmetic.

Finally, we note that Tarski [270] discovered in 1930 a decision method for real arithmetic with both addition and multiplication. The method has been improved several times, and apparently the most recent and most efficient (worst case $2^{2^{kn}}$) method is that of Collins [62].

Our main purpose here is to survey the use of quantitative methods for solving the decision problem in various arithmetics. We will present methods for the $\forall, \exists, \forall, \exists$ and $\exists\forall$ fragments of real arithmetic and then modify the methods for integer arithmetic.

5.5.2 Quantitative Methods for Presburger Real Arithmetic

Williams [293] has pointed out that linear programming can be useful for solving decision problems in Presburger real arithmetic. Here we treat the various fragments separately. Since there are no uninterpreted predicates in Presburger arithmetic, a formula is satisfiable if and only if it is valid. This means that the $\exists$ and $\forall$ fragments present essentially the same problem, as we see below.

The $\exists$ Fragment. Suppose we want to test whether the formula

$$\neg \forall x \forall y \forall z [(2x < y \lor x + y \geq 5 \lor x > 6) \land (y + z < 2 \lor y + z \geq 1)] \quad (5.26)$$

is valid. In prenex form the formula is

$$\exists x \exists y \exists z \neg [(2x < y \lor x + y \geq 5 \lor x > 6) \land (y + z < 2 \lor y + z \geq 1)]$$

so that (5.26) indeed belongs to the $\exists$ fragment. We next rewrite the expression in brackets (the *matrix*) in disjunctive normal form:

$$\exists x \exists y \exists z [(2x \geq y \land \neg(x + y = 5) \land x \leq 6) \lor (y + z \geq 2 \land y + z < 1)]$$

An equivalent formula gives each disjunct its own quantifiers:

$$\exists x \exists y (2x \geq y \land \neg(x+y = 5) \land x \leq 6) \lor \exists y \exists z (y + z \geq 2 \land y + z < 1) \quad (5.27)$$

Thus (5.26) is valid if and only if either disjunct of (5.27) is true.

To check whether the first disjunct of (5.27) is true, we first note that $\neg(x + y = 5)$ is equivalent to $(x + y \leq 5 - \epsilon) \lor (x + y \geq 5 + \epsilon)$, where ϵ is an arbitrarily small positive number. Now we need only determine whether at least one of the following systems of inequalities has a real solution.

$$\begin{array}{ll} 2x - y \geq 0 & 2x - y \geq 0 \\ x + y \leq 5 - \epsilon & x + y \geq 5 + \epsilon \\ x \leq 6 & x \leq 6 \end{array} \quad (5.28)$$

This can be done by using a linear programming routine to check for feasibility. Since both linear programs are feasible in this case, the first disjunct of (5.27) is true and (5.26) is valid. If the first disjunct had been false, we would have checked the second disjunct in similar fashion.

The $\forall$ Fragment (Bledsoe Real Arithmetic). It is equally easy to deal with the $\forall$ fragment: simply negate the formula, so as to produce a formula in the $\exists$ fragment. Then the original formula is valid if and only if none of the resulting linear programs are feasible. Thus linear programming routines solve problems traditionally solved by the SUP and INF procedures.

5.5 THE LOGIC OF ARITHMETIC

Conversely, these procedures can be used to solve linear programs [261], but they are grossly inefficient for large examples.

The $\forall \exists$ Fragment. When $\forall$ quantifiers precede $\exists$ quantifiers, we can combine linear programming with a variable elimination method to obtain a decision procedure. Suppose we wish to check the validity of

$$\forall x \exists y \exists z [(x \geq 2y + 2z \wedge 2x \geq -3y - 6 \wedge x \geq y - z) \vee \neg(x \geq 0)] \quad (5.29)$$

The formula is already in prenex form, and the matrix already in disjunctive normal form. We next eliminate all occurrences of $<, >, =$ and $\neg$ by using ϵ's and disjunctions, as in the previous example. This yields

$$\forall x \exists y \exists z [(x \geq 2y + 2z \wedge x \geq -y - 2 \wedge x \geq y - z) \vee x \leq -\epsilon]$$

which is equivalent to

$$\forall x [\exists y \exists z (x \geq 2y + 2z \wedge 2x \geq -3y - 6 \wedge x \geq y - z) \vee x \leq -\epsilon] \quad (5.30)$$

The next step is to eliminate the existentially quantified variables in both disjuncts of (5.30). Since the second disjunct has none, we focus on the first. We use *Fourier-Motzkin elimination*, which goes as follows. To eliminate y we first "solve" each inequality for y:

$$\begin{array}{rl} y & \leq (1/2)x - z \\ -(2/3)x - 2 \leq y & \\ y & \leq x + z. \end{array} \quad (5.31)$$

To say that some y satisfies these relations is to say that we can pair the expressions on the left with those on the right to form new valid inequalities that exclude y:

$$\begin{array}{rl} -(2/3)x - 2 & \leq (1/2)x - z \\ -(2/3)x - 2 & \leq x + z \end{array} \quad (5.32)$$

Simplifying, this becomes

$$\begin{array}{rl} -(7/6)x + z & \leq 1 \\ -(5/3)x - z & \leq 2 \end{array}$$

We now eliminate z in the same fashion to obtain

$$x \geq -24/17$$

We conclude that there exist y and z satisfying the first disjunct of (5.30) if and only if $x \geq -24/17$.

Since y and z do not occur in the second disjunct of (5.30), we can now write (5.30) with y and z eliminated:

$$\forall x(x \geq -24/17 \ \lor \ x \leq -\epsilon)$$

This formula belongs to Bledsoe real arithmetic and can be evaluated as described above. In this case the formula is obviously valid, so that (5.29) is valid.

We remark in passing that Fourier-Motzkin elimination is a classical method for projecting a polyhedron. In this case we projected the polyhedron described by the inequalities in the first disjunct of (5.30), which sits in (x, y, z)-space, onto a 1-dimensional space, the x axis. The result is a ray described by $x \geq -24/17$.

The $\exists\forall$ Fragment. Prenex formulas in this form can be negated to produce a formula in $\forall\exists$ form, which can be treated as above. If the latter formula is false, the original formula is valid. Formulas in neither the $\forall\exists$ nor the $\exists\forall$ fragment, such as $\exists x \forall y \exists z(x + y \geq z)$, require Skolem functions.

5.5.3 Quantitative Methods for Presburger (Integer) Arithmetic

When variables and constants are restricted to integers, the above methods can be adapted to check for validity.

The $\exists$ Fragment. Merely use an integer rather than a linear programming routine to check the systems (5.28) for feasibility, since we now require x and y to be integers. Also, $x + y \leq 5 - \epsilon$ can be replaced with $x + y \leq 4$, and similarly for $x + y \geq 5 + \epsilon$.

The $\forall\exists$ Fragment. We can use Williams's modification of Fourier elimination to eliminate variables [290]. Note that since y must now be an integer, (5.31) is no longer equivalent to (5.32). Rather, it is equivalent to

$$-4x - 12 \leq \ 6y \ \leq 3x - 6z \qquad (5.33)$$
$$-2x - 6 \leq \ 3y \ \leq 3x + 3z \qquad (5.34)$$

We have multiplied the inequalities by the necessary constants to maintain integer coefficients. Williams points out that (5.34) is equivalent to the following disjunction:

$$\bigvee_{i=0}^{5} [-4x - 12 + 1 \leq 3x - 6z \ \land \ -4x - 12 + i \equiv 0 \ (\text{mod } 6)] \qquad (5.35)$$

5.5 THE LOGIC OF ARITHMETIC

Since the congruence has no solution for $i = 1, 3, 5$, (5.35) simplifies to

$$[-7x + 6z \leq 12 \ \wedge \ x \equiv 0 \ (\text{mod } 3)]$$
$$\vee \ [-7x + 6z \leq 8 \ \wedge \ x \equiv 1 \ (\text{mod } 3)] \quad (5.36)$$
$$\vee \ [-7x + 6z \leq 10 \ \wedge \ x \equiv 2 \ (\text{mod } 3)]$$

Similarly, (5.34) is equivalent to

$$[-5x - 3z \leq 6 \ \wedge \ x \equiv 0 \ (\text{mod } 3)]$$
$$\vee \ [-5x - 3z \leq 5 \ \wedge \ x \equiv 1 \ (\text{mod } 3)] \quad (5.37)$$
$$\vee \ [-5x - 3z \leq 4 \ \wedge \ x \equiv 2 \ (\text{mod } 3)]$$

The conjunction of (5.36) and (5.37) simplifies to

$$[-7x + 6z \leq 12 \ \wedge \ -5x - 3z \leq 6 \ \wedge \ x \equiv 0 \ (\text{mod } 3)]$$
$$\vee \ [-7x + 6z \leq 8 \ \wedge \ -5x - 3z \leq 5 \ \wedge \ x \equiv 1 \ (\text{mod } 3)] \quad (5.38)$$
$$\vee \ [-7x + 6z \leq 10 \ \wedge \ -5x - 3z \leq 4 \ \wedge \ x \equiv 2 \ (\text{mod } 3)]$$

Having eliminated y, we now eliminate z from each disjunct of (5.38). From the first disjunct we get

$$-10x - 12 \leq 6z \leq 7x + 12 \ \wedge \ x \equiv 0 \ (\text{mod } 3) \quad (5.39)$$

We can expand this as in (5.35), except that now we must combine the congruence $x \equiv 0 \ (\text{mod } 3)$ with each congruence in the expansion. In general, a set of congruences is either inconsistent or can be written as a single congruence, using the Chinese remainder theorem. In this case, we get mostly inconsistencies, and (5.39) is equivalent to

$$-17x \leq 24 \ \wedge \ x \equiv 0 \ (\text{mod } 3)$$

Similar treatment of the other two disjuncts yields that (5.29) is equivalent to,

$$\forall x[(17x \geq -24 \ \wedge \ x \equiv 0 \ (\text{mod } 3))$$
$$\vee \ (17x \geq -16 \ \wedge \ x \equiv 1 \ (\text{mod } 3)) \quad (5.40)$$
$$\vee \ (17x \geq -16 \ \wedge \ x \equiv 2 \ (\text{mod } 3)) \ \vee \ x \leq -1].$$

This formula is in the $\forall$ fragment and can be treated as above. To apply integer programming we must remove the congruences, but any congruence $A \equiv a \ (\text{mod } b)$ can be removed by adding the constraint $A = bw + a$, where w is a new integer variable. We can see on inspection that (5.40) is true, and we conclude that (5.29) is valid even for integers.

The $\exists \forall$ Fragment. Negate and check for falsehood, as above.

Chapter 6

Nonclassical and Many-Valued Logics

Propositional, predicate, and to a lesser extent probabilistic and many-valued logics can be considered classical. The focus in this book has been on exploring the use of optimization methods for solving inference problems in classical logics. However, modern usage of logic formalisms in specific application domains has resulted in new logics that require alternate inference mechanisms. It would be natural to end the book with some discussion of how the mathematical programming of logic can be extended to encompass some of these new inference problems. This is what we set out to accomplish in this chapter. We describe mathematical programming embeddings of inference in nonmonotonic, many-valued, modal, and partially interpreted logics.

The presentation is largely focused on formulation of inference in these logics as mathematical programs. There is little discussion of the specialization of mathematical programming methods to actually solve the inference problems. This ommission is due mostly to the fact that such specializations are yet to be worked out. This chapter is therefore an invitation to mathematical programmers to take a closer look at these new inference problems.

Constraint logic programming (CLP) began as a natural merger of two declarative paradigms: constraint solving and logic programming. This combination helps make CLP programs both expressive and flexible, and in some cases more efficient than other kinds of programs. Of particular interest here is the variety of CLP denoted CLP($\Re$), in which constraints are linear inequalities on real-valued variables. Thus CLP($\Re$) brings together the techniques of linear programming and logic programming in a declarative setting. In Section 6.4 we will see that the infinite dimensional linear programming embedding of definite logic programs (see Section 5.4) naturally extends to an embedding of inference in CLP($\Re$). This embedding requires some technical extensions of the compactness theorems proved in the last chapter to accommodate real-valued variables that are unbounded.

Modal logic can be described briefly as the logic of necessity and possibility. Modalities have been found to be very useful in recent times for the temporal specification of systems (e.g., see [214]) using temporal logics. We simply observe in Section 6.3 that inference in propositional modal logic can be reformulated as inference in first-order or predicate logic. From the results of the previous chapter, it follows that integer programming methods can be used for solving inference in modal logics.

Many-valued logics have been proposed and studied by many ancient civilizations (e.g., the Saad Nyaya system of seven-valued logic used by Jain monks). More modern uses have been the use of 3- and 4-valued logics in the semantics of default reasoning. Hardware verification of switch-level models of transistor circuits using an integer-programming-based theorem prover for propositional many-valued logic has been reported by Hähnle and Kernig [122]. The many values of "truth" correspond to different levels of voltage in the circuit. In Section 6.2 we will see a simple embedding of deduction in propositional many-valued logics as mixed integer programs.

We begin with nonmonotonic logics, which have been popularized by research in default reasoning in artificial intelligence. The notion of default reasoning arises from wanting to conclude that a statement is true when it cannot be proved false. This "closed world assumption" is usually expressed through nonmonotonic negation in logic programs. While most of the effort has been on the semantics of various model-theoretic nuances of negation-as-failure schemes for logic programming, the task of actually devising computational strategies for inference in these formalisms has remained largely ignored. Bell et al. [18] have recently argued that integer programming methods may be in fact the most natural way of implementing nonmonotonic reasoning. It is this work that we review in the next section.

6.1 Nonmonotonic Logics

The ability to reason with temporarily held beliefs, in the absence of contrary knowledge, has been the motivation for the study of nonmonotonic reasoning systems. Also known as default reasoning, nonmonotonic reasoning within the frameworks of rule systems and logic programming has

6.1 NONMONOTONIC LOGICS

been an active area of research. A large part of the work has concentrated on semantic theories for nonmonotonic logics. A recent unifying theme has been to show that the notion of stable models of normal logic programs may be used as a basis to reformulate theories and extensions of default logic and truth maintenance systems [206].

There is a very simple integer programming scheme for computing the stable models of a normal logic program. In fact, this scheme also serves as the most straightforward and constructive definition of stable models. These ideas are due to Bell et al. [18] and will be the basis of our presentation here.

A normal logic program is a rule-based system with clauses (rules) of the form

$$(L_1 \wedge L_2 \wedge \cdots \wedge L_k) \to A_o$$

where A_o is an atomic predicate and L_i are literals (atomic predicates and their negations). For the moment let us assume that the predicates can have variables as arguments but no function symbols. This assumption implies that the Herbrand universe is finite and that complete instantiation would result in a ground system of rules that is finite. So if $\{x_1, x_2, \ldots, x_n\}$ denotes the set of ground predicates, we could as well consider this as a propositional rule-based system with the caveat that the consequent in every rule is a single atom and the antecedent can contain negative literals (hence any CNF formula can be expressed as a normal logic program).

A *minimal* model of a normal logic program is a Herbrand model that is minimal in the sense that the atoms set to true have no subset that forms a model. One simple way of finding a minimal model would be to solve the integer program

$$\begin{aligned} \min \quad & \sum_{i=1}^{n} x_i \\ \text{s.t.} \quad & Ax \geq b \\ & x \in \{0,1\}^n \end{aligned} \qquad (6.1)$$

where the constraints are the usual integer programming constraints of satisfiability. Starting with a minimal model M_1 it is an easy matter to generate another solving the integer program again with an added constraint

$$\sum_{i \in P(M_1)} x_i \leq k_1 - 1$$

where $P(M_1)$ denotes the set of subscripts i for which x_i is set to true in M_1, and k_1 denotes the cardinality of $P(M_1)$. It is easy to see that if we have an integer programming technique to generate all optimal solutions, it could be adapted to a systematic procedure for generating all minimal models of a logic program.

The reason for generating all minimal models of a logic program is that we can now sieve the stable models from this collection, defined as follows. Given a model M of a normal logic program, let the *Gelfond-Lifschitz transform* of the logic program with respect to M be formed by (a) selecting the rules of the logic program in which every atom that occurs negated in the antecedent is false in M, and (b) deleting all negated atoms from the antecedents of these rules. A minimal model M of a normal logic program is *stable* if it is an optimal solution to the integer program (6.1) defined on the Gelfond-Lipschitz transform of the logic program with respect to M.

In effect then we use integer programming to generate all minimal models and then sieve out the stable models by solving, for each minimal model, another integer program. Notice that the latter is actually finding the least model of a Horn formula, and so by the observations of Section 2.3 in Chapter 2 we know that a linear programming technique suffices. In fact, Horn resolution would clearly be the best technique for implementing the sieve.

As an example, consider the normal logic program,

1. Joe will visit us if he gets a day off, unless the train tickets are sold out.
2. Joe will go fishing if he gets a day off, unless it rains or he visits us.
3. The train tickets will sell out if it rains.
4. Joe will get a day off unless it rains.

If we do not know whether the tickets are sold out, rule 1 allows us to infer by default that Joe will visit if he gets a day off. If we later learn, however, that he cannot get a ticket, we must retract this inference. The reasoning is nonmonotonic in the sense it does not always move forward; we must sometimes backtrack in this fashion. Rules 2 and 4 are similarly interpreted. The rules can be written symbolically as,

$$\begin{aligned} (D \wedge \neg S) &\to V \\ (D \wedge \neg R \wedge \neg V) &\to F \\ R &\to S \\ \neg R &\to D \end{aligned} \tag{6.2}$$

To find minimal models we solve the integer programming problem,

$$\begin{aligned} \min \quad & d + f + r + s + v \\ \text{s.t.} \quad & -d + s + v \geq 0 \\ & -d + f + r + v \geq 0 \\ & -r + s \geq 0 \\ & d + r \geq 1 \\ & d, f, r, s, v \in \{0, 1\} \end{aligned} \tag{6.3}$$

6.2 MANY-VALUED LOGICS

where 0-1 variable d corresponds to proposition D, and so forth. One optimal solution of (6.3) is $(d, f, r, s, v) = (0, 0, 1, 1, 0)$, which is therefore a minimal model. In this model it rains, the tickets sell out, and Joe spends the day at work. This model is not stable, however. The Gelfond-Lifschitz transform contains the rules that are enforced because their "unless" clauses are inoperative. In this case only rule 3 is enforced, $R \rightarrow S$. The remaining rules in (6.2) contain at least one false atom that is negated in the antecedent. The integer programming problem that corresponds to the transform is

$$\begin{aligned} \min \quad & d + f + r + s + v \\ \text{s.t.} \quad & -r + s \geq 0 \\ & d, f, r, s, v \in \{0, 1\} \end{aligned}$$

The model $(d, f, r, s, v) = (0, 0, 1, 1, 0)$ is not stable because it is not optimal in this problem. Intuitively, the one rule that we actually use does not require both rain and a booked train. In fact, it requires neither.

A second minimal model is obtained solving (6.3) with the additional constraint $r + s \leq 1$, which yields $(d, f, r, s, v) = (1, 0, 0, 0, 1)$. Here Joe gets the day off and visits us in sunny weather. Because rules 1 and 4 are enforced, the Gelfond-Lifschitz transform contains the rules $D \rightarrow F$ and $\rightarrow D$. The corresponding optimization problem is,

$$\begin{aligned} \min \quad & d + f + r + s + v \\ \text{s.t.} \quad & -d + v \geq 0 \\ & d \geq 1 \\ & d, f, r, s, v \in \{0, 1\} \end{aligned} \quad (6.4)$$

Because positive unit resolution fixes D and V to true, the model $(d, f, r, s, v) = (1, 0, 0, 0, 1)$ is optimal in (6.4) and is stable.

A third minimal model $(d, f, r, s, v) = (1, 1, 0, 1, 0)$ is obtained by solving (6.3) with the additional constraints $r + s \leq 1$, $d + v \leq 1$. Joe gets the day off and goes fishing because the train is booked. The Gelfond-Lifschitz transform contains the rules $D \rightarrow F$ and $\rightarrow D$. Because its optimal solution does not require the tickets to sell out, the third minimal model is not stable.

6.2 Many-Valued Logics

Since two-valued logic took its origin from certain semantic assumptions about truth and falsity, the classical approach to many-valued logics has been to start with a semantic interpretation for the many values. But the modern view [208] is that many-valued logics can be organized into formal deductive systems based on consistent syntactic rules of deductive

312   CHAPTER 6   NONCLASSICAL AND MANY-VALUED LOGICS

inference. As has been the constant theme in this book, we shall thoroughly mix the syntactics and semantics by specifying inference in propositional many-valued logic as embeddings that manifest themselves as mixed integer programming problems. This description is based on the work of Reiner Hähnle [123]. We will use the framework of 3-valued Lukasiewicz logic (cf. [236]) to illustrate Hähnle's mathematical programming embedding of inference in many-valued logics.

Our focus will be on finite but many-valued propositional logic. The extension to infinite values appears straightforward but cumbersome to describe. We assume that the set of *truth values* N is finite and consists of equidistant rational numbers between 0 and 1. So if the cardinality of N is a finite number n, the set N is given by $\{0, 1/(n-1), \cdots, (n-2)/(n-1), 1\}$. A *connective* can be k-ary and is interpreted by the function $f : N^k \to N$. The truth value $v(.)$ of a formula is recursively defined by assigning values in N for atomic propositions and by setting

$$v(F(P_1, P_2, \ldots, P_k)) = f(v(P_1), v(P_2), \ldots, v(P_k)).$$

For any $S \subseteq N$, we define a many-valued proposition P to be *S-satisfiable* if there is a valuation such that $v(P) \in S$, or equivalently, such that $v(.)$ is an S-model for P. P is an *S-tautology* if all valuations are S-models for P.

A clever device for talking about many-valued logics within the vernacular of 2-valued logic is the notion of a *signed formula*, notated SP, where P is a proposition and its "sign" S is a set of truth values. We say that a signed formula SP is satisfiable if P is S-satisfiable. Since we have embedded the truth values in the interval $[0, 1]$, we use the notation $\boxed{\leq t}$ to denote the truth values $N \cap [0, t]$ and $\boxed{\geq t}$ to denote the truth values $N \cap [t, 1]$. Thus $\boxed{\leq t} P$ denotes a signed formula for which $v(P) \in \boxed{\leq t}$ in order for $v(.)$ to be a model for P.

Having many values of truth affects the basic semantics of propositional logic since, as we just saw, the definition and meaning of connectives and of satisfiability are affected. Let us first take a closer look at connectives. We can still use $\{\neg, \vee, \wedge, \to\}$ as unary and binary connectives operating on atomic propositions and recursively on formulas to build up a set of well-formed formulas. However, their effects on truth valuations have to be suitably interpreted.

- Negation $\neg$ can be viewed as one of several kinds of unary connectives and needs to be specified by some rule. For example, in Lukasiewicz logics, one interpretation of $\boxed{\leq t} \neg P$ is simply $\boxed{\geq 1-t} P$.

6.2 MANY-VALUED LOGICS

- Conjunction ∧ and disjunction ∨ are usually interpreted as min and max operators, respectively, with the natural ordering induced by the embedding of N in $\Re$.

- Implication → is a connective that, like negation, is open to several interpretations. In 3-valued Lukasiewicz logic, implication $R \to C$ follows the truth table given by the (3×3) matrix below, where the rows correspond to the truth values $\{0, \frac{1}{2}, 1\}$ of R and the columns to those of C:

$$\begin{pmatrix} 1 & 1 & 1 \\ \frac{1}{2} & 1 & 1 \\ 0 & \frac{1}{2} & 1 \end{pmatrix}$$

We will assume that all connectives in our many-valued logic satisfy the following condition: the function f corresponding to each k-ary connective has a graph

$$\text{graph}(f) = \{(v_0, v_1, \ldots v_k) \mid v_0 = f(v_1, \ldots, v_k)\}$$

that is MILP representable (mixed integer linear program representable, a concept that was defined in Chapter 4). This is a fairly mild assumption that is met by most reasonable interpretations of connectives. For instance, the Lukasiewicz implication for 3-valued logic that we defined above is easily seen to be MILP representable. Let v_1 and v_2 denote the truth values of R and C, respectively. The optimal value of the following mathematical program is the truth value of $R \to C$:

$$\begin{aligned} \max \quad & v_0 \\ \text{s.t.} \quad & v_0 \leq 1 - v_1 + v_2 \\ & v_0 \in \{0, \tfrac{1}{2}, 1\} \end{aligned}$$

Note that this is not exactly a 0-1 integer linear program but can be easily converted to one. In fact, if the v_i were allowed to take values in the continuum $[0, 1]$ (infinite many-valued logic), this connective is representable as a linear program. Connectives defined by nonlinear conditions and by conditions involving open sets would not meet the assumption of MILP representability.

We are finally ready to describe Hähnle's embedding of inference in many-valued propositional logic as a mixed integer linear program. Continuing with Lukasiewicz implication for 3-valued propositional logic, let us see how this embedding might work for proving that $\boxed{\geq 1}\, (p \to (q \to p))$ is a tautology in this logic.

To formulate the appropriate mixed integer linear program we note that the formula $\boxed{\leq t}\, (a \to b)$ reduces to the disjunction of two conditions.

314   CHAPTER 6   NONCLASSICAL AND MANY-VALUED LOGICS

Either (i) $t = 1$, or (ii) $\boxed{\geq v_a}\, a$ and $\boxed{\leq v_b}\, b$ with $t = 1 - v_a + v_b$. Condition (i) leads to a trivially satisfiable formula since all valuations have to be no larger than 1. Note that the "or" above is not exclusive since $t = 1$ can occur in both branches. A similar reduction can be formulated for formulas of the form $\boxed{\geq t}\,(a \to b)$, but we do not need it with this example. The idea is to repeatedly apply these reductions until the signed formulas are reduced to a simple form involving just atomic propositions; Hähnle [123] calls this a *constraint tableau*. But now we have several disjunctions with which to contend. These we can handle with appropriately chosen 0-1 variables.

The constraint tableau for $p \to (q \to p)$ appears in Figure 6.1. Here (v_p, v_q, v_{qp}) are the truth values of the propositions $(p, q, q \to p)$. The Lukasiewicz valuation requires that $t = 1 - v_p + v_{qp}$ and $v_{qp} = 1 - v_q + v_p$. The 0-1 variables z_1, z_2 indicate which branches are taken. Thus we write constraints $t \geq z_1$ and $t \geq 1 - v_p + v_{qp} - z_1$ to represent the first branch, and similarly for the second branch. This results in the mathematical program:

$$\begin{aligned}
\min \quad & t \\
\text{s.t.} \quad & t \geq z_1 \\
& t \geq 1 - v_p + v_{qp} - z_1 \\
& v_{qp} \geq z_2 \\
& v_{qp} \geq 1 - v_q + v_p - z_2 \\
& z_1, z_2 \in \{0, 1\} \\
& v_p, v_q, v_{qp}, t \in \{0, \tfrac{1}{2}, 1\}
\end{aligned}$$

Because the optimal value of this program is 1, $\boxed{\leq t}\,(p \to (q \to p))$ is a tautology. In this case the optimal value can be deduced on inspection, because if $t < 1$, constraint 1 implies that $z_1 = 0$, which means by constraint 2 that $v_p > v_{qp}$. Thus $v_{qp} < 1$, and by constraint 3 $z_2 = 0$. Because $z_1 = z_2 = 0$, constraints 2 and 4 combine to yield $t \geq 1$, which contradicts the assumption that $t < 1$.

6.3   Modal Logics

Modal logic extends classical logic by introducing new quantifiers over formulas. They are called modalities and are generally represented by the modal operator □ and its dual operator ◇. In one type of modal logic, □ϕ may indicate that proposition ϕ is necessarily true, whereas ◇ϕ says that ϕ is possibly true. In temporal logic □ can mean "at all times" and ◇ "at some time." The modalities can also be viewed as programs or single atomic actions within programs (defined in Section 6.4.1 below). In general the set of modalities is $\{\Box_a\}_{a \in \Sigma}$ (and their duals $\{\Diamond_a\}_{a \in \Sigma}$) where Σ is a set of symbols.

6.3 MODAL LOGICS

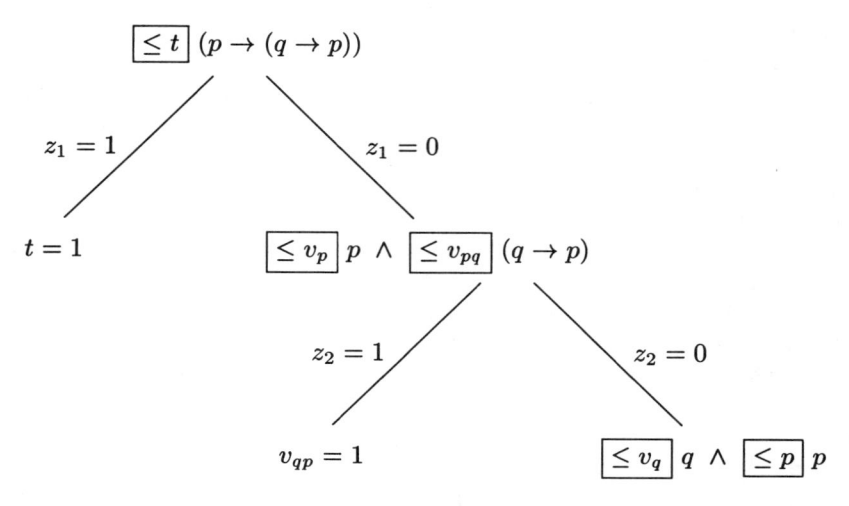

```
                    [≤t] (p → (q → p))
                   /                  \
           z₁=1  /                      \  z₁=0
                /                        \
            t=1                [≤vₚ] p ∧ [≤vₚq] (q → p)
                              /                    \
                       z₂=1 /                        \ z₂=0
                           /                          \
                      vqp = 1                  [≤vq] q ∧ [≤p] p
```

Figure 6.1: Constraint tableau for a satisfiability problem in 3-valued logic.

A model $M = (W, T, \sigma)$ for a modal formula ϕ is given by a set of *worlds* W, a family of transition functions $T = \{t_a : W \to 2^W\}_{a \in \Sigma}$ labeled by the symbols in Σ, and a valuation $\sigma : V \to 2^W$ that associates with each atomic proposition the set of worlds in which it is true. For a given modality a represented by operators $\Box_a$ and $\Diamond_a$, the set $t_a(w)$ can be regarded as containing the worlds that are possible relative to world w; that is, the worlds that are possible if w is actual, or possible "in" w. The formula $\Diamond_a \phi$ is true in w if and only if ϕ is true in some world that is possible relative to w, and $\Box_a$ is true in w if and only if ϕ is true in all worlds that are possible relative to w.

Given a model $M = (W, T, \sigma)$ and a world w in W, let $M, w \models \phi$ indicate that ϕ is true in world w of model M. The truth value of a modal formula is defined by induction on the structure of the formula as follows:

- $M, w \models x$ iff $w \in \sigma(x)$,
- $M, w \models \phi \land \psi$ iff both $M, w \models \phi$ and $M, w \models \psi$,
- $M, w \models \phi \lor \psi$ iff either $M, w \models \phi$ or $M, w \models \psi$,
- $M, w \models \neg \phi$ iff not $M, w \models \phi$,
- $M, w \models \Box_a \phi$ iff $M, v \models \phi$ for all $v \in t_a(w)$,
- $M, w \models \Diamond_a \phi$ iff $M, v \models \phi$ for some $v \in t_a(w)$.

A formula ϕ is true in a model M, written $M \models \phi$, if ϕ is true in all worlds of M. A formula ϕ is satisfiable iff it is true in some model.

So far, we have defined a logic language that extends propositional logic, and we have defined a notion of satisfiability for formulas in that language. We now show how modal logic is formulated as a special case of predicate logic. Because $\Box\phi$ is interpreted as stating that ϕ is true in all worlds that are possible relative to the actual world, this statement can be encoded in predicate logic by quantifying over possible worlds, and similarly for $\Diamond\phi$. To make this more precise, associate every atomic proposition x with a predicate $X(w)$ that is true when x is true in world w. Let $R_a(w, v)$ be the *alternativeness relation* for modality a, indicating that v is possible relative to w. The following rules allow one to construct a first-order formula $\mathbf{f}(\phi)$ to represent any modal formula ϕ:

$$\begin{aligned}
\mathbf{f}(x) &= X(w) \\
\mathbf{f}(\neg\phi) &= \neg\mathbf{f}(\phi) \\
\mathbf{f}(\phi \wedge \psi) &= \mathbf{f}(\phi) \wedge \mathbf{f}(\psi) \\
\mathbf{f}(\phi \vee \psi) &= \mathbf{f}(\phi) \vee \mathbf{f}(\psi) \\
\mathbf{f}(\Box_a \phi) &= \forall v (R_a(w, v) \rightarrow \mathbf{f}(\phi)[v/w]) \\
\mathbf{f}(\Diamond_a \phi) &= \exists v (R_a(w, v) \wedge \mathbf{f}(\phi)[v/w])
\end{aligned}$$

The notation $\mathbf{f}(\phi)[v/w]$ denotes the result of substituting v for every occurrence of w in the formula $\mathbf{f}(\phi)$. Clearly, the first-order formula $\forall x \mathbf{f}(\phi)$ is satisfiable iff ϕ has a model.

We can now combine the above formulation with the embedding of inference in first-order logic as infinite dimensional integer programs (see Section 5.4 of the previous chapter) to obtain a mathematical programming embedding of inference in modal logic. This formulation however, strikes us as being unsatisfactory since we have reduced a finite problem (inference in modal logic) to an infinite one (inference in predicate logic). This suggests an interesting research problem, which is to obtain a finite embedding of inference in modal logic as a mathematical program (see [29] for some preliminary results in this direction).

6.4 Constraint Logic Programming

In a formal logical system, predicates are uninterpreted. The predicates P and Q in the formula $P(x) \vee Q(x)$, for example, have no prior meanings attached. The system itself can define a predicate in terms of others,

6.4 CONSTRAINT LOGIC PROGRAMMING

however, by asserting formulas that relate them. For example, the formula

$$(\neg R(x) \vee P(x) \vee Q(x)) \wedge (R(x) \vee \neg P(x)) \wedge (R(x) \vee \neg Q(x))$$

in effect defines the predicate R to be the disjunction of P and Q. Similarly, the Peano axioms of arithmetic can be used to define a sum predicate $S(x, y, z)$, interpreted as saying $x + y = z$, in terms of a primitive successor function $s(x)$, interpreted as $x + 1$. One such axiom might be

$$S(x, y, z) \rightarrow S(s(x), y, s(z))$$

In principle, this and the other Peano axioms allow a logical inference algorithm to compute the sum of 2 and 3, for instance, by deducing the proposition

$$S(s(s(0)), s(s(s(0))), s(s(s(s(s(0)))))))$$

where 0 is a primitive symbol that represents zero.

It is impractical, however, to manipulate arithmetic and other complex predicates purely with logical inference methods. It is much more efficient to use specialized methods that are designed for these predicates. For this reason, most early implementations of logic programming (viz., PROLOG) used "built-in predicates" that have some prior meaning, such as the arithmetic sum. There are also other reasons for including built-in predicates, emanating from programming ease. Of course, this meant that the theoretical framework, in particular the Herbrand interpretation, cannot be used to analyze the semantics of such programs.

The constraint logic programming (CLP) scheme [158-160] was proposed in the mid-1980s by Jaffar, Lassez, and Maher to address this conflict between theory and practice of logic programming. CLP works with partially interpreted Horn formulas, where some of the predicates have specific interpretations as constraints that have useful expressive power and efficient solution methods. In CLP, compactness properties of constraint domains are used to generalize Herbrand's theorem in a richer setting.

Thus constraint logic programming began as a natural merger of two declarative paradigms: constraint solving and logic programming. This combination helps make CLP programs both expressive and flexible, and in some cases more efficient than other kinds of programs. We apply our embedding technique to a particular kind of CLP known as CLP($\Re$) to bring out its inherent mathematical programming nature. Constraints in CLP($\Re$) are linear inequalities on real-valued variables. Thus CLP($\Re$) brings together the techniques of linear programming and logic programming in a declarative programming language setting.

6.4.1 Some Definitions

If Σ is a signature (set of predicate and function symbols), a Σ-structure $\mathcal{M}$ consists of a set D and an assignment of functions and relations on D to the symbols of Σ that respects the arities of the symbols. A Σ-theory $\mathcal{T}$ is a collection of closed Σ-formulas. A model of a Σ-theory $\mathcal{T}$ is a Σ-structure $\mathcal{M}$ such that all formulas of $\mathcal{T}$ evaluate to true under the interpretation provided by $\mathcal{M}$. A primitive constraint has the form $p(t_1, \ldots, t_n)$, where $t_1, \ldots, t_n$ are Σ-terms and $p \in \Sigma$. A constraint (first-order) formula is built from the primitive constraints in the usual way using logical connectives and quantifiers [158, 160].

In constraint logic programming there is also a signature Π consisting of the uninterpreted predicates that are defined by a logic program. A CLP atom has the form $p(t_1, \ldots, t_n)$, where $t_1, \ldots, t_n$ are terms and $p \in \Pi$. A program P is of the form $p(\bar{x}) \leftarrow C, \bar{q}(\bar{y})$, where $p(\bar{x})$ is an atom, $\bar{q}(\bar{y})$ is a finite sequence of atoms in the body of the program, and C is a conjunction of constraints.[1] A goal G is a conjunction of constraints and atoms. A rule of the form $p(\bar{x}) \leftarrow C$ is called a *fact*.

We assume that programs and goals are in the following standard form.

- All arguments in atoms are variables and each variable occurs in at most one atom. This involves no loss of generality since a rule such as

$$p(\bar{t}) \leftarrow C, q(\bar{s})$$

can be replaced by the rule

$$p(\bar{x}) \leftarrow \bar{x} = \bar{t}, \bar{y} = \bar{s}, C, q(\bar{y})$$

- All rules defining the same predicate have the same head, and no two rules have any other variables in common (this is simply a matter of renaming).

For any signature Σ, let $\mathcal{M}$ be a Σ-structure (the domain of computation) and $\mathcal{L}$ be a class of Σ-formulas (the constraints). We call the pair $(\mathcal{M}, \mathcal{L})$ a constraint domain. We also make the following assumptions.

- The binary predicate symbol "=" is contained in Σ and interpreted as identity in $\mathcal{M}$.

- There are constraints *True* and *False* in $\mathcal{L}$ which are, respectively, true and false in $\mathcal{M}$.

---

[1]To conform to the CLP literature, an implication $(A \wedge B) \rightarrow C$ is written $C \leftarrow A, B$ throughout the rest of this chapter.

6.4 CONSTRAINT LOGIC PROGRAMMING

- The class of constraints in $\mathcal{L}$ is closed under variable renaming, conjunction, and existential quantification.

A valuation σ is a mapping from variables to the domain D. An $\mathcal{M}$-interpretation of a formula is an interpretation of the formula with the same domain as that of $\mathcal{M}$ and the same interpretation for the symbols in Σ as $\mathcal{M}$. It can be represented as a subset of $\mathcal{B}_\mathcal{M}$, where $\mathcal{B}_\mathcal{M} = \{p(\bar{d}) \mid p \in \Pi, \bar{d} \in D_k \text{ for some k}\}$. An $\mathcal{M}$-model of a closed formula is an $\mathcal{M}$-interpretation that is a model of the formula. The usual logical semantics are based on the $\mathcal{M}$-models of P.

Let Σ contain the constants 0 and 1, the binary function symbols $+$ and $*$, and the binary predicate symbols $=, <,$ and $\leq$. Let D be the set of real numbers and let $\mathcal{M}$ interpret the symbols of Σ as usual (i.e., $+$ is interpreted as addition, etc.). Let $\mathcal{L}$ be the constraints generated by the primitive constraints. Then $\Re = (\mathcal{M}, \mathcal{L})$ is the constraint domain of arithmetic over the real numbers. For our purpose *we will consider only the function symbol $+$ and the predicate symbol $\geq$ in Σ*. A typical rule in CLP($\Re$) will look like $p(x) \leftarrow (2x+3y \geq 2), (4y \geq 3), q(y)$, where $x, y \in \Re$. When we associate the variables in a rule in CLP($\Re$) with values over the reals $\Re$, we obtain a ground instance of that rule. It is easy to see that *the ground instances of a rule in CLP($\Re$) are uncountable*.

6.4.2 The Embedding

In order to illustrate the formulation, let us assume that a rule R_i in a given CLP($\Re$) is of the form

$$p(x) \leftarrow \tilde{c}_1, \tilde{c}_2, q_1(y_1), q_2(y_2)$$

where $\tilde{c}_1$ and $\tilde{c}_2$ are primitive constraints of the form $f(x, y) \geq 0$ and $g(x, y) \geq 0$, respectively, and $f(x, y), g(x, y)$ are linear functions of x, y. The q_i are atoms. We associate with R_i a linear (clausal) inequality lc(R_i) as follows:

$$v_{p(x)} + \sum_{i=1}^{2}(1 - u_{\tilde{c}_i}) + \sum_{i=1}^{2}(1 - v_{q(y_i)}) \geq 1$$

This clausal inequality can be rewritten as

$$v_{p(x)} - \sum_{i=1}^{2} u_{\tilde{c}_i} - \sum_{i=1}^{2} v_{q(y_i)} \geq (1 - k)$$

where k ($= 4$ in this example) is the total number of primitive constraints and atoms in the body of the rule.

320 CHAPTER 6   NONCLASSICAL AND MANY-VALUED LOGICS

We also associate the linear equalities

$$f(x,y) + (1 - u_{\tilde{c}_1})M \geq 0$$
$$g(x,y) + (1 - u_{\tilde{c}_2})M \geq 0$$

with the rule R_i, where M is an arbitrary large number. Note that a constraint must be solvable if the corresponding u variable is to take value 1. Also, if a particular value of u is feasible, then so are all smaller values of u (as far as these inequalities are concerned). We rewrite these inequalities as

$$f(x,y) - Mu_{\tilde{c}_1} \geq -M$$
$$g(x,y) - Mu_{\tilde{c}_2} \geq -M$$

respectively and denote them as $\text{le}(R_i)$.

Given a CLP($\Re$) program $\mathcal{P}$ we construct a 0-1 mixed integer program $\mathcal{F}_\mathcal{P}\{0,1\}$ as follows:

1. For each rule R in $\mathcal{P}$, the linear inequality $\text{lc}(R)$ is in $\mathcal{F}_\mathcal{P}\{0,1\}$.

2. For each rule R in $\mathcal{P}$, the inequality $\text{le}(R)$ corresponding to the constraints appearing in $\mathcal{P}$ is in $\mathcal{F}_\mathcal{P}\{0,1\}$.

3. For any atom $p(\bar{x})$ appearing in $F_\mathcal{P}\{0,1\}$ the restriction $v_{p(\bar{x})} \in \{0,1\}$ is in $\mathcal{F}_\mathcal{P}\{0,1\}$.

4. For every primitive constraint $\tilde{c}$ appearing in $\mathcal{P}$, the restriction $u_{\tilde{c}} \in \{0,1\}$ is in $\mathcal{F}_\mathcal{P}\{0,1\}$.

5. For every variable x appearing in $\mathcal{P}$, the restriction $x \in \Re$ is in $\mathcal{F}_\mathcal{P}\{0,1\}$.

When we replace restriction 3 by $v_{p(\bar{x})} \in [0,1]$ and restriction 4 by $u_{\tilde{c}} \in [0,1]$, we get $\mathcal{F}_\mathcal{P}[0,1]$.

The inequalities in $\text{lc}(R)$ have to hold for universally quantified variables x taking values in the reals. If we ground these inequalities by fully instantiating the variables over $\Re$, we say that we have *ground the formulation* $\mathcal{F}_\mathcal{P}\{0,1\}$. This ground formulation of $\mathcal{F}_\mathcal{P}\{0,1\}$ is an infinite 0-1 mixed integer program. We now require a compactness theory for such a mixed integer program before we can even attempt a semidecision procedure for solving it. The compactness theorems we developed in Section 5.4 are not adequate for this task since they assumed that the real-valued variables are bounded. We require a compactness theory that makes no such assumption. Since mixed integer programs generalize linear programs, we will first look at compactness in infinite linear programs and then in infinite mixed integer programs.

6.4 CONSTRAINT LOGIC PROGRAMMING

6.4.3 Infinite Linear Programs

The compactness theorem (Theorem 61) applies to infinite linear programs that arise as the relaxation of 0-1 programs. In such programs, all the variables are bound by the interval $[0,1]$. If we permit variables to take arbitrary values in $\Re$, compactness can be obtained only under certain assumptions on the recession cones of the underlying convex sets.

Let I, L denote possibly uncountable index sets. Let $\Re_i$ = a replica of $\Re$ for $i \in L$ and $\Re^I = \pi_{i \in L} \Re_i$ with the product topology. Assume I to be well ordered and write $x \in \Re^I$ as $x = [x_\alpha, \alpha \in I]$.

For finite $J \subset I$, denote by $x_J = \Re^{|J|}$ the appropriately ordered $|J|$-tuple $(x_\alpha, \alpha \in I)$. Also, define $||x||_\infty = \sup_\alpha |x_\alpha|$ (possibly $+\infty$).

For each $i \in L$, we have a constraint C_i of the type

$$A_i x_{J(i)} \geq b_i$$

where $J(i) \subset I$ is finite, $A_i \in \Re^{|J(i)| \times |J(i)|}$, and $b_i \in \Re^{|J(i)|}$. Without loss of generality, let $\cup_{i \in L} J(i) = I$.

Let θ be the zero vector in $\Re^I$ and θ_n the zero vector in $\Re^n$. We assume that

$$\bigcap_{i \in L} \{x \mid A_i x_{J(i)} \geq \theta_{|J(i)|}\} = \{\theta\} \qquad (6.5)$$

This says that the convex sets defined by C_i for $i \in L$ have no common direction of recession.

Theorem 63 *There is a finite $M \subset L$ such that for $J = \bigcup_{i \in M} J(i)$*

$$\bigcap_{i \in M} \left\{ x_J \in \Re^{|J|} \mid A_i x_{J(i)} \geq \theta_{|J(i)|} \right\} = \{\theta_J\} \qquad (6.6)$$

Proof: From (6.5), we have

$$\bigcap_{i \in L} \{x \mid A_i x_{J(i)} \geq \theta_{|J(i)|}, \ ||x||_\infty = 1\} = \emptyset \qquad (6.7)$$

Each set above is compact (being a closed subset of $\{x \mid ||x||_\infty = 1\}$, which is compact by Tychonoff's theorem). Thus by the finite intersection property of families of compact sets, there is a finite $M \subset L$ such that

$$\bigcap_{i \in M} \{x \mid A_i x_{J(i)} \geq \theta_{|J(i)|}, \ ||x||_\infty = 1\} = \emptyset$$

Hence

$$\bigcap_{i \in M} \left\{ x_J \in \Re^{|J|} \ \middle| \ A_i x_{J(i)} \geq \theta_{|J(i)|}, \ \max_{\alpha \in J} |x_\alpha| = 1 \right\} = \emptyset \qquad (6.8)$$

322 CHAPTER 6 NONCLASSICAL AND MANY-VALUED LOGICS

Suppose there is an $\bar{x} \in \Re^{|J|}$ such that

$$A_i \bar{x}_{J(i)} \geq \theta_{|J(i)|}, \quad \text{for all } i \in M$$
$$\bar{x}_{J(i)} \neq \theta_{|J(i)|}, \quad \text{for at least one } i \in M$$

Then if $a = \max_{\alpha \in J} |x_\alpha|$, we have $a > 0$. Define $\tilde{x} \in \Re^{|J|}$ by

$$\tilde{x}_\alpha = \begin{cases} x_\alpha/a, & \text{if } \alpha \in J \\ 0, & \text{if } \alpha \notin J \end{cases}$$

Then $\max_{\alpha \in J} |x_\alpha| = 1$ and $A_i \tilde{x}_{J(i)} \geq \theta_{|J(i)|}$ for all $i \in M$. That is, $\tilde{x}$ belongs to the left-hand side of (6.8), a contradiction. Therefore no such $\bar{x}$ can exist. In other words, (6.6) holds. □

By abuse of notation, let C_i denote the closed convex subset of $\Re^I$ for which $A_i x_{J(i)} \geq b_i$.

Theorem 64 *Under the assumption (6.5), if $\bigcap_{i \in L} C_i = \emptyset$, then there exists a finite $K \subset L$ such that $\bigcap_{i \in K} C_i = \emptyset$.*

Proof: Let $M, J(i)$ for $i \in M$, and J be defined as in the preceding theorem. Let

$$C'_i = \{x^J \in \Re^{|J|} \mid x \in C_i\}$$

denote the projection of C_i to $\Re^{|J|}$ under the map $x \to x_J$. It suffices to show that there is a finite $K \subset L$ for which $\bigcap_{i \in K} C'_i = \emptyset$. Define $\bar{C}_i = C'_i \cap \bigcap_{j \in M} C'_j$, $i \in L$. By the preceding theorem, C'_j, $j \in M$ do not have a common direction of recession and therefore $\bigcap_{j \in M} C'_j$ is bounded (cf. Rockafellar [245], pp. 60–61). It is also closed and therefore compact. Thus $\bar{C}_i$, $i \in L$ are compact and $\bigcap_{i \in L} \bar{C}_i = \emptyset$. By the finite intersection property of families of compact sets, it follows that there exists a finite $T \subset L$ such that $\bigcap_{i \in T} \bar{C}_i = \emptyset$. Let $K = T \cup M$. Then $\bigcap_{i \in K} C'_i = \emptyset$. □

An example shows that one cannot relax the assumption (6.5) that the sets C_i have no common recession direction. Let $I = \{1\}$ and $C_i = \{x_1 \mid x_1 \geq i\}$ for $i = 1, 2, \ldots$. Then $\bigcap_i C_i = \emptyset$, but no finite intersection is empty. A second example shows that the assumption is needed even when the components of A_i, b_i remain bounded as i increases. Let $I = \{1, 2\}$ and define $C_i, i = 0, 1, \ldots$, by

$$C_i = \begin{cases} \{(x_1, x_2) \mid x \leq 0\}, & \text{if } i = 0 \\ \{(x_1, x_2) \mid x_1 + \frac{1}{i} x_2 \geq 1\}, & \text{if } i \geq 1 \end{cases}$$

Again, $\bigcap_i C_i = \emptyset$ but no finite intersection is empty.

6.4 CONSTRAINT LOGIC PROGRAMMING

6.4.4 Infinite 0-1 Mixed Integer Programs

In the context of partially interpreted logics, we will need compactness results for infinite linear programs have 0-1 variables along with real-valued variables with no explicit bounds on them. The compactness theorem (Theorem 64) can be extended to this case as well. The addition of 0-1 variables causes no difficulty to compactness since they are bounded. The assumption that the constraint regions have no common recession direction is modified to the assumption that the projections of the constraint regions onto the space of real-valued variables have no common recession direction.

Let K denote a possibly uncountable index set. Let $\{0,1\}_i$ denote a replica of $\{0,1\}$ for $i \in L$ and let $\{0,1\}^K = \prod_{i \in L}\{0,1\}_i$ with the product topology. Assume K to be well ordered and write $u \in \{0,1\}^K$ as $u = (u_\beta, \beta \in K)$. For finite $T \subset K$, denote by $u_T \in \{0,1\}^{|T|}$ the appropriately ordered $|K|$-tuple $(u_\beta, \beta \in K)$.

For each $i \in L$, we have a constraint C_i of the type

$$A_i x_{J(i)} + B_i u_{T(i)} \geq h_i$$

where $J(i) \subset I$ is finite, $T(i) \subset K$ is finite, $A_i \in \Re^{|M(i)|} \times \Re^{|J(i)|}$, $B_i \in \Re^{|M(i)|} \times \Re^{|T(i)|}$, $h_i \in \Re^{|M(i)|}$, and $M(i)$ is finite. Without loss of generality, let $\cup_{i \in L} J(i) = I$ and $\cup_{i \in L} T(i) = K$.

Let $\theta = $ the zero vector in $\Re^I$ and θ_n the zero vector in $\Re^n$. We assume that

$$\bigcap_{i \in L} \mathcal{P}_x \{x \mid A_i x_{J(i)} + B_i u_{K(i)} \geq \theta_{|M(i)|}\} = \{\theta\} \qquad (6.9)$$

Here $\mathcal{P}_x$ denotes the projection operator that projects a given set in x, u-space onto x-space. Note that each $C_i, i \in L$ represents a union of convex sets. The assumption is that the x-projections of these sets have no common direction of recession.

The compactness results are now derived exactly as they were for the case of infinite linear programs. Note that convexity of the constraint regions was never used in the compactness proofs in that case. The result for the case of infinite 0-1 mixed integer programs is summarized by the theorem below.

Theorem 65 *Under the assumption (6.9), if $\bigcap_{i \in L} C_i = \emptyset$, then there exists a finite $N \subset L$ such that $\bigcap_{i \in N} C_i = \emptyset$.*

Since the case of infinite linear programs is a special case of the infinite 0-1 mixed integer programs (where all the u variables are bound to 0 or 1), assumption (6.9) is again needed.

324 CHAPTER 6   NONCLASSICAL AND MANY-VALUED LOGICS

Finally, we apply the compactness theorem to the embedding of CLP($\Re$) that we obtained earlier in this section. Under suitable assumptions about the common recession directions on the real-valued constraints, we can invoke Theorem 65 to show compactness of the mathematical programming embedding of CLP($\Re$). We get a stronger result by noting that the Horn clause structure of rules in CLP($\Re$) allows us to solve inference by optimization over a linear programming relaxation to obtain a least model. The infinite linear program

$$\begin{array}{ll} \inf & \sum v_{p(.)} + \sum u_{\bar{c}_i} \\ \text{s.t.} & u, v, x \text{ satisfy ground } \mathcal{F}_\mathcal{P}[0,1] \end{array}$$

has a minimum v, u solution that is guaranteed to be 0-1 valued. The proof of this result is completely analogous to that of Theorem 62 proved in the previous chapter for the case of definite logic programs.

Appendix
Linear Programming

This appendix presents a very brief review of some of the basic concepts of the geometry of linear inequalities and the cornerstones of the geometry and duality of linear programming. We also highlight some key results on the computational complexity of linear programming. This is really no substitute for a proper introduction to the subject (such as [47, 249]) and can serve at best as a refresher for a reader whose working knowledge of linear programming is a little rusty. We begin with linear homogeneous inequalities. This involves the geometry of (convex) polyhedral cones.

Polyhedral Cones

A homogeneous linear equation $ax = 0$ in n variables defines a *hyperplane* of dimension $n - 1$ that contains the origin and is therefore a linear subspace. A homogeneous linear inequality $ax \leq 0$ defines a *half-space* on one "side" of the hyperplane, defined by converting the inequality into an equation. A system $Ax \leq 0$ of linear homogeneous inequalities therefore defines an object that is the intersection of finitely many half-spaces, each of which contains the origin in its boundary. A simple example of such an object is the nonnegative orthant. Clearly, the objects in this class resemble cones with the apex defined at the origin and with a prismatic boundary surface. We call them *convex polyhedral cones*.

A convex polyhedral cone is a set of the form

$$\mathcal{K} = \{x \mid Ax \leq 0\}$$

Here A is assumed to be an $m \times n$ matrix of real numbers. A set is *convex* if it contains the line segment connecting any pair of points in the set. A convex set is called a *convex cone* if it contains all nonnegative, scalar multiples of points in it. A convex set is *polyhedral* if it is represented by a finite system of linear inequalities. Since we shall deal exclusively with cones that are both convex and polyhedral, we shall refer to them simply as cones.

The representation of a cone as the solution to a finite system of homogeneous linear inequalities is sometimes referred to as the "constraint" or "implicit" description. It is implicit because it takes an algorithm to generate points in the cone. An "explicit" or "edge" description can also be derived for any cone.

Theorem 66 *Every cone* $\mathcal{K} = \{x \mid Ax \leq 0\}$ *has an "edge" representation of the following form:* $\mathcal{K} = \{x \mid x = \sum_{j=1}^{L} e^j \mu_j, \ \mu_j \geq 0 \ \forall j\}$, *where each distinct edge of $\mathcal{K}$ is represented by a point e^j.*

Thus for any cone we have two representations:

- *Constraint representation*: $\mathcal{K} = \{x \mid Ax \leq 0\}$.

- *Edge representation*: $\mathcal{K} = \{x \mid x = E\mu, \ \mu \geq 0\}$.

The matrix E is a representation of the edges of $\mathcal{K}$. Each column of E contains the coordinates of a point on a distinct edge. Since positive scaling of the columns is permitted, we fix the representation by scaling each column so that the last nonzero entry is either 1 or -1. This scaled matrix E is called the *canonical matrix of generators* of the cone $\mathcal{K}$.

Every point in a cone can be expressed as a positive combination of the columns of E. Since the number of columns of E can be huge, the edge representation does not seem very useful. Fortunately, the following tidy result, due to Carathéodory, helps us out.

Theorem 67 ([37]) *For any cone $\mathcal{K}$, every point in $\mathcal{K}$ can be expressed as the positive combination of at most d edge points, where d is the dimension of $\mathcal{K}$.*

Convex Polyhedra

The transition from cones to polyhedra may be conceived of, algebraically, as a process of dehomogenization. This is to be expected, of course, since polyhedra are represented by systems of (possibly inhomogeneous) linear inequalities, and cones by systems of homogeneous linear inequalities. Geometrically, this process of dehomogenization corresponds to realizing that a polyhedron is the Minkowski or set sum of a cone and a polytope (bounded polyhedron). Just as cones in $\Re^n$ are generated by taking *positive* linear combinations of a finite set of points in $\Re^n$, polytopes are generated by taking *convex* linear combinations of a finite set of (generator) points.

Given K points $\{x^1, x^2, \ldots, x^K\}$ in $\Re^n$ the *convex hull* of these points is given by

$$conv(\{x^i\}) = \left\{ x \mid x = \sum_{i=1}^{K} \alpha_i x^i, \sum_{i=1}^{K} \alpha_i = 1, \alpha \geq 0 \right\}$$

That is, the convex hull of a set of points in $\Re^n$ is the object generated in $\Re^n$ by taking all convex linear combinations of the given set of points. Clearly, the convex hull of a finite list of points is always bounded.

Theorem 68 ([289]) *A subset of $\Re^n$ is a polytope if and only if it is the convex hull of a finite set of points.*

An *extreme point* of a convex set S is a point $x \in S$ satisfying

$$[x = \alpha \bar{x} + (1-\alpha)\tilde{x};\ \bar{x}, \tilde{x} \in S;\ \alpha \in (0,1)] \rightarrow (x = \bar{x} = \tilde{x})$$

Equivalently, an extreme point of a convex set S is one that cannot be expressed as a convex linear combination of some other points in S. When S is a polyhedron, extreme points of S correspond to the geometric notion of corner points. This correspondence is formalized in the corollary below. Let $extr(P)$ be the set of extreme points of P.

Corollary 11 *A polytope P is the convex hull of its extreme points, so that $P = conv(extr(P))$.*

Now we go on to discuss the representation of (possibly unbounded) convex polyhedra.

Theorem 69 *Any convex polyhedron P represented by a linear inequality system $\{x \mid Ax \leq b\}$ can be also represented as the Minkowski or set addition of a convex cone R and a convex polytope Q.*

$$P = Q + R = \{x + r \mid x \in Q,\ r \in R\}$$

where

$$Q = \left\{ x \mid x = \sum_{i=1}^{K} \alpha_i x^i,\ \sum_{i=1}^{K} \alpha_i = 1,\ \alpha \geq 0 \right\}$$

$$R = \left\{ r \mid r = \sum_{j=1}^{L} \mu_j r^j,\ \mu \geq 0 \right\}$$

and where $x^i \in P$ for $i = 1, \ldots, K$ and $r^j \in R$ for $j = 1, \ldots, L$.

If this representation is minimal, the vectors r^j are the *extreme rays* of P, and $\{x^i\}$ the extreme points. It follows from the statement of the theorem that P is nonempty if and only if the polytope Q is nonempty. We proceed now to discuss the representations of R and Q.

The cone R associated with the polyhedron P is called the *recession or characteristic cone* of P. A hyperplane representation of R is also readily available. It is easy to show that

$$R = \{r \mid Ar \leq 0\}$$

An obvious consequence is that P equals the polytope Q if and only if $R = \{0\}$.

The polytope Q associated with the polyhedron P is the convex hull of a finite collection $\{x^i\}$ of points in P. It is not difficult to see that the minimal set $\{x^i\}$ is precisely the set of extreme points of P.

A polyhedron is said to be pointed if it does not contain a line extending to infinity in both directions. A nonempty pointed polyhedron P, it follows, must have at least one extreme point. The set of generators of a polyhedron (extreme rays and extreme points) can be a huge set. As in the case of cones, we are helped by a compact representation theorem.

Theorem 70 *Every point in a convex polyhedron P can be expressed as the positive combination of at most $d+1$ extreme points and extreme rays of P, where d is the dimension of P.*

The *affine hull* of P is given by

$$\mathit{aff}(P) = \left\{ x \,\middle|\, x = \sum_{x^i \in extr(P)} \alpha_i x^i,\ \sum_i \alpha_i = 1 \right\}$$

Thus $\mathit{aff}(P)$ is the smallest affine set that contains P. A hyperplane representation of $\mathit{aff}(P)$ is also possible. Let $(A^=, b^=)$ be the largest set of rows of (A, b) for which all $x \in P$ satisfy $A^= x = b^=$. Let (A^+, b^+) be the remaining rows of (A, b), so that all $x \in P$ satisfy $A^+ x \leq b^+$. It follows that P must contain at least one point $\bar{x}$ satisfying $A^= \bar{x} = b^=$ and $A^+ \bar{x} < b^+$.

Lemma 34 $\mathit{aff}(P) = \{x \mid A^= x = b^=\}$.

The *dimension* of a polyhedron P in $\Re^n$ is defined to be the dimension of the affine hull of P, which equals the maximum number of affinely independent points in $\mathit{aff}(P)$, minus one. P is said to be full dimensional if its dimension equals n or, equivalently, if the affine hull of P is all of $\Re^n$.

A *supporting hyperplane* of the polyhedron P is a hyperplane H,
$$H = \{x \mid b^T x = z^*\}$$
satisfying
$$b^T x \leq z^*, \text{ for all } x \in P$$
$$b^T x = z^*, \text{ for some } x \in P$$

A supporting hyperplane H of P is one that touches P such that all of P is contained in a half-space of H. Note that a supporting plane can touch P at more than one point.

A *face* of a nonempty polyhedron P is a subset of P that is either P itself or is the intersection of P with a supporting hyperplane of P. It follows that a face of P is itself a nonempty polyhedron. A face whose dimension is one less than the dimension of P is called a *facet*. A face of dimension one is called an *edge* (note that extreme rays of P are in correspondence with the unbounded edges of P). A face of dimension zero is called a *vertex* of P (the vertices of P are precisely the extreme points of P). Two vertices of P are said to be adjacent if they are both contained in an edge of P. Two facets are said to be adjacent if they both contain a common face of dimension one less than that of a facet. Many interesting aspects of the facial structure of polyhedra can be derived from the following representation lemma.

Lemma 35 *F is a face of $P = \{x \mid Ax \leq b\}$ if and only if F is nonempty and $F = P \cap \{x \mid \tilde{A}x = \tilde{b}\}$, where $\tilde{A}x \leq \tilde{b}$ is a subsystem of $Ax \leq b$.*

As a consequence of the lemma, we have an algebraic characterization of extreme points of polyhedra.

Theorem 71 *Given a polyhedron $P = \{x \mid Ax \leq b\}$, x^i is an extreme point of P if and only if it is a face of P satisfying $\tilde{A}x^i = \tilde{b}$, where $(\tilde{A}, \tilde{b})$ is a submatrix of (A, b) and the rank of $\tilde{A}$ equals n.*

In the context of the simplex method of linear programming, we are often faced with polyhedra of the form $\{x \in \Re^n \mid Ax = b, x \geq 0\}$. Note that such a polyhedron is always pointed since it is contained in the nonnegative orthant. An extreme point of a polyhedron in this form is obtained by a partition of A given by $(B|N)$, where B is an $m \times m$ nonsingular "basis" matrix. The extreme point is specified by a corresponding partition of the x variables into x_B and x_N, the *basic* and *nonbasic variables*. The extreme point coordinates are given by $x_B = B^{-1}b \geq 0$ and $x_N = 0$ and such a representation is called a *basic feasible solution*.

Now we come to Farkas lemma, which says that a linear inequality system has a solution if and only if a related (polynomial size) linear inequality system has no solution. This lemma is representative of a large body of theorems in mathematical programming known as *theorems of the alternative*.

Lemma 36 ([92]) *Exactly one of the alternatives*

I. $Ax \leq b$ *for some* $x \in \Re^n$

II. $A^T y = 0, b^T y < 0, y \geq 0$ *for some* $y \in \Re^m$

is true for any given matrices $A \in \Re^{m \times n}, b \in \Re^m$.

Optimization and Dual Linear Programs

The two fundamental problems of linear programming (which are polynomially equivalent) are:

- *Solvability:* This is the problem of checking if a system of linear constraints on real (rational) variables is solvable or not. Geometrically, we have to check if a polyhedron, defined by such constraints, is nonempty.

- *Optimization:* This is the problem of optimizing a linear objective function over a polyhedron described by a system of linear constraints.

Building on polarity in cones and polyhedra, duality in linear programming is a fundamental concept which is related to both the complexity of linear programming and to the design of algorithms for solvability and optimization. We will encounter the solvability version of duality (called the Farkas lemma) while discussing the Fourier-Motzkin elimination technique below. Here we will state the main duality results for optimization. If we take the *primal* linear program to be

$$\begin{array}{ll} \text{minimize} & cx \\ \text{subject to} & Ax \geq b \end{array}$$

there is an associated *dual* linear program

$$\begin{array}{ll} \text{maximize} & b^T y \\ \text{subject to} & A^T y = c^T \\ & y \geq 0 \end{array}$$

where the first problem optimizes over $x \in \Re^n$ and the second over $y \in \Re^m$. The two problems satisfy

1. For any $\hat{x}$ and $\hat{y}$ feasible in the primal and dual (i.e., they satisfy the respective constraints), we have $c\hat{x} \geq b^T \hat{y}$ (*weak duality*).

2. The primal has a finite optimal solution if and only if the dual does.

3. x^* and y^* are a pair of optimal solutions for the primal and dual, respectively, if and only if x^* and y^* are feasible in the primal and dual (i.e., they satisfy the respective constraints) and $cx^* = b^T y^*$ (*strong duality*).

4. x^* and y^* are a pair of optimal solutions for the primal and dual, respectively, if and only if x^* and y^* are feasible in the primal and dual (i.e., they satisfy the respective constraints) and $(Ax^* - b)^T y^* = 0$ (*complementary slackness*).

The strong duality condition above gives us a good stopping criterion for optimization algorithms. The complementary slackness condition, on the other hand, gives us a constructive tool for moving from dual to primal solutions and vice versa. The weak duality condition gives us a technique for obtaining lower bounds for minimization problems and upper bounds for maximization problems.

Note that the properties above have been stated for linear programs in a particular form. The reader should be able to check that if, for example the primal is of the form

$$\begin{aligned} \min \quad & cx \\ \text{s.t.} \quad & Ax \geq b \\ & x \geq 0 \end{aligned}$$

then the corresponding dual will have the form

$$\begin{aligned} \max \quad & b^T y \\ \text{s.t.} \quad & A^T y \leq c^T \\ & y \geq 0 \end{aligned}$$

The tricks needed for deriving the duals of various forms are (a) any equation can be written as two inequalities, (b) an unrestricted variable can be substituted by the difference of two nonnegatively constrained variables, and (c) an inequality can be treated as an equality by adding a non-negatively constrained variable to the lesser side. Using these tricks, the reader could also check that dual construction in linear programming is involutory (i.e., the dual of the dual is the primal).

Complexity of Linear Programming

We have seen that every extreme point of a polyhedron $P = \{x \mid Ax = b,\ x \geq 0\}$ is the solution of an $m \times m$ linear system whose coefficients come from (A, b). Therefore, if a linear inequality system is solvable (the polyhedron is nonempty), we can always guess a polynomial length string representing an extreme point and check its membership in P in polynomial time.

Lemma 37 *The language $\mathcal{L}_I = \{(A, b) \mid \exists x \geq 0 \text{ such that } Ax = b\}$ belongs to NP.*

Lemma 36 (the Farkas lemma) tells us that insolubility of a linear inequality system is always witnessed by the solubility of a "dual" polynomial size system. Thus we have the following:

Lemma 38 *The language $\mathcal{L}_I$ is in $NP \cap coNP$.*

That $\mathcal{L}_I$ can be recognized in polynomial time, follows from algorithms for linear programming that we now discuss. The history of algorithms for linear programming is now at least 175 years old. In 1823, L.B.J. Fourier [95] gave us a classical variable elimination method for solving systems of linear inequalities. It leads to simple proofs of the duality principle of linear programming and many other insights (cf. [42]). The simplex method of linear programming [75] uses the vertex-edge structure of a convex polyhedron to execute a local search. The simplex method has been finely honed over five decades now and continues to be the workhorse of linear programming practice. However, the classical method of Fourier and the simplex method of Dantzig are both exponential-time and so do not reveal the computational complexity of linear programming.

In 1979, L. G. Khachiyan [183] showed that the ellipsoid method of convex optimization specializes to a polynomial-time algorithm for linear programming, and we have the following theorem:

Theorem 72 *The language $\mathcal{L}_I$ is in P.*

N. K. Karmarkar's [178] breakthrough in 1984 made practical the theoretical demonstrations of tractability of linear programming that were provided via the ellipsoid method.

That $\mathcal{L}_I$ is P-complete was established in an interesting short paper by D. Dobkin et al. [85]. They used a reduction of the satisfiability of Horn propositions to linear programming, an idea that we have found very useful in designing "optimization methods for logical inference."

Bibliography

[1] Ahu, A. V., J. E. Hopcroft, and J. D. Ullman, *The Design and Analysis of Algorithms*, Addison-Wesley (Reading MA, 1974).

[2] Andersen, K. A., and J. N. Hooker, Bayesian logic, *Decision Support Systems* **11** (1994) 191–210.

[3] Andersen, K. A., and J. N. Hooker, A linear programming framework for logics of uncertainty, *Decision Support Systems* **16** (1996) 39–53.

[4] Araque, G. J. R., and V. Chandru, Some facets of satisfiability, Technical Report CC-91-13, Institute for Interdisciplinary Engineering Studies, Purdue University, West Lafayette, IN (1991).

[5] Arvind, V., and S. Biswas, $O(n^2)$ algorithm for the satisfiability problem of a subset of propositonal sentences in CNF that includes all Horn sentences, *Information Processing Letters* **24** (1987) 67–69.

[6] Aspvall, B., Recognizing disguised NR(1) instance of the satisfiability problem, *Journal of Algorithms* **1** (1980) 97–103.

[7] Aspvall, B., M. F. Plass, and R. E. Tarjan, A linear time algorithm for testing the truth of certain quantified boolean formulas, *Information Processing Letters* **8** (1979) 121–123.

[8] Balas, E., S. Ceria, and G. Cornuejols, Mixed 0-1 programming by lift and project in a branch-and-cut framework, *Management Science* **42** (1996) 1129–1246.

[9] Balas, E., and R. Jeroslow, Canonical cuts on the unit hypercube, *SIAM Journal of Applied Mathematics* **23** (1972) 61–69.

[10] Balas, E., and C. H. Martin, Pivot and complement—a heuristic for 0-1 programming, *Management Science* **26** (1980) 86–96.

[11] Balas, E., and S.M. Ng, On the set covering polytope I: all facets with coefficients in $\{0,1,2\}$, *Mathematical Programming B*, **43**, (1989), II: lifting the facets with coefficients in $\{0,1,2\}$, *Mathematical Programming B*, **45**, (1989), 1–20.

[12] Balas, E., and J. Xue, Minimum weighted coloring of triangulated graphs with application to maximum weight vertex packing and clique finding in arbitrary graphs, *SIAM Journal on Computing* **20** (1991) 209–221.

[13] Balas, E., and C. S. Yu, Finding a maximum clique in an arbitrary graph, *SIAM Journal on Computing* **15** (1986) 1054–1068.

[14] Barnes, E. R., A variation on Karmarkar's algorithm for solving linear programming problems, *Mathematical Programming* **36** (1986) 174–182.

[15] Barnett, J. A., Computational methods for a mathematical theory of evidence, Information Sciences Institute, University of Southern California (not dated).

[16] Barth, P., *Logic-Based 0-1 Constraint Programming*, Kluwer (Dordrecht, 1996).

[17] Beale, E. M. L., Nonlinear programming using a general mathematical programming system, in H. J. Greenberg, ed., *Design and Implementation of Optimization Software*, Sijthoff and Noordhoff (Groningen, Netherlands, 1978).

[18] Bell, C., A. Nerode, R. T. Ng, and V. S. Subrahmanian, Mixed integer programming methods for computing nonmonotonic deductive databases, *Journal of the ACM* **41** (1994) 1178–1215.

[19] Benders, J. F., Partitioning procedures for solving mixed variables programming problems, *Numerische Mathematik* **4** (1962) 238–252.

[20] Bibel, W., *Automated Theorem Proving*, 2nd ed., Vieweg (Braunschweig, 1987).

[21] Billionnet, A., and A. Sutter, An efficient algorithm for the 3-satisfiability problem, *Operations Research Letters* **12** (1992) 29–36.

[22] Blair, C., Two rules for deducing valid inequalities for 0-1 problems, *SIAM Journal of Applied Mathematics* **31** (1976) 614–617.

[23] Blair, C., R. G. Jeroslow, and J. K. Lowe, Some results and experiments in programming techniques for propositional logic, *Computers and Operations Research* **13** (1988) 633–645.

[24] Blake, A., *Canonical Expressions in Boolean Algebra*, Ph.D. thesis, University of Chicago (1937).

[25] Bledsoe, W. W., The sup-inf method in Presburger arithmetic, Memo ATP-18, Mathematics Department, University of Texas, Austin, TX (1974).

[26] Bledsoe, W. W., A new method for proving certain Presburger formulas. In *Advance Papers, 4th International Joint Conference on Artificial Intelligence*, Tibilisi, Georgia, USSR (1975) 15–21.

[27] Boole, G., *An Investigation of the Laws of Thought, on Which Are Founded the Mathematical Theories of Logic and Probabilities*, Dover Publications (New York, 1951). Original work published 1854.

[28] Boole, G., *Studies in Logic and Probability*, edited by R. Rhees, Watts and Co. (London) and Open Court Publishing Company (La Salle IL, 1952).

[29] Borkar, V. S., V. Chandru, D. Micciancio, and S.K. Mitter, Mathematical programming embeddings of logic, Technical Report IISC-CSA-98-02, Department of Computer Science & Automation, Indian Institute of Science, Bangalore, India (1998).

[30] Boros, E., Y. Crama, and P. L. Hammer, Polynomial time inference of all valid implications for Horn and related formulae, *Annals of Mathematics and Artificial Intelligence* **1** (1990) 21–32.

[31] Boros, E., Y. Crama, P. L. Hammer, and M. Saks, A complexity index for satisfiability problems, *SIAM Journal on Computing* **23** (1994) 45–49.

[32] Boros, E., P. L. Hammer, and J. N. Hooker, Predicting cause-effect relationships from incomplete discrete observations, *SIAM Journal on Discrete Mathematics* **7** (1994) 531–543.

[33] Boros, E., P. L. Hammer, and J. N. Hooker, Boolean regression, *Annals of Operations Research* **58** (1995) 201–226.

[34] Boros, E., P. L. Hammer, and X. Sun, Reconition of q-Horn formulas in linear time, *Discrete Applied Mathematics* **55** (1994) 1–13.

[61] Chvátal, V., and E. Szemerédi, Many hard problems for resolution, *Journal of the ACM* **35** (1988) 759–768.

[62] Collins, G. E., Quantifier elimination for real closed fields by cylindrical algebraic decomposition, in G. Goos and J. Hartmanis, eds., *Automata Theory and Formal Languages 2nd GI Conference*, Lecture Notes in Computer Science **33**, Springer (New York, 1975) 134–183.

[63] Conforti, M., and G. Cornuéjols, A class of logical inference problems soluble by linear programming, *Journal of the ACM* **42** (1995) 1107–1113.

[64] Conforti, M., G. Cornuèjols, A. Kapoor, and K. Vuskovic, Recognizing balanced 0, +or−1 matrices, in *Proceedings, Fifth Annual SIAM/ACM Symposium on Discrete Algorithms* (1994) 103–111. r

[65] Conforti, M., G. Cornuèjols, M. R. Rao, Decomposition of balanced 0,1 matrices, Parts I–VII, manuscripts, Carnegie Mellon University, Pittsburgh, PA 15213 USA, 1991.xx

[66] Cook, S. A., The complexity of theorem-proving procedures, i *Proceedings of the Third Annual ACM Symposium on the Theory of Computing* (1971) 151–158.

[67] Cook, S. A., Feasibly constructive proofs and the propositional calculus, *Proceedings of the Seventh Annual ACM Symposium on the Theory of Computing* (1975) 83–97.

[68] Cook, W., C. R. Coullard, and Gy. Turán, On the complexity of cutting-plane proofs, *Discrete Applied Mathematics* **18** (1987) 25–38.

[69] Cooper, D. C. Theorem proving in arithmetic without multiplication, in B. Meltzer and D.Mitchie, eds., *Machine Intelligence* **7**, American Elsevier (New York, 1972) 43–59.

[70] Cottle, R. W., and A. F. Veinott, Polyhedral sets having a least element, *Mathematical Programming* **3** (1972) 238–249.

[71] Cornuéjols, G., and A. Sassano, On the 0,1 facets of the set covering polytope, *Mathematical Programming B* **45** (1989) 45–56.

[72] Crama, Y., P. L. Hammer, and T. Ibaraki, Cause-effect relationships and partially defined Boolean functions, *Annals of Operations Research* **16** (1988) 299–325.

[73] Crama, Y., P. Hansen, and B. Jaumard, The basic algorithm for pseudo-boolean programming revisited, *Discrete Applied Mathematics* **29** (1990) 171–185.

[74] Crawford, J. M., and L. D. Auton, Experimental results on the crossover point in satisfiability problems, in *Proceedings of the Eleventh National Conference on Artificial Intelligence, AAAI93* (1993) 21–27.

[75] Dantzig, G. B., Maximization of a linear function of variables subject to linear inequalities, in C. Koopmans, ed., *Activity Analysis of Production and Allocation*, Wiley (New York, 1951) 339–347.

[76] Davis. M., and H. Putnam, A computing procedure for quantification theory, *Journal of the ACM* **7** (1960) 201–215.

[77] Davis, R., and D. B. Lenat, *Knowledge-Based Systems in Artificial Intelligence*, McGraw-Hill (New York, 1982).

[78] Davis, R., B. Buchanan, and E. Shortliffe, Production rules as a representation for a knowledge-based consultation program, *Artificial Intelligence* **8** (1977) 15–45.

[79] Dempster, A. P., Upper and lower probabilities induced by a multivalued mapping, *Annals of Mathematical Statistics* **38** (1967) 325–339.

[80] Dempster, A. P., A generalization of Bayesian inference, *Journal of the Royal Statistical Society (Series B)* **30** (1968) 205–247.

[81] Dhar, V., and N. Ranganathan, Integer programming vs. expert systems: an experimental comparison, *Communications of the ACM* **33** (1990) 323–336.

[82] Dijkstra, E. W., A note on two problems in connexion with graphs, *Numerische Mathematik* **1** (1959) 269–271.

[83] Dikin, I. I., Iterative solution of problems of linear and quadratic programming, *Doklady Akademiia Nauk SSSR* **174** (1967) 747–748. English translation in *Soviet Mathematics—Doklady* **8** (1967) 674–675.

[84] Dikin, I. I., On the convergence of an iterative process, *Upravlyaemye Systemi* **12** (1974) 54–60.

[85] Dobkin, D., R. J. Lipton, and S. Reiss, Linear programming is log-space hard for P, *Information Processing Letters* **8** (1979) 96–97.

[86] Douanya Ngueste, G.-B., P. Hansen, and B. Jaumard, Probabilistic satisfiability and decomposition, in *Proceedings, Symbolic and Quantititaive Approaches to Reasoning and Uncertainty, European Conference, ECSQARU 75*, Springer (Berlin, 1995) 151–161.

[87] Dowling, W. F., and J. H. Gallier, Linear-time algorithms for testing the satisfiability of propositional Horn formulae, *Journal of Logic Programming* **1** (1984) 267–284.

[88] Dubois, D., and H. Prade, The principle of minimum specificity as a basis for evidential reasoning, *Uncertainty in Knowledge-Based Systems, Lecture Notes in Computer Science* **286** (1986) 75–84.

[89] Dubois, D., and H. Prade, A tentative comparison of numerical approximate reasoning methodologies, *International Journal Man-Machine Studies* **27** (1987) 149–183.

[90] Dyson, F. J., *Infinite in All Directions: Gifford Lectures at Aberdeen, Scotland, April-November 1985*, Harper and Row (New York, 1988).

[91] Dugat, V., and S. Sandri, Complexity of hierarchical trees in evidence theory, *ORSA Journal on Computing* **6** (1994) 37–49.

[92] Farkas, Gy., A Fourier-féle mechanikai elv alkalmazásai (in Hungarian), *Mathematikai és Természettudományi Értesitö* **12** (1894) 457–472.

[93] Feo, T., M. Resende, and S. Smith, A greedy randomized adaptive search procedure for maximum independent set, *Operations Research* **42** (1994) 860–878.

[94] Fletcher, R., *Practical Methods of Optimization*, Wiley (New York, 1987).

[95] Fourier, L. B. J., reported in: Analyse des travaux de l'Académie Royale des Sciences, pendant l'année 1824, Partie mathématique, *Histoire de l'Académie Royale des Sciences de l'Institut de France* **7** (1827) xlvii–lv. (Partial English translation in: D.A. Kohler, Translation of a report by Fourier on his work on linear inequalities, *Opsearch* **10** (1973) 38–42.)

[96] Franco, J., On the performance of algorithms for the satisfiability problem, *Information Processing Letters* **23** (1986) 103–106.

[97] Franco, J., and M. Paul, Probabilistic analysis of the Davis-Putnam procedure for solving the satisfiability problem, *Discrete Applied Mathematics* **5** (1983) 77–87.

[98] Franco, J., J. M. Plotkin, and J. W. Rosenthal, Correction to probabilistic analysis of the Davis-Putnam procedure for solving the satisfiability problem, *Discrete Applied Mathematics* **17** (1987) 295–299.

[99] Franco, J., and Yuan Chuan Ho, Probabilistic performance of a heuristic for the satisfiability problem, *Discrete Applied Mathematics* **22** (1988) 35–51.

[100] Funabiki, N., Y. Takefuji, and Kuo-Chun Lee, A neural network model for finding a near-maximum clique, *Journal of Parallel and Distributed Computing* **14** (1992) 340–344.

[101] Gallaire, H., and J. Minker, *Logic and Data Bases*, Plenum Press (New York, 1978).

[102] Gallo, G., G. Longo, S. Nguyen, and S. Pallottino, Directed hypergraphs and applications, Technical Report, Dip. di Informatica, Università di Pisa, Italy (April 1990).

[103] Gallo, G., and G. Rago, First order satisfiability: Partial instantiation and hypergraph algorithms, Manuscript, Dip. di Informatica, Università di Pisa (1993).

[104] Gallo, G., and S. Pallottino, Shortest path methods: a unifying approach, *Mathematical Programming Study* **26** (1986) 38–64.

[105] Gallo, G., and G. Rago, A hypergraph approach to logical inference for datalog formulae, Manuscript, Dip. di Informatica, Università di Pisa, Italy (September 1990).

[106] Gallo, G., and M. Scutellà, Polynomially soluble satifiability problems, *Information Processing Letters* **29** (1988) 221–227.

[107] Gallo, G., and G. Urbani, Algorithms for testing the satisfiability of propositional formulae, *Journal of Logic Programming* **7** (1989) 45–61.

[108] Garey, M. R., and D. S. Johnson, *Computers and Intractability: A Guide to the Theory of NP-Completeness*, W. H. Freeman (San Francisco, 1979).

[109] Geiger, D., and J. Pearl, On the logic of causal models, *Uncertainty in Artificial Intelligence* **4**, North-Holland (Amsterdam, 1990) 3–14.

[110] Gendreau, M., P. Soriano, and L. Salvail, Solving the maximum clique problem using a tabu search approach, *Annals of Operations Research* **41** (1993) 385–403.

[111] Genesereth, M. R., and N. J. Nilsson, *Logical Foundations of Artificial Intelligence*, Morgan Kaufmann (Los Altos CA, 1987).

[112] Gent, I. P., and T. Walsh, The SAT phase transition, in A. G. Cohn, ed., *Proceedings of the Eleventh European Conference on Artificial Intelligence, ECAI94*, Wiley (New York, 1994) 105–109.

[113] Georgakopolous, G., D. Kavvadias, and C. H. Papadimitriou, Probabilistic satisfiability, *Journal of Complexity* **4** (1988) 1–11.

[114] Gill, P. E., W. Murray, and M. H. Wright, *Practical Optimization*, Academic Press (London, 1981).

[115] Glover, F., Tabu search–Part I, *ORSA Journal on Computing* **1** (1989) 190–206.

[116] Glover, F., and H. J. Greenberg, Logical testing for rule-based management, *Annals of Operations Research* **12** (1988) 199–215.

[117] Gödel, K., Die Vollständigkeit der Axiome des logischen Functionkalküs, *Monatsch. Math. Phys.* **37** (1930) 349–360. Reprinted in J. van Heijenoort, *From Frege to Gödel*, Harvard University Press (Cambridge MA, 1967) 596–616.

[118] Gorden, J., and E. H. Shortliffe, The Dempster-Shafer theory of evidence, in B. G. Buchanan and E. H. Shortliffe, eds., *Rule-Based Expert Systems: The MYCIN Experiments of the Stanford Heuristic Programming Project*, Addison-Wesley (Reading MA, 1984).

[119] Grosof, B. N., An inequality paradigm for probabilistic reasoning, in J. F. Lemmer and L. N. Kanal, eds., *Uncertainty in Artificial Intelligence* **1**, North-Holland (Amsterdam, 1986).

[120] Grosof, B. N., Non-monotonicity in probabilistic knowledge, in J. F. Lemmer and L. N. Kanal, eds., *Uncertainty in Artificial Intelligence* **2**, North-Holland (Amsterdam, 1986).

[121] Haddawy, P., and A. M. Frisch, Modal logics and higher-order probability, *Uncertainty in Artificial Intelligence* **4**, North-Holland (Amsterdam, 1990) 133–148.

[122] Hähnle, R., and W. Kernig, Verification of switch-level designs with many-valued logic, in *Proceedings, 4th International Conference on Logic Programming and Automated Reasoning LPAR'93*, Springer (Berlin, 1993) 158–169.

[123] Hähnle, R., Many-valued logic and mixed integer programming, *Annals of Mathematics and Artificial Intelligence* **12** (1994) 231–263.

[124] Hailperin, T., *Boole's Logic and Probability*, Studies in Logic and the Foundations of Mathematics **85**, North-Holland (Amsterdam, 1976).

[125] Hailperin, T., Probability logic, *Notre Dame Journal of Formal Logic* **25** (1984) 198–212.

[126] Hailperin, T., *Boole's Logic and Probability*, 2nd ed., Studies in Logic and the Foundations of Mathematics **85**, North-Holland (Amsterdam, 1986).

[127] Hadjiconstantinou, E., and G. Mitra, Tools for reformulating logical forms into zero-one mixed integer programs, *European Journal of Operations Research* **72** (1994) 262–276.

[128] Haken, A., The intractability of resolution, *Theoretical Computer Science* **39** (1985) 297–308.

[129] Hammer, P., and S. Rudeanu, *Boolean Methods in Operations Research and Related Areas*, Springer (Berlin, 1968).

[130] Hansen, P., and B. Jaumard, Algorithms for the maximum satisfiability problem, *Computing* **44** (1990) 279–303.

[131] Hansen, P., B. Jaumard, and G. Plateau, An extension of nested satisfiability, RUTCOR Technical Report 29-93, Rutgers University, New Brunswick, NJ (1993).

[132] Harche, F., J. N. Hooker, and G. Thompson, A computational study of satisfiability algorithms for propositional logic, *ORSA Journal on Computing* **6** (1994) 423–435.

[133] Harche, F., and G. L. Thompson, The column subtraction algorithm: an exact method for solving the weighted set covering problem, *Computers and Operations Research* **21** (1990) 689–705.

[134] Hayes-Roth, F., D. A. Waterman, and D. B. Lenat, *Building Expert Systems*, Addison-Wesley (Reading MA, 1983).

[135] Henrion, M., An introduction to algorithms for inference in belief nets, *Uncertainty in Artificial Intelligence* **5**, North-Holland (Amsterdam, 1990).

[136] Henrion, M., R. D. Shachter, L. N. Kanal, J. F. Lemmer, eds., *Uncertainty in Artificial Intelligence* **5**, North-Holland (Amsterdam, 1990).

[137] Herbrand, J., Recherches sur la théorie de la démonstration, *Travaux de la Société des sciences et des lettres de Varsovie, Cl. III, math.-phys.* **33** (1930) 33–160.

[138] Hooker, J. N., A mathematical programming model for probabilistic logic, working paper 05-88-89, Graduate School of Industrial Administration, Carnegie Mellon University, Pittsburgh, PA (July 1988).

[139] Hooker, J. N., Resolution vs. cutting plane solution of inference problems: some computational experience, *Operations Research Letters* **7** (1988) 1–7.

[140] Hooker, J. N., Generalized resolution and cutting planes, *Annals of Operations Research* **12** (1988) 217–239.

[141] Hooker, J. N., A quantitative approach to logical inference, *Decision Support Systems* **4** (1988) 45–69.

[142] Hooker, J. N., Input proofs and rank one cutting planes, *ORSA Journal on Computing* **1** (1989) 137–145.

[143] Hooker, J. N., Solving the incremental satisfiability problem, *Journal of Logic Programming* **15** (1993) 177–186.

[144] Hooker, J. N., Generalized resolution for 0-1 linear inequalities, *Annals of Mathematics and Artificial Intelligence* **6** (1992) 271–286.

[145] Hooker, J. N., Logical inference and polyhedral projection, Computer Science Logic Conference 1991, *Lecture Notes in Computer Science* **626** (1992) 184–200.

[146] Hooker, J. N., New methods for computing inferences in first order logic, *Annals of Operations Research* **43** (1993) 479–492.

[147] Hooker, J. N., Resolution and the integrality of satisfiability problems, *Mathematical Programming* **74** (1996) 1–10.

[148] Hooker, J. N., Constraint satisfaction methods for generating valid cuts, in D. L. Woodruff, ed., *Advances in Computational and Stochastic Optimization, Logic Programming and Heuristic Search*, Kluwer (Dordrecht, 1997) 1–30.

[149] Hooker, J. N. and C. Fedjki, Branch-and-cut solution of inference problems in propositional logic, *Annals of Mathematics and Artificial Intelligence* **1** (1990) 123–140.

[150] Hooker, J. N., and G. Rago, Partial instantiation methods for logic programming, *Journal of Automated Reasoning* (to appear).

[151] Hooker, J. N., and V. Vinay, Branching rules for satisfiability, *Journal of Automated Reasoning* **15** (1995) 359–383.

[152] Hooker, J. N., and Hong Yan, Verifying logic circuits by Benders decomposition, in V. Saraswat and P. Van Hentenryck, eds., *Principles and Practice of Constraint Programming: The Newport Papers*, MIT Press (Cambridge MA, 1995) 267–288.

[153] Hooker, J. N., and Hong Yan, Tight representation of logical constraints as cardinality rules, *Mathematical Programming* (to appear).

[154] Hooker, J. N., Hong Yan, I. Grossmann, and R. Raman, Logic cuts for processing networks with fixed charges, *Computers and Operations Research* **21** (1994) 265–279.

[155] Hopcroft, J. E., and R. E. Tarjan, Efficient planarity testing, *Journal of the ACM* **21** (1974) 549–568.

[156] Howard, R. A., and J. E. Matheson, Influence diagrams, in R. A. Howard and J. E. Matheson, eds., *The Principles and Applications of Decision Analysis* **2**, Strategic Decision Group (Menlo Park CA, 1981).

[157] Hu, S.-T., *Threshold Logic*, University of California Press (Berkeley, 1965).

[158] Jaffar, J., and J.-L. Lassez, Constraint logic programming, Technical Report 86/73, Department of Computer Science, Monash University, 1986.

[159] Jaffar, J., and J.-L. Lassez, Constraint logic programming, in *Proc. 14th symposium on Principles of programming Languages* Munich (Jan. 1987) 111–119.

[160] Jaffar, J., and M. J. Maher, *Constraint Logic Programming: a survey*, Journal of Logic Programming **19/20** (1994) 503–581.

[161] Jaumard, B., P. Hansen, and M. P. Aragaö, Column generation methods for probabilistic logic, *ORSA Journal on Computing* **3** (1991) 135–148.

[162] Jaumard, B., and B. Simeone, On the complexity of the maximum satisfiability problem, *Information Processing Letters* **26** (1987–88) 1–4.

[163] Jeavons, P., D. Cohen, and M. Gyssens, A test for tractability, in E. C. Freuder, ed., *Principles and Practice of Constraint Programming - CP96*, Lecture Notes in Computer Science **1118** (1996) 267–281.

[164] Jensen, F. V., S. L. Lauritzen, and K. G. Olesen, Bayesian updating in causal probabilistic networks by local computations, *Computational Statistics Quarterly* **4** (1990) 269–282.

[165] Jeroslow, R. E., Representability in mixed integer programming, I: Characterization results, *Discrete Applied Mathematics* **17** (1987) 223–243.

[166] Jeroslow, R. E., Computation-oriented reductions of predicate to propositional logic, *Decision Support Systems* **4** (1988) 183–197.

[167] Jeroslow, R. E., On monotone chaining procedures of the CF type, *Decision Support Systems* **4** (1988) 179–182.

[168] Jeroslow, R. E., *Logic-Based Decision Support: Mixed Integer Model Formulation*, Annals of Discrete Mathematics **40**, North-Holland (Amsterdam, 1989).

[169] Jeroslow, R. E., and J. K. Lowe, Modeling with integer variables, *Mathematical Programming Studies* **22** (1984) 167–184.

[170] Jeroslow, R. G., R. K. Martin, R. L. Rardin, and J. Wang, Gain-free Leontief flow problems, Manuscript, Graduate School of Business (Martin's address), University of Chicago, Chicago, IL USA (1989).

[171] Jeroslow, R. E., and J. Wang, Solving propositional satisfiability problems, *Annals of Mathematics and Artificial Intelligence* **1** (1990) 167–188.

[172] Johnson, D. S., and M. A. Trick, eds., *Cliques, Coloring and Satisfiability*, DIMACS Series in Discrete Mathematics and Theoretical Computer Science **26**, American Mathematical Society (Providence RI, 1996).

[173] Kagan, V., A. Nerode, and V. S. Subrahmanian, Computing definite logic programs by partial instantiation, *Annals of Pure and Applied Logic* **67** (1994) 161–182.

[174] Kamath, A. P., N. K. Karmarkar, K. G. Ramakrishnan, and M. G. C. Resende, Computational experience with an interior point algorithm on the satisfiability problem, *Annals of Operations Research* **25** (1990) 43–45; also in R. Kannan and W. R. Pulleyblank, eds., *Integer Programming and Combinatorial Optimization*, University of Waterloo Press (Waterloo, Ontario, 1990) 333–349.

[175] Kamath, A. P., N. K. Karmarkar, K. G. Ramakrishnan, and M. G. C. Resende, An interior point algorithm to solve computationally difficult set covering problems, Manuscript, AT&T Bell Laboratories, Murray Hill, NJ (1991).

[176] Kamath, A. P., N. K. Karmarkar, K. G. Ramakrishnan, and M. G. C. Resende, An interior point approach to Boolean vector function synthesis, *Proceedings*, 36th IEEE Midwest Symposium on Circuits and Systems, vol. 1 (1993) 185–189.

[177] Kanal, L. N., T. S. Levitt, and J. F. Lemmer, *Uncertainty in Artificial Intelligence 3*, North-Holland (Amsterdam, 1989).

[178] Karmarkar, N., A new polynomial-time algorithm for linear programming, *Combinatorica* **4** (1984) 373–395.

[179] Karmarkar, N., M. G. C. Resende, and K. G. Ramakrishnan, An interior point algorithm for zero-one integer programming, presented at the 13th International Symposium on Mathematical Programming, Mathematical Programming Society, Tokyo (1988).

[180] Karmarkar, N., M. G. C. Resende, and K. G. Ramakrishnan, An interior point approach to the maximum independent set problem in dense random graphs, Technical Report, Mathematical Sciences Research Center, AT&T Bell Laboratories, Murray Hill, NJ 07974 USA (1989).

[181] Karp, R. M., Reducibility among combinatorial problems, in R. E. Miller and J. W. Thatcher, eds., *Complexity of Computer Computations*, Plenum Press (New York, 1972) 85–103.

[182] Kavvadias, D., and C. H. Papadimitriou, A linear programming approach to reasoning about probabilities, *Annals of Mathematics and Artificial Intelligence* **1** (1990) 189–206.

[183] Khachiyan, L.G., A polynomial algorithm in linear programming, *Soviet Mathematics–Doklady* **20** (1979) 191–194.

[184] Klir, G. J., Is there more to uncertainty than some probability theorists might have us believe? *International Journal of General Systems* **15** (1989) 347–378.

[185] Kneale, W., and M. Kneale, *The Development of Logic*, Oxford University Press (London, 1962).

[186] Knuth, D. E., Nested satisfiability, *Acta Informatica* **28** (1990) 1–6.

[187] Konolige, K. G., An information-theoretic approach to to subjective Bayesian inference in rule-based systems, working paper, SRI International (Menlo Park, CA, 1982).

[188] Kowalski, R., *Logic for Problem Solving*, Elsevier North-Holland (New York, 1979).

[189] Kopf, R., and G. Ruhe, A computational study of the weighted independent set problem for general graphs, *Foundations of Control Engineering* **12** (1987) 167–180.

[190] Larrabee, T., and Y. Tsujii, Evidence for a satisfiability threshold for random 3cnf formulas, in H. Hirsh et al., eds., *Proceedings of the Spring Symposium on Artificial Intelligence and NP-Hard Problems*, Stanford, CA (1993) 112–118.

[191] Lauritzen, S. L., and D. J. Spiegelhalter, Local computations with probabilities on graphical structures and their application to expert systems, *Journal of the Royal Statistical Society* **B 50** (1988) 157–224.

[192] Lemmer, J. F., and S. W. Barth, Efficient minimum information updating for Bayesian inferencing in expert systems, *Proceedings of the National Conference on Artificial Intelligence*, Pittsburgh, PA, 1982. Morgan Kaufmann (Los Altos CA, 1982) 424–427.

[193] Lemmer, J. F., and L. N. Kanal, eds., *Uncertainty in Artificial Intelligence* **1**, North-Holland (Amsterdam, 1986).

[194] Lemmer, J. F., and L. N. Kanal, eds., *Uncertainty in Artificial Intelligence* **2**, North-Holland (Amsterdam, 1988).

[195] Lewis, H., Complexity results for classes of quantification formulas, *Journal of Computer and Systems Sciences* **21** (1980) 317–353.

[196] Lewis, P. M., and C. L. Coates, *Threshold Logic*, Wiley (New York, 1967).

[197] Lewis, H. R., Renaming a set of clauses as a Horn set, *Journal of the ACM* **25** (1978) 134–135.

[198] Lloyd, J. W., *Foundations of Logic Programming*, 2nd ed., Springer-Verlag (New York, 1995).

[199] Loveland, D. W., *Automated Theorem Proving: A Logical Basis*, North-Holland (Amsterdam, 1978).

[200] Lowe, J. K., Modeling with integer variables, Ph.D. thesis, Georgia Institute of Technology (1984).

[201] Löwenheim, L., Über das Auflösungsproblem in logischen Klassenkalkul, *Sitzungsberichte der Berlinen Mathematischen Gesellschaft* **7** (1908) 89–94.

[202] Löwenheim, L., Über die Auflösung von Gleichungen im logischen Gebietkalkul, *Mathematische Annalen* **68** (1910) 169–207.

[203] Löwenheim, L., Über Transformationen im Gebietkalkul, *Mathematische Annalen* **73** (1910) 245–272.

[204] Mamdani, E. H., Application of fuzzy logic to approximate reasoning using linguistic systems, *IEEE Transactions on Computers* **26** (1977) 1182–1191.

[205] Mannila, H., and K. Mehlhorn, A fast algorithm for renaming a set of clauses as a Horn set, *Information Processing Letters* **21** (1985) 261–272.

[206] Marek, W., A. Nerode, and J. B. Remmel, The stable models of a predicate logic program, *Journal of Logic Programming* **21** (1994) 129–154.

[207] Marriott, K., and P. J. Stuckey, *Programming with Constraints: An Introduction*, MIT Press (Cambridge MA, 1998).

[208] Martin, N. M., *Systems of logic*, Cambridge University Press (Cambridge, England, 1989).

[209] Mayer, J., I. Mitterreicher, and F. J. Radermacher, Running time experiments on some algorithms for solving propositional satisfiability problems, *Annals of Operations Research* **55** (1995) 135–178.

[210] McLeish, M., Probabilistic logic: some comments and possible use for nonmonotonic reasoning, in J. F. Lemmer and L. N. Kanal, eds., *Uncertainty in Artificial Intelligence* **2**, North-Holland (Amsterdam, 1986).

[211] McLeish, M., Nilsson's probabilistic entailment extended to Dempster-Shafer theory, in *Uncertainty in Artificial Intelligence* **3**, North-Holland (Amsterdam, 1989) 23–34.

[212] McNaughton, R., A theorem about infinite-valued sentential logic, *Journal of Symbolic Logic* **16** (1951) 1–13.

[213] Mendelson, E., *Introduction to Mathematical Logic*, 2nd ed., Van Nostrand (Princeton NJ, 1979).

[214] Milner, R., *Communication and Concurrency*, Prentice Hall, London, 1989.

[215] Minker, J., ed., *Foundations of Deductive Databases and Logic Programming*, Morgan Kaufmann (Los Altos, CA, 1988).

[216] Minoux, M., *Mathematical Programming: Theory and Algorithms*, Wiley (New York, 1986).

[217] Minoux, M., LTUR: a simplified linear-time unit resolution algorithm for Horn formulae and computer implementation, *Information Processing Letters* **29** (1988) 1–12.

[218] Mitchell, D., B. Selman, and H. Levesque, Hard and easy distributions of SAT problems, in *Proceedings, Tenth National Conference on Artificial Intelligence, AAAI92*, MIT Press (Cambridge MA, 1992) 459–465.

[219] Mundici, D., Normal forms in infinite-valued logic: the case of one variable, *Proceedings, Computer Science Logic 91* Springer (Berlin, 1992) 272–277.

[220] Munkres, J.R., *Topology: A First Course*, Prentice-Hall (Englewood Cliffs NJ, 1975).

[221] Muroga, S., *Threshold Logic and Its Applications*, Wiley-Interscience (New York, 1971).

[222] Nemhauser, G. L., and L. A. Wolsey, *Integer and Combinatorial Optimization*, Wiley-Interscience (New York, 1988).

[223] Nilsson, N. J., Probabilistic logic, *Artificial Intelligence* **28** (1986) 71–87.

[224] Nobili, P., and A. Sassano, Facets and lifting procedures for the set covering polytope, *Mathematical Programming* **45** (1989), 111–137.

[225] Ohyanagi, T., M. Yamamoto, and A. Ohuchi, An algorithm for the beta clause satisfiability problem in propositional logic, *Transactions of the Institute of Electrical Engineers of Japan* **114-C** (1994) 786–804.

[226] Oliver, R. M., and J. Q. Smith, *Influence Diagrams, Belief Nets and Decision Analysis*, Wiley (Chichester, UK, 1990).

[227] Oppen, D., A $2^{2^{2^n}}$ upper bound on the complexity of Presburger arithmetic, *Journal of Computer and Systems Science* (1975).

[228] Paaß, G., Probabilistic logic, in P. Smets et al., eds., *Non-standard Logics for Automated Reasoning*, Academic Press (New York, 1988) 213–251.

[229] Padberg, M. W., and G. Rinaldi, A branch-and-cut algorithm for the solution of large-scale symmetric traveling salesman problems, Technical Report, Stern School of Business, New York University, New York (1989).

[230] Pardalos, P., and G. Rodgers, A branch and bound algorithm for the maximum clique problem, *Computers and Operations Research* **19** (1992) 363–375.

[231] Patrizi, G., The equivalence of an LCP to a parametric linear program with a scalar parameter, *European Journal of Operational Research* **51** (1991) 367–386.

[232] Patrizi, G., and C. Spera, Solving in polynomial time some hard combinatorial problems, manuscript, Dip. di Statistica, Probabilità e Statistiche Applicate, Università di Roma "La Sapienza," not dated.

[233] Pearl, J., Fusion, propagation and structuring in belief networks, *Artificial Intelligence* **29** (1986) 241–288.

[234] Pearl, J., *Probabilistic Reasoning in Intelligent Systems: Networks of Plausible Inference*, Morgan Kaufmann (San Mateo CA, 1988).

[235] Plaisted, J. A., Complete problems in the first-order predicate calculus, *Journal of Computer and Systems Sciences* **29** (1984) 8–35.

[236] Pogorzelski, W. A., The deduction theorem for Lukasiewicz many-valued propositional calculi, *Studia Logica*, **15** (1964) 7–23.

[237] Presburger, M., Über die Vollständigkeit eines gewissen Systems der Arithmetik ganzer Zählen in welchem die Addition als einzige Operation hervortritt, *Sprawozdanie z I Kongresu Matematykow Krajow Slowcanskich Warszawa*, Warsaw, Poland (1929) 92–101.

[238] Pretolani, D., Ph.D. Thesis, University of Pisa (1995).

[239] Purdom, P., A survey of average time analyses of satisfiability algorithms, *Journal of Information Processing* **13** (1990) 449–455.

[240] Purdom, P. W., and C. A. Brown, Polynomial-average-time satisfiability algorithms, *Information Science* **41** (1987) 23–42.

[241] Quine, W. V., The problem of simplifying truth functions, *American Mathematical Monthly* **59** (1952) 521–531.

[242] Quine, W. V., A way to simplify truth functions, *American Mathematical Monthly* **62** (1955) 627–631.

[243] Reiter, R., Nonmonotonic reasoning, *Annual Reviews of Computer Science* **2** (1987) 147–186.

[244] Robinson, J. A., A machine-oriented logic based on the resolution principle, *Journal of the ACM* **12** (1965) 23–41.

[245] Rockafeller, R. T., *Convex Analysis*, Princeton University Press (Princeton NJ, 1970).

[246] Sassano, A., On the facial structure of the set covering polytope, *Mathematical Programming* **44**, (1989), 181–202.

[247] Schaefer, T. J., The complexity of satisfiability problems, *Proceedings of the 10th ACM Symposium on Theory of Computing (STOC)*, (1978) 216–226.

[248] Schlipf, J. S., F. S. Annexstein, J. V. Franco, and R. P. Swaminathan, On finding solutions for extended Horn formulas, *Information Processing Letters* **54** (1995) 133–137.

[249] Schrijver, A., *Theory of Linear and Integer Programming*, Wiley (New York, 1986).

[250] Shachter, R. D., Evaluating influence diagrams, *Operations Research* **34** (1986) 871–82.

[251] Shachter, R. D., T. S. Levitt, L. N. Kanal, and J. F. Lemmer, eds., *Uncertainty in Artificial Intelligence* **4**, North-Holland (Amsterdam, 1990).

[252] Shafer, G., *A Mathematical Theory of Evidence,* Princeton University Press (Princeton NJ, 1976).

[253] Shastri, L., *Semantic Networks: An Evidential Formalization and Its Connectionist Realization,* Pitman (San Mateo CA, 1988).

[254] Shenoy, P. P., A valuation-based language for expert systems, *International Journal for Approximate Reasoning* **3** (1989) 383–411.

[255] Shenoy, P. P., Consistency in valuation-based systems, *ORSA Journal on Computing* **6** (1994) 281–291.

[256] Shenoy, P. P., Valuation-based systems for propositional logic, *Methodologies for Intelligent Systems* **5**, Proceedings of the Fifth International Symposium (North Holland, Amsterdam, 1990) 305–312.

[257] Shenoy, P. P., Valuation-based systems for discrete optimization, Working Paper No. 221, School of Business, University of Kansas, Lawrence, KA (1990).

[258] Sherali, H. D., and W. P. Adams, A hierarchy of relaxations and convex hull characterizations for mixed-integer zero-one programming problems, *Discrete Applied Mathematics* **52** (1994) 83–186.

[259] Sherali, H. D., M. S. Bazaraa and C. M. Shetty, *Nonlinear Programming: Theory and Algorithms,* 2nd ed., Wiley (New York, 1992).

[260] Shindo, M., and E. Tomita, A simple algorithm for finding a maximum clique and its worst-case time complexity, *Systems and Computers in Japan* **21** (1990) 1–13.

[261] Shostack, R. E., On the SUP-INF method for proving Presburger formulas, *Journal of the ACM* **26** (1979) 351–360.

[262] Shostack, R. E., A practical decision procedure for arithmetic with function symbols, *Journal of the ACM* **26** (1979) 351–360.

[263] Sombé, L., *Reasoning under Incomplete Information in Artificial Intelligence,* Wiley (New York, 1990).

[264] Spera, C., Computational results for solving large general satisfiability problems, Technical Report, Centro di Calcolo Elettronico, Università degli Studi di Siena, Italy (1990).

[265] Spiegelhalter, D. J., and S. L. Lauritzen, Sequential updating of conditional probabilities on directed graphical structures, *Networks* **20** (1990) 579–605.

[266] Sterling, L., and E. Shapiro, *The Art of Prolog: Advanced Programming Techniques*, MIT Press (Cambridge MA, 1986).

[267] Swaminathan, R. P., and D. K. Wagner, The arborescence-realization problem, *Discrete Applied Mathematics* **59** (1995) 267–283.

[268] Tarjan, R. E., Depth-first search and linear graph algorithms, *SIMM Journal of Computing* **1** (1972) 146–160.

[269] Tarjan, R. E., and A. E. Trojanowski, Fast algorithm for finding a maximum independent set in a graph, *SIAM Journal on Computing* **3** (1977) 537–546.

[270] Tarski, *A Decision Method for Elementary Algebra and Geometry*, University of California Press (Berkeley, 1951).

[271] Triantaphyllou, E., and A. L. Soyster, A relationship between CNF and DNF systems derivable from examples, *ORSA Journal on Computing* **7** (1995) 283–285.

[272] Triantaphyllou, E., and A. L. Soyster, On the minimum number of logical clauses inferred from examples, *Computers and Operations Research* **23** (1996) 783–799.

[273] Triantaphyllou, E., and A. L. Soyster, An approach to guided learning of boolean functions, *Mathematical and Computer Modeling* **23** (1996) 69–86.

[274] Triantaphyllou, E., A. L. Soyster, and S. R. T. Kumara, Generating logical expressions from positive and negative examples via a branch-and-bound approach, Manuscript, Industrial and Management Systems Engineering, Pennsylvania State University, University Park, PA (1991).

[275] Trillas, E., and L. Valverde, On mode and implication in approximate reasoning, in M. M. Gupta, A. Kandel, W. Bandler and J. B. Kiszka, eds., *Approximate Reasoning in Expert Systems*, North-Holland (Amsterdam, 1985) 157–166.

[276] Truemper, K., *Effective Logic Computation*, Wiley (New York, 1998).

[277] Truemper, K., and F. J. Radermacher, Analyse der Leistungsfähigkeit eines neues Systems zur Auswertung aussagenlogischer Probleme, Report FAW-TR-90003, Forschungsinstitut für anwendungsorientierte Wissensverarbeitung, Ulm, Germany (1991).

[278] Tseitin, G. S., On the complexity of derivations in the propositional calculus, in A. O. Slisenko, ed., *Structures in Constructive Mathematics and Mathematical Logic, Part II* (translated from Russian, 1968) 115–125.

[279] Turing, A., On computable numbers, with an application to the Entscheidungsproblem, *Proceedings, London Mathematical Society* **42** (1936) 230–265.

[280] Urquhart, A., Hard examples for resolution, *Journal of the ACM* **34** (1987) 209–219.

[281] Ursic, S., Generalizing fuzzy logic probabilistic inferences, in J. F. Lemmer and L. N. Kanal, eds., *Uncertainty in Artificial Intelligence 2*, North-Holland (Amsterdam, 1988) 337–364.

[282] Tsang, E., *Foundations of Constraint Satisfaction*, Academic Press (London, 1993).

[283] Van der Gaag, L., Computing probability intervals under independency constraints, Manuscript, Computer Science Department, Utrecht University, Utrecht, Netherlands (not dated).

[284] Van der Gaag, L., Probability-based models for plausible reasoning, Ph.D. Thesis, University of Amsterdam (1990).

[285] Van der Gaag, L., On evidence absorption for belief networks, *International Journal of Approximate Reasoning* **15** (1996) 165–286.

[286] Van Hentenryck, P. *Constraint Satisfaction in Logic Programming*, MIT Press (Cambridge MA, 1989).

[287] Van Hentenryck, P., and V. Saraswat (1996). Strategic directions in constraint programming, *ACM Computing Surveys* **28** (1996) 701–726.

[288] Vanderbei, R. J., M. S. Meketon, and B. A. Freedman, A modification of Karmarkar's linear programming algorithm, *Algorithmica* **1** (1986) 395–407.

[289] Weyl, H., Elemetere Theorie der konvexen polyerer, *Comm. Math. Helv.* **I** (1935) 3–18 (English translation in *Annals of Mathematics Studies* **24** (1950) xx-xx).

[290] Williams, H. P., Fourier-Motzkin elimination extension to integer programming problems, *Journal of Combinatorial Theory* **21** (1976) 118–123.

[291] Williams, H. P., Logical problems and integer programming, *Bulletin of the Institute of Mathematics and its Implications* **13** (1977) 18–20.

[292] Williams, H. P., Fourier's method of linear programming and its dual, *American Mathematical Monthly* **93** (1986) 681–695.

[293] Williams, H. P., Linear and integer programming applied to the propositional calculus, *International Journal of Systems Research and Information Science* **2** (1987) 81–100.

[294] Williams, H. P., *Model Building in Mathematical Programming*, Wiley (Chichester, 1988).

[295] Williams, H. P., Computational logic and integer programming: connections between the methods of logic, AI and OR, Technical Report, University of Southampton, United Kingdom (1991).

[296] Williams, H. P., Logic applied to integer programming and integer programming applied to logic, *European Journal of Operations Research* **81** (1995) 605–616.

[297] Wilson, J. M., Compact normal forms in propositional logic and integer programming formulations, *Computers and Operations Research* **90** (1990) 309–314.

[298] Wise, B. P., and M. Henrion, A framework for comparing uncertain inference systems to probability, in *Proceedings of Workshop on Uncertainty and Probability in Artificial Intelligence*, AAAI (Los Angeles, 1985).

[299] Wolfe, P., Methods of nonlinear programming, in R. L. Graves and P. Wolfe, eds., *Recent Advances in Mathematical Programming* (1963).

[300] Yager, R. R., Approximate reasoning as a basis for rule-based expert systems, *IEEE Transactions on Systems, Man and Cybernetics* **14** (1984) 636–643.

[301] Yamada, Y., E. Tomita, and H. Takahashi, A randomized algorithm for finding a near-maximum clique and its experimental evaluations, *Systems and Computers in Japan* **25** (1994) 1–7.

[302] Yamasaki, S., and S. Doshita, The satisfiability problem for a class consisting of Horn sentences and some non-Horn sentences in propositional logic, *Information and Control* **59** (1993) 1–12.

[303] Zadeh, L. A., Fuzzy sets, *Information and Control* **8** (1965) 338–353.

[304] Zadeh, L. A., A theory of approximate reasoning, *Machine Intelligence* **9** (1979) 149–194.

Index

2-satisfiability 200

absorption 18
    generalized 154
affine dimension 161
affine hull 328
affine scaling method 174
alternativeness relation 316
ancestral set 232
antecedent 22
arborescence-chain pattern 73
Aristotle 3
arithmetic, logic of 4,269
artificial intelligence 5,6
artificial variable 25
atomic proposition 13,24
automated theorem proving 6

backtracking 88,102,121
    limited 87
backward chaining 23
balanced clause set 78
balanced matrix 78
B-arc 288
barrier function 175
basic probability function 236
basic probability number 236
basic solution 125,131
basic solution 329
basic variable 329
basis matrix 329
Bayesian inference 218
Bayesian inference 7

Bayesian logic 204,219
Bayesian network 7,204,219–220
belief function 238,239,244
    incomplete 244,248
belief net 205
belief net 7
Benders cut 181
Benders decomposition 177,181,
    224,225
B-graph 288
    cycle in 289
    path in 289
Billionnet-Sutter method 123
binary tree 101
Bledsoe real arithmetic 301
blocked valuation 278
Boole, G. 2,4,203
boolean function 153,177,183,186
bound variable 269
bounded convex representability
    262
bounded resolution 123
B-path 289
    shortest 290
branch-and-bound method 117,
    201
branch-and-cut method 133,144
branching algorithm 87,94,101,
    103,116,174
branching rule 103
breadth-first search 104

canonical cut 159

INDEX

canonical matrix of generators 326
Carathéodory's theorem 326
causal reasoning 205
Chinese remainder theorem 305
Church, A. 9
Chvátal cut 134,149,153,196
circuit cut 184
clause 17,24
    definite 31
    Horn 22,31
clause set 162
closed world assumption 308
CLP 307
CNF 13,17
column generation 204–205,211, 224
column subtraction method 131
combining evidence 239,246
compactness 275
compactness theorem 296,297, 317,321
complementary slackness 331
complexity of inference 9
conditional independence 219, 221
conditional probability 221
cone
    canonical matrix of generators of 326
    constraint representation of 326
    convex 326
    convex polyhedral 325
    edge representation of 326
    recession 328
confidence factor 204,250,262
confidence function 250
conjunction 16
conjunctive normal form 13,17, 149,179,200
    conversion to 17,20
consensus 100,123

consequent 22
conservative confidence function 258,263
consistency 89
constant, in first-order logic 269
constraint logic programming 5, 300,307,317
constraint programming 7
constraint representation of a cone 326
constraint satisfaction 5
convex cone 326
convex hull 327
convex polyhedral cone 325
convex set 325
    polyhedral 326
cutting plane 120,133,134,149, 151–152,174
cutting plane proof 149,161
cycle, in B-graph 289

data base 7
datalog formula 268,287
datalog formula, hypergraph model of 290
Davis-Putnam-Loveland procedure 9,98,99,101,104,110, 114,276
De Morgan's laws 16
decision tree 7
default logic 7,204
default reasoning 308,308
definite clause 31
degeneracy 120
degree of an inequality 154
Dempster's combination rule 236, 240,246
    modified 248
Dempster-Shafer theory 7,204, 235,246
depth-first search 104
determining set 186

diagonal sum 151,155,158
dimension of polyhedron 328
disguised proposition 41
disjunction 16
disjunctive normal form 17,186
disjunctive programming 259
distributive law 17
DNF 17
domain 270
DPL procedure 101
dual cost 210
dual linear program 330
dual variables for a Horn proposition 37
duality
    in linear programming 330
    duality, strong 331
    duality, weak 330
dynamic programming 200

edge of a polyhedron 329
edge representation of a cone 326
ellipsoid method 332
entropy 216
erasure 61
evidential reasoning 239,246
exclusive OR 187
existential quantifier 269
expert system 6,31,189,201
extended ancestral set 232
extended clause 159
extended Horn proposition 64,68
extended Horn rule base 73
extended nested satisfiability problem 61
extended resolution 149
extended star-chain 64,72
extended unit resolution 65,72
extension 191
extreme point 327
extreme ray 328

face of a polyhedron 329
facet cut 153,154,160
facet of a polyhedron 329
Farkas lemma 329,332
firing a rule 252
first-order logic 9,269
    completeness of 271
    undecidability of 268,270
    with equality 270
fixed charge 259
fixed density model 113
fixed probability model 113
forbidden subgraph 50
forward chaining 22
Fourier-Motzkin elimination 190, 194,196,303,332
fractional programming 218,250
frame of discernment 236
free variable 269
Frege, G. 3
functions, in first order logic 269
fuzzy logic 7,205

Gödel, K. 3
Gallo-Urbani method 122
generalization, in first-order logic 269
generalized Horn proposition 94
generalized resolution 151–152, 154
generations, in a network 229
ground instance 269
ground level 269
ground resolution 98

half-space 325
head of hyperarc 288
Helly's theorem 297
H-equivalence 33
Herbranch ground instance 275
Herbranch universe 275
Herbrand base 275

INDEX

Herbrand extension 275,295,298
Herbrand interpretation 275
Herbrand universe 309
Herbrand's theorem 274,298,317
Hilbert, D. 4
homogeneous linear equation 325
homogeneous linear inequality 325
Horn clause 9,22,31,193
Horn equivalence 33
Horn polytope 34
    least element of 35
Horn proposition 32,317,332
    dual variables for 37
    graph representation 32
Horn relaxation 31,122
hyperarc 288
    head of 288
    layer of 288
    tail of 288
hyperedge 61
hypergraph 61,288
hypergraph model of propositional logic 288
hyperplane 325
    supporting 329

implication graph 42
incremental satisfiability 106,189, 276
independence of events 219
independence, in Dempster-Shafer theory 247
independent propositions 221
individual, in first-order logic 269
inductive inference 201
inference engine 6
inference, complexity of 9
infinite dimensional programming 296,316,320

influence diagram 7,204
initialized rule system 254
input proof 137
input refutation 136
input resolution 136
integer programming 24
integral polytope 76,81
interior point method 174–175,201
interpretation, for first-order logic 270

Jeroslow, R. 2,20,37,38,103,117, 121,201,251,267,275,278, 282,288
Jeroslow-Wang method 121

K-resolution 191
Karmarkar's method 332

layer of hyperarc 288
least element of a Horn polytope 35
least element of a polyhedron 34, 65
least element property 310
Leibniz, G. W. 4
Levenberg-Marquardt algorithm 174
lifting 151
linear complementarity 199
linear program 330
    dual 330
    primal 330
linear programming 5,24,124,317, 325
    complexity of 332
    ellipsoid method for 332
    Fourier's method for 332
    Karmarkar's method for 332
    P-completeness of 332
    relaxation 27
    simplex method for 332

linear relaxation 118,127
literal 16
logic circuit 98,100,177
logic circuit verification 6,177,179, 308
logic cut 133,152
logic gate 177,186
logic model 6,7
logic of arithmetic 4
logic programming 5,308,317
logic
    Bayesian 204,219
    default 204
    default 7
    first-order 9
    fuzzy 205
    fuzzy 7
    Lukasiewicz 312
    many-valued 308,311
    modal 205,308,314
    modal 7
    multivalued 201
    nonmonotonic 204,308
    nonmonotonic 7
    possibilistic 205
    predicate 98
    probabilistic 2,7,9,77
    probabilistic 203,205,238–239,245
    propositional 13
logical projection 190–191
Löwenheim, L. 98
Lukasiewicz logic 312

many-sorted logic 270
many-valued logic 308,311
Markov tree 200,220
master problem 181
material conditional 206
mathematical programming 8,23
matrix of a formula 271
maximum clique problem 90

maximum satisfiability problem 58
median operator 44
median set 44
MILP representability 261,313
minimal element property 299
minimal model 33,309
Minkowski sum 326–327
mixed integer programming 133, 153,251,259,263,313
modal logic 7,205,308,314
model 15
    for modal logic 315
    in the logical sense 206
    logic 6,7
    minimal 309
    minimal 33
    stable 309–310
modus ponens 22
modus tollens 22
monotone constraint set 82
monotone literal 140
monotone set of inequalities 152, 155
monotone variable fixing 102
moral graph 235
multivalued logic 201

negation 16
negation-as-failure 308
nested proposition 55,59
    recognition of 57
nested satisfiability problem, extended 61
nonbasic variable 329
nonlinear programming 173–174
nonmonotonic logic 7,204,308
normal form 16
NP-completeness 9,97

partial instantiation 267,275
    dual approach 268,284

INDEX

primal approach 268,278
partition variable 57
path tree 255
path, in B-graph 289
Phase I
    problem 26
    simplex method 126
phase transition 116
pigeonhole problem 97,110,150, 160–161
pivot and complement method 128
point probability value 216
pointed polyhedron 328
polyhedral convex set 326
polyhedron 326
    dimension of 328
    edge of 329
    face of 329
    facet of 329
    least element of 34
    pointed 328
    projection of 2
    vertex of 329
polytope 327
    Horn 34
    integral 76,81
positive unit resolution 31,37
positively homogeneous function 262
possibilistic logic 205
possible world 205,206
potential function 175
predicate 269
predicate logic 98,267
prenex form 271
Presburger arithmetic 301,304
Presburger real arithmetic 301–302
primal linear program 330
prime cover 241
prime implication 19,99,151,161

probabilistic logic 2,7,9,77,203, 205,238–239,245–246
probabilistic satisfiability 208, 246
probability interval, width of 217
product topology 297
projection 151,181,191,216
projection
    of a polyhedron 2,190,193
    logical 98,190–191
projection cut 193,195
projective scaling method 174
PROLOG 6
propagation 220
    atomic 13,24
    disguised 41
    extended Horn 64,68
    first-order Horn 299
    generalized Horn 94
    Horn 317,332
    Horn 32
    nested 55
    Q-Horn 53
    quadratic 41–42
    renamable extended Horn 70
    renamable Horn 41,68,88, 89
    split Horn 96
propositional logic 13
    hypergraph model for 288
propositions, nested 59
pseudo-boolean optimization 214

Q-Horn proposition 53
quadratic proposition 41–42
quantified logic 269
quantifier 269
query 295
Quine, W. V. 98–99

random problem 9,113
rank one cut 133–134,136,143,147
recession cone 328
recession direction 260
recognition
    of nested proposition 57
    of renamable Horn proposition 45
reduction 154,196
regular resolution 112,140,192
relaxation 117
renamable extended Horn proposition 70
renamable Horn proposition 41, 68,88–89
    recognition 45
representability 204
    in mixed integer programming 259,261
resolution 9,18,98-99,113
    bounded 123
    completeness of 100
    extended 149
    extended unit 65,72
    generalized 151-152,154
    ground 98
    Horn 31
    input 136
    K- 191
    positive unit 31,37
    regular 112,140
    regular 193
    unit K- 193,197
    unit 19,23,28,31,63,69,101, 120,136,182
resolvent 19,99
root node 101
rooted arborescence 64,68
rounding theorem 65,69
rule 22,250–251,308
rule-based system 22
Russell, B. 4

sampling heuristic 117
satisfaction set 191
satisfiability problem 13,15,24,97
satisfiability problems, hierarchy of 91
satisfiability
    in predicate logic 270
    incremental 106,189,276
    probabilistic 208,246
    quadratic (2-SAT) 41,200
Schönfinkel-Bernays form 9,272
search tree 101
semantics 13
semidecidability 271,275
sensitivity analysis 209
separating cut 134
separation algorithm 135,143, 146
sequential linear programming 227
sequential quadratic programming 224,226
set-covering problem 241
set-covering problem 82,100,149, 161
set-packing problem 90,158
signature 318
signed formula 312
simple support function 241
simplex method 124,329,332
    phase I 126
singly connected network 220,234
Skolem function 273
Skolem normal form 271
split Horn proposition 96
splitting 99
stable model 309–310
standardizing variables apart 271
strong component 42
strong duality 331
subproblem in Benders decomposition 181

INDEX

SUP-INF method 5,301
supporting hyperplane 329
syntactics 13

tableau method 101,123
tail of hyperarc 288
Tarski, A. 4
tautology 15
tautology checking 177–178
threshold function 153
tight formulation 200
topological order 42
total unimodularity 77
trust region 175
Tseitin, G. 20,109,111,149
Turing, A. 9
Tychonoff theorem 297
typed logic 270
undominated cut 151,154,156

unit K-resolution 193,197
    complexity of 198
unit proof 136,190
unit refutation 136
unit resolution 19,23,28,31,63,69,
    101,120,136,182
unit resolution property 65,71,78,
    82
universal quantifier 269

validity, in predicate logic 270
vertex of a polyhedron 329
VLSI design 177

weak duality 330
well formed formula 13
Whitehead, A. N. 4
Williams, H. P. 2,135,190,196,200,
    302,304

WILEY-INTERSCIENCE
SERIES IN DISCRETE MATHEMATICS AND OPTIMIZATION

ADVISORY EDITORS

RONALD L. GRAHAM
AT & T Laboratories, Florham Park, New Jersey, U.S.A.

JAN KAREL LENSTRA
*Department of Mathematics and Computer Science,
Eindhoven University of Technology, Eindhoven, The Netherlands*

ROBERT E. TARJAN
*Princeton University, New Jersey, and
NEC Research Institute, Princeton, New Jersey, U.S.A.*

AARTS AND KORST • Simulated Annealing and Boltzmann Machines: A Stochastic Approach to Combinatorial Optimization and Neural Computing
AARTS AND LENSTRA • Local Search in Combinatorial Optimization
ALON, SPENCER, AND ERDŐS • The Probabilistic Method
ANDERSON AND NASH • Linear Programming in Infinite-Dimensional Spaces: Theory and Application
AZENCOTT • Simulated Annealing: Parallelization Techniques
BARTHÉLEMY AND GUÉNOCHE • Trees and Proximity Representations
BAZARRA, JARVIS, AND SHERALI • Linear Programming and Network Flows
CHANDRU AND HOOKER • Optimization Methods for Logical Inference
CHONG AND ZAK • An Introduction to Optimization
COFFMAN AND LUEKER • Probabilistic Analysis of Packing and Partitioning Algorithms
COOK, CUNNINGHAM, PULLEYBLANK, AND SCHRIJVER • Combinatorial Optimization
DASKIN • Network and Discrete Location: Modes, Algorithms and Applications
DINITZ AND STINSON • Contemporary Design Theory: A Collection of Surveys
ERICKSON • Introduction to Combinatorics
GLOVER, KLINGHAM, AND PHILLIPS • Network Models in Optimization and Their Practical Problems
GOLSHTEIN AND TRETYAKOV • Modified Lagrangians and Monotone Maps in Optimization
GONDRAN AND MINOUX • Graphs and Algorithms *(Translated by S. Vajdā)*
GRAHAM, ROTHSCHILD, AND SPENCER • Ramsey Theory, Second Edition
GROSS AND TUCKER • Topological Graph Theory
HALL • Combinatorial Theory, Second Edition
JENSEN AND TOFT • Graph Coloring Problems
KAPLAN • Maxima and Minima with Applications: Practical Optimization and Duality
LAWLER, LENSTRA, RINNOOY KAN, AND SHMOYS, Editors • The Traveling Salesman Problem: A Guided Tour of Combinatorial Optimization
LAYWINE AND MULLEN • Discrete Mathematics Using Latin Squares
LEVITIN • Perturbation Theory in Mathematical Programming Applications
MAHMOUD • Evolution of Random Search Trees
MARTELLO AND TOTH • Knapsack Problems: Algorithms and Computer Implementations
McALOON AND TRETKOFF • Optimization and Computational Logic
MINC • Nonnegative Matrices
MINOUX • Mathematical Programming: Theory and Algorithms *(Translated by S. Vajdā)*
MIRCHANDANI AND FRANCIS, Editors • Discrete Location Theory
NEMHAUSER AND WOLSEY • Integer and Combinatorial Optimization
NEMIROVSKY AND YUDIN • Problem Complexity and Method Efficiency in Optimization *(Translated by E. R. Dawson)*
PACH AND AGARWAL • Combinatorial Geometry
PLESS • Introduction to the Theory of Error-Correcting Codes, Third Edition

ROOS AND VIAL • Ph. Theory and Algorithms for Linear Optimization: An Interior Point Approach
SCHEINERMAN AND ULLMAN • Fractional Graph Theory: A Rational Approach to the Theory of Graphs
SCHRIJVER • Theory of Linear and Integer Programming
TOMESCU • Problems in Combinatorics and Graph Theory *(Translated by R. A. Melter)*
TUCKER • Applied Combinatorics, Second Edition
WOLSEY • Integer Programming
YE • Interior Point Algorithms: Theory and Analysis